Applications of Stem Cells and Derived Exosomes in Neurodegenerative Disorders

Sadaf Jahan • Arif Jamal Siddiqui
Editors

Applications of Stem Cells and Derived Exosomes in Neurodegenerative Disorders

Springer

Editors
Sadaf Jahan
Department of Medical Laboratory Sciences
College of Applied Medical Sciences, Majmaah University
Al-Majmaah, Saudi Arabia

Arif Jamal Siddiqui
Department of Biology
College of Science, University of Hail
Hail, Saudi Arabia

ISBN 978-981-99-3847-6 ISBN 978-981-99-3848-3 (eBook)
https://doi.org/10.1007/978-981-99-3848-3

This Springer imprint is published by the registered company Springer Nature Singapore Pte Ltd.
The registered company address is: 152 Beach Road, #21-01/04 Gateway East, Singapore 189721, Singapore

Preface

This book is based on therapeutic strategies against neurodegenerative disorders (NDDs). Currently, NDDs have become a severe burden for patients and community healthcare because of their relatively long duration and expensive treatment costs. There are many strategies for therapy, but they have less efficacy and more side effects. The stem cells-based therapy is promising. Stem cells can be differentiated into the desired cells *in vitro* as well as *in vivo*. Apart from stem cells, stem cells-derived molecules are also found to be more effective tools against NDDs. Therefore, in the chapters, we provided information on the treatment strategies for different types of neurodegenerative disorders with the help of stem cells and derived molecules like exosomes. So we investigated the causes of NDDs and the history of disease in the book. Apart from that, we have shed light on the role of stem cells and their mechanistic approach against the NDDs. The role of exosomes in the treatment of disease is also discussed in the book. We are very grateful to our authors, who have taken time off their busy schedules to contribute one or more chapters to this book. Dr. Bhavik Sawhney at Springer Nature has been the enthusiastic initiator of this project, and without his continuing encouragement, one or the other editor would probably have given up along the way. Many thanks are due to supportive staff in the background, particularly the copy editors, who, as always, have done an outstanding job polishing the chapters.

There are numerous people who deserve thanks. I, Sadaf Jahan, would like to begin by thanking the Almighty and my late grandma, Mrs. Akhtari Begam, for their unceasing support. My parents, Mr. Furkan Ali and Mrs. Farhat Sultana, my brothers, Mr. Shoeb Ali and Mr. Suhail Ali, and all family members have been a major source of inspiration for me throughout my life. My husband Muhammad Alam and my son Azlaan Muhammad Umar are great blessing and support to fulfill my dreams.

I, Arif Siddiqui, express my special thanks to my parents Mr. Hakimuddin Siddiqui and Mrs. Shahnaz Begum, wife Ruhi Parveen, son Ariz Siddiqui, and all family members for their everlasting love, enthusiasm for science, and encouragement to pursue every task successfully.

Al-Majmaah, Saudi Arabia — Sadaf Jahan
Hail, Saudi Arabia — Arif Jamal Siddiqui

Contents

Editors and Contributors

About the Editors

Sadaf Jahan is currently working as Assistant professor in the Department of Medical Laboratory Sciences, College of Applied Medical Sciences, Majmaah University, Al-Majmaah, Saudi Arabia. She has earlier served as Assistant Professor (Microbiology) in the Department of Microbiology; Chinmaya Degree College, H.N.B. Garhwal University, Haridwar, India (2018–2019). Previously, she has also worked as a National Post-Doctoral Fellowship through DST-Science and Engineering Research Board (SERB) at CSIR-Central Drug Research Institute, Lucknow, India, between 2016 and 2018. In 2009, she qualified a National Level Exam (UGC CSIR NET-JRF) in Life Sciences and pursued a PhD in Biological Sciences in 2010 at CSIR-Indian Institute of Toxicology Research (IITR), Lucknow, India, through ***AcSIR***-Academy of Scientific and Innovative Research, New Delhi, India, and completed her PhD in 2015. Her research experience and interest are broadly focused on developing and validating in vitro model systems to study developmental neurotoxicity of environmental chemicals and drugs using chemical-induced differentiating neural cells, derived from human umbilical cord blood stem cells. She is an editorial board member of various journals. She has served as a referee for a number of international journals. She has more than 10 years of research experience in neurosciences and stem cells.

She has also published more than 85 research articles in peer-reviewed international journals and authored or co-authored many book chapters. She is a member of many international scientific societies and organizations, importantly International Placenta Stem Cell Society (IPLASS), International Society for Neurochemistry (ISN), American Society for Neurochemistry, Life Member of Indian Academy of Neuroscience (IAN), India, Life Member of Society for Free Radical Research (SFRR), India. She has reviewed more than 100 manuscripts for reputed journals.

She is running several projects as PI (principal investigator) and Co-PI obtain from Deanship of Research, Majmaah University as well as other Saudi university. Since 2020 to present, she is also a member of Quality Assurance Committee at the Department of Medical Laboratory Sciences, College of applied Medical Sciences, Majmaah University, Al-Mjamaah, Saudi Arabia.

Arif Jamal Siddiqui is an Associate Professor and Principal Investigator at the Department of Biology, College of Science, University of Hail, Saudi Arabia. He received his PhD from CSIR-Central Drug Research Institute, Lucknow, India, in 2015. From 2015 to 2018, he worked as a Post-Doctoral Fellow at Texas Tech University Health Sciences Center, Lubbock, Texas, United States. He has more than 12 years of experience in research, teaching, and administration. In his professional work, he has received research grants as a Principal Investigator from various renowned organizations. He has successfully published more than 100 publications in internationally recognized peer-reviewed prestigious journals, published several book chapters for internationally renowned publishers and presented many articles and posters in various conferences/workshops worldwide. He is associate editor of 3 Biotech (Springer Nature), as well as academic editor and editorial board member of various other journals. His research experience and interest are to study host–parasites interaction, role of stem cells in neuro-disorder diseases, immune response (specific T, B, MQ, dendritic cells and specific memory T, B-related study) during parasite interaction, drug and vaccine development against parasites infection. Presently, he is working on parasite diversity, novel role of stem cells in various neuro-related diseases, natural bioactive compound role in cancer, immune behaviors and responses, diagnosis and vaccine development of *Schistosoma mansoni*, *Schistosoma japonicum*, *Schistosoma haematobium* (Trematode worm), and *Plasmodium* parasites (Malaria). Furthermore, he is a member of The Indian Science Congress Association, India, and the Annals of Parasitology, Poland. He has reviewed more than 180 manuscripts and he has edited more than 100 manuscripts. He is running several projects as PI (principal investigator) and Co-PI obtain from Deanship of Research, University of Hail as well as other Saudi university.

List of Contributors

Shaista Ahmed Centre of Translational and Clinical Research, School of Chemical and Lifesciences, Jamia Hamdard, New Delhi, India

Sahar Aldosari Department of Medical Laboratory Sciences, College of Applied Medical Sciences, Majmaah University, Al-Majmaah, Saudi Arabia

Athanasios Alexiou Novel Global Community Educational Foundation, Hebersham, NSW, Australia

AFNP Med Austria, Wien, Austria

Shaheen Ali Department of Molecular Medicine, SIST, Jamia Hamdard, New Delhi, India

Medicinal Chemistry and Molecular Modeling Lab, Department of Pharmaceutical Chemistry, SPER, Jamia Hamdard, New Delhi, India

Saeed Alsharani Department of Pharmacology and Toxicology, College of Pharmacy, Jazan University, Jazan, Saudi Arabia

Adil Alshoaibi Department of Medical Laboratory Sciences, College of Applied Medical Sciences, Al- Majmaah, Saudi Arabia

Department of Physics, College of Science, King Faisal University, Al-Ahsa, Saudi Arabia

Deepika Arora Department of Biotechnology, School of Engineering and Technology, Shobhit Institute of Engineering & Technology (Deemed-to-be-University), Meerut, India

Research & Development, Regrow Biosciences Private Limited., Lonavala, Maval, Pune, India

Riadh Badraoui Laboratory of General Biology, Department of Biology, University of Ha'il, Ha'il City, Saudi Arabia

Laboratory of induced and developed diseases LR19ES12, Medicine Faculty of Sfax. University of Sfax, Sfax, Tunisia

Section of Histology–Cytology, Medicine Faculty of Tunis El Manar University, Tunis, Tunisia

Insaf Bahrinil Department of Biology, College of Science, University of Ha'il, Ha'il City, Saudi Arabia

Hmed Ben-Nasr Laboratory of Pharmacology. Medicine faculty of Sfax, Sfax, Tunisia

Department of Life Sciences, Faculty of Sciences of Gafsa, Gafsa, Tunisia

Akanksha Bhardwaj ICAR-Indian Agricultural Research Institute (ICAR-IARI), Delhi, India

Bibhas Bhunia Department of Biotechnology, School of Engineering and Technology, Shobhit Institute of Engineering & Technology (Deemed-to-be-University), Meerut, India

Sakshi Chaudhary Department of Biotechnology, School of Engineering and Technology, Shobhit Institute of Engineering & Technology (Deemed-to-be-University), Meerut, India

Mohammed Faruq Genomics and Molecular Medicine Division, CSIR-Institute of Genomics and Integrated Biology, New Delhi, India

Academy of Scientific and Innovative Research (AcSIR), Ghaziabad, India

Shruti Gairola National Institute of Biologicals (NIB), Noida, Uttar Pradesh, India

Anindita Ghosh Genomics and Molecular Medicine Division, CSIR-Institute of Genomics and Integrated Biology, New Delhi, India

Divya Goel Genomics and Molecular Medicine Division, CSIR-Institute of Genomics and Integrated Biology, New Delhi, India

Department of Pharmacology, SPER, Jamia Hamdard, New Delhi, India

Saweza Hashmi Department of Environmental Science, Faculty of Engineering, Jamia Millia Islamia, Delhi, India

Sadaf Jahan Department of Medical Laboratory Sciences, College of Applied Medical Sciences, Majmaah University, Al-Majmaah, Saudi Arabia

Andleeb Khan Department of Pharmacology and Toxicology, College of Pharmacy, Jazan University, Jazan, Saudi Arabia

Johra Khan Department of Medical Laboratory Sciences, College of Applied Medical Sciences, Majmaah University, Al-Majmaah, Saudi Arabia

Dipak Kumar Department of Zoology, KKM College, Munger University, Munger, India

Ayush Madan Department of Biotechnology, School of Engineering and Technology, Shobhit Institute of Engineering & Technology (Deemed-to-be-University), Meerut, India

Sidharth Mehan Neuropharmacology Division, Department of Pharmacology, ISF College of Pharmacy, Moga, Punjab, India

Rabab Syeda Mirza Department of Biotechnology, School of Chemical and Life Sciences, Jamia Hamdard University, New Delhi, India

Shouvik Mukherjee Department of Biotechnology, School of Chemical and Life Sciences, Jamia Hamdard University, New Delhi, India

Ghazala Muteeb Department of Nursing, College of Applied Medical Science, King Faisal University, Al-Ahsa, Saudi Arabia

Uzma Nayeem Department of Molecular Medicine, School of Interdisciplinary Science and Technology, Jamia Hamdard, New Delhi, India

Chetan Singh Rajpurohit Huffington Center on Aging, Baylor College of Medicine, One Baylor Plaza (Lab), Houston, USA

Gulzar Rather Department of Biomedical Engineering, Sathyabama Institute of Science & Technology deemed to be University, Chennai, Tamil Nadu, India

Mashoque Ahmed Rather Department of Molecular Pharmacology & Physiology, Morsani College of Medicine, University of South Florida, Tampa, USA

Nimisha Rawat Department of Biotechnology, School of Chemical and Life Sciences, Jamia Hamdard University, Delhi, India

Kalpeeta Roy CliniRx Research Pvt. Ltd, New Delhi, India

Mongi Saoudi Laboratory of Animal Ecophysiology, Faculty of Sciences of Sfax, University of Sfax, Sfax, Tunisia

Arif Jamal Siddiqui Department of Biology, College of Science, University of Hail, Hail City, Saudi Arabia

Tanisha Singh School of Medicine, University of Pittsburgh, Pittsburgh, USA

I. Irem Tatli Department of Pharmaceutical Botany, Faculty of Pharmacy, Hacettepe University, Ankara, Turkey

Deepanshi Thakur Department of Biotechnology, School of Chemical and Life Sciences, Jamia Hamdard University, Delhi, India

Qamar Zia Department of Medical Laboratory Sciences, College of Applied Medical Sciences, Majmaah University, Al- Majmaah, Saudi Arabia

Health and Basic Sciences Research Center, Majmaah University, Al-Majmaah, Saudi Arabia

History, Origin and Types of Neurological Disorders

1

Shouvik Mukherjee, Shaheen Ali, Saweza Hashmi, and Sadaf Jahan

Abstract

Neurological disorders generally affect the brain and spinal cord as well as the nerves found throughout the body. These disorders might result in symptoms such as muscle weakness, paralysis, seizures, loss of sensation, confusion, pain or even altered levels of consciousness. The existance of such diseases dates back to the history of mankind. However, the study of such diseases gained momentum a few decades back only. There may be several causes of neurological disorders such as genetic defects, congenital abnormalities, malnutrition, spinal cord injury, brain damage, nerve injury, disorders of the central or peripheral nervous system such as epilepsy, Alzheimer's disease, dementias, stroke, migraine, Parkinson's disease, multiple sclerosis, traumatic disorders, brain tumours, neuronal infections, etc. According to an estimate, around six million people die due to stroke every year, out of which 80% cases are observed in low-income countries.

S. Mukherjee
Department of Biotechnology, School of Chemical and Life Sciences, Jamia Hamdard, New Delhi, India

S. Ali
Department of Molecular Medicine, SIST, Jamia Hamdard, New Delhi, India

Medicinal Chemistry and Molecular Modeling Lab, Department of Pharmaceutical Chemistry, SPER, Jamia Hamdard, New Delhi, India

S. Hashmi
Department of Environmental Science, Faculty of Engineering, Jamia Milia Islamia, New Delhi, India

S. Jahan (✉)
Department of Medical Laboratory Sciences, College of Applied Medical Sciences, Majmaah University, Al-Majmaah, Saudi Arabia
e-mail: s.jahan@mu.edu.sa

S. Jahan, A. J. Siddiqui (eds.), *Applications of Stem Cells and Derived Exosomes in Neurodegenerative Disorders*, https://doi.org/10.1007/978-981-99-3848-3_1

More than 50 million people have epilepsy, 47.5 million people suffer from dementia and more than 10% have migraine worldwide. This chapter discusses the history, origin and types of neurological disorders, their prevalence over the ages and remedies available for their management.

Keywords

Neurology · Neurological disorders · Nervous system · Diagnosis · Treatment

1.1 Introduction

Neurological disorders have affected a large population throughout the world. According to the World Health Organization (WHO), these are the diseases of the peripheral and central nervous system. Alternatively, any disorder in the spinal cord, brain, cranial nerves, peripheral nerves, autonomic nervous system, muscles and neuromuscular junction are neurological disorders including Alzheimer's disease (AD), epilepsy, cerebrovascular disease like stroke, headaches, migraine, multiple sclerosis, Parkinson's disease, brain tumours, traumatic disorders, etc. It is reported by the WHO that around six million people die due to stroke every year. Around 50 million of people have epilepsy. According to an estimate, there are 47.5 million people suffering from dementia worldwide and 7.7 million new cases are reported each year. AD is the most frequent cause of dementia and may be blamed for 60 to 70% of the cases. Indeed, the prevalence of migraine is above 10% worldwide (Mental Health: Neurological Disorders 2016) (Fig. 1.1).

The causes of the neurological disorders may vary on genetic and environment factors. Congenital abnormalities, lifestyle and environmental factors may cause such disorders. According to the US National Library of Medicine, there are more than 600 neurological diseases. Whatever type of neurologic disease may be, all are caused by some or other kinds of abnormalities in the nervous system. Depending on the type or site of onset and severity, it determines the level of damage to communication, hearing, vision, movement or cognition (Neurological Disorders 2022).

1.2 History of Neurology and Neurological Disorders

The term 'neurology' is derived from a combination of two words: 'neuron' or the string, which means nerve, and 'logia,' which means study of. Neurology, a branch of medicine, deals with the structures, diseases and functions of the nervous system (ACGME 2013. In 1700 BC, Edwin Smith Papyrus, an Egyptian, wrote a treatise concerning trauma surgery, and it contains various treatments including those of neurological disorders. It has description for meninges, external surface of the brain and cerebrospinal fluid as well as intracranial pulsations. The work also mentions that some bodily functions can be impaired by brain injuries or by injuries to the cervical spine. Sumerians in 4000 BC, in a sculpture, observed about paraplegia of a lion with an arrow on the back. During the Vedic period, in a literature of medicine in

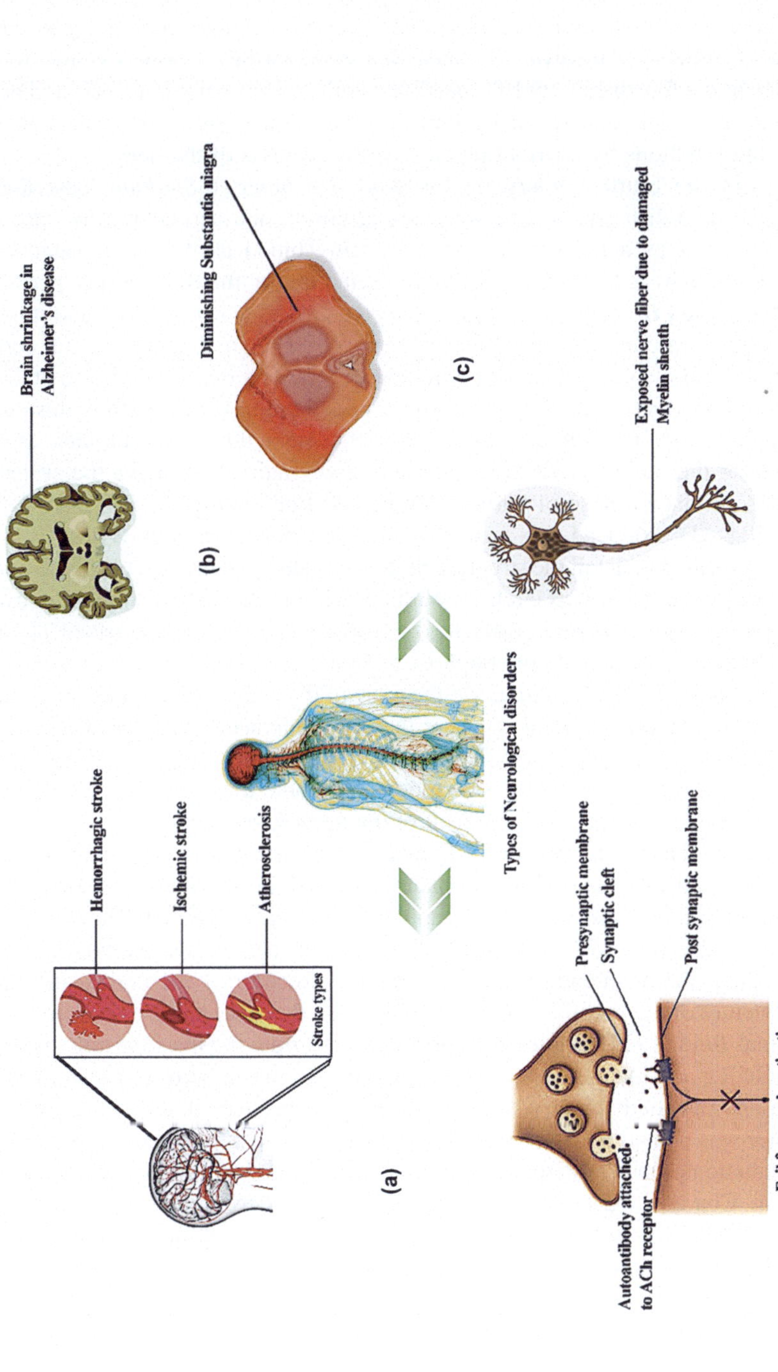

Fig. 1.1 Different types of neurological disorders: **(a)** stroke, **(b)** Alzheimer's disease, **(c)** Parkinson's disease, **(d)** multiple sclerosis and **(e)** myasthenia gravis

ancient India, *Charaka Samhita* discussed various symptoms and possible treatments of epilepsy. However, it stated that epilepsy has a sacred cause until Hippocrates in the fourth century BC stated that epilepsy has a natural cause and not a sacred cause. It was also stated by him that the brain is the seat of all mental processes and functions by distributing air to different parts of the body.

Aristotle in the fourth century, in his work *De motu Animalium*, described meninges and also distinguished between the cerebrum and cerebellum. He stated that the brain acts as a radiator for the circulating blood as it is overheated by seething heart. Andreas Vesalius, a Greek physician during the Renaissance period, wrote a book called *De Humani Corporis Fabrica Libri Septem* in 1543, and it has detailed images depicting the ventricles, pituitary gland, cranial nerves, spinal cord and meninges and also an image of peripheral nerves. His observation also stated that the brain consists of a pair of seven nerves called brain nerves, each with some specialized function. He did not support the statement that the ventricles were responsible for the brain functions arguing that many animals do have the system of ventricles similar to those of humans but has no true intelligence. Leonardo da Vinci in 1504 AD produced wax cast of human ventricles. Thomas Willis, in the year 1664, described in detail the structure of the brainstem, ventricles, cerebrum and cerebral hemispheres as well as circle of Willis. His book *On Epilepsy and the use of Guaiac wood in that affection: Likewise of tinnitus of the Ears* described about epilepsy. He also described about apoplexy and paralysis. René Descartes in 1648 proposed the theory of Dualism to tackle the issue of the brain's relation to the mind. He asserted that the pineal gland is where the mind interacted with the body after recording brain mechanism responsible for circulating cerebrospinal fluid. He also stated that every activity recorded is a response to some external stimuli, and the responses too carried under a specific path of the nervous system.

Then, Steven Hales and Robert Whytt demonstrated spinal segmental reflex action, and around that time, Felice Fontana and J Aldini demonstrated electrical stimulation in animals as well as decapitated criminals. Samuel Thomas von Sömmering in the year 1778 described all 12 cranial nerves. Francesco Gennari and Felix Vicq d'Azyr demonstrated the laminar structure of the cortex and the internal structure of the brain. Domenico Cotugno, on the other hand, discovered the cerebrospinal fluid. J E Purkinje described about neurons for the first time, while Matthew Baillie and Jean Cruveilhier illustrated lesions in stroke. Later in the nineteenth and twentieth century, several things were discovered, terms like blood-brain barrier was coined by M Lewandowsky in 1900, and 5 years later, the phrase parasympathetic nervous system was introduced. Nevertheless, the contributions of Jean-Martin Charcot are remarkable, leading to consider him as the father of neurology. Neurology in the recent years has developed a lot, and treatment and diagnosis for several neurological disorders have remarkably improved, helping the world population to find a better solution for such disorders (Neurological Disorders 2022; Gardner-Thorpe 2000; Kumar et al. 2011; Tyler et al. 2003; Mishra et al. 2013; Brown 2019).

1.3 Biology of Organs Related to Neurological Disorders

The organs related to the neurological disorders are the organs of the central and peripheral nervous system. The central nervous system consists of the brain and spinal cord, while the peripheral nervous system consists of nerves. A neuron consists of the axon, dendrite, cell body and nucleus. The axon is protected by a myelin sheath (oligodendrocytes). A long, tail-like structure is an axon. The axon hillock, a specialized junction, is where it connects to the cell body. Fibrous roots known as dendrites extend out from the cell body. Dendrites, like antennae, pick up and interpret messages sent by the axons of other neurons. Dendritic trees, or many sets of dendrites on a single neuron, are possible. Types of neurons include unipolar (single axon at the same side), pyramidal (several dendrites forming a pyramidal shape), multipolar (single axon and symmetrical dendrite), bipolar (one axon and dendrite) and Purkinje (multiple dendrites).

1.3.1 Brain

The brain functions in a unique way. It has billions of nerve cells. All kinds of thoughts, moods, beliefs and behaviours arise from the brain. The brain is the location of cognition and intellect as well as the body's control centre (Maiese 2023a, b, c). The brain is made up of 60% fat, while the rest (40%) is a mixture of protein, carbohydrate, salts and water. Both the brain and the spinal cord consist of the two things called the grey matter and white matter. The grey matter consists of neuron somas and is present on the outer part of the brain, while the white matter consists of axons wrapped in myelin which is present inside. The grey matter is responsible for processing and interpreting data, while the white matter oversees transmitting data to other regions of the nervous system (Anatomy of the Brain 2023).

The brain can be divided into three main sections: the cerebrum, cerebellum and brainstem. The cerebrum comprises the largest part of the brain and is divided into left and right hemispheres. It is responsible for speech, reasoning, emotions, learning and interpretation of touch, vision and hearing. The cerebellum, which is located under the cerebrum, is responsible for muscle movements and maintains posture and balance. The brainstem works as a relay centre that connects the cerebrum and the cerebellum directly to the spinal cord. It performs many functions such as autonomic activities, e.g. breathing, temperature, heartbeat, wake and sleep cycles, sneezing, digestion, coughing, vomiting and swallowing. Different parts of the brain have different functions which we shall study in the next subsection. Lobes of the brain fissures pass through the hemisphere of the brain which divide the brain into two hemispheres. Each of these hemispheres has four lobes having different functions (Mayfield Brain and Spine 2018). The functions of these lobes are summarized below (Mayfield Brain and Spine 2018) (Fig. 1.2).

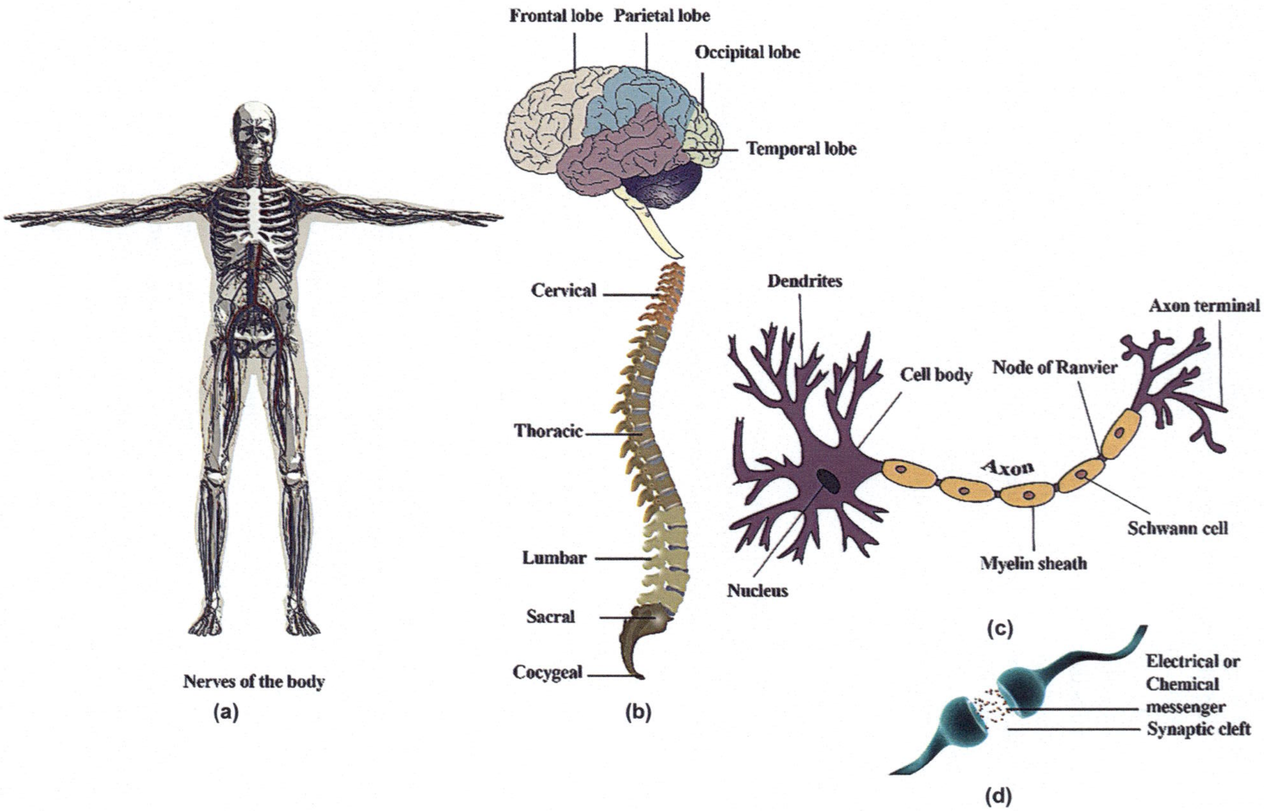

Fig. 1.2 The nervous system and its components: (**a**) nerves, (**b**) brain and spinal cord (**c**). Neuron present in the brain and spinal cord (**d**) Synaptic cleft between neurons responsible for signal transfer

Frontal	Parietal	Occipital	Temporal
Personality Behaviour Emotions Judgement Problem solving Broca's area for speaking and writing Motor strip area that helps for body movement Intelligence Concentration Self-awareness	Interpretation of word and language Sensory strip that helps to sense temperature, pain, touch Interprets signals from vision, hearing, motor, memory and sensory Spatial perception Visual perception	Interpretation of vision based on light, colour, movement	Wernicke's area present in the lobe helps to understand language To sequence as well as organise Memory and hearing

1.3.2 Spinal Cord

The spinal cord in humans begins at the bottom of the brainstem and ends in the lower back by forming a cone-like structure called conus medullaris. The human spinal cord is around 18 inches in length consisting of 33 bones called vertebrae. The coccyx is made up of four tiny vertebrae that are fused together. The rest of the spinal cord has cervical vertebrae (C1 to C7- located in the neck), thoracic vertebrae (T1 to T12—attached to the ribcage), lumbar vertebrae (L1 to L5—located in the lower back) and sacral vertebrae (S1 to S5—located in the pelvis). The discs in between the vertebral bodies, except for the cervical vertebrae 1 and 2, act as the spine's support system. The nucleus pulposus, a softer substance, is enclosed by a stiff outer layer (annulus fibrosus) in these oval-shaped discs. The spinal bones are cushioned by these discs. Ligaments that are joined to the vertebrae also act as structural supports (A Neurosurgeon's Overview of the Anatomy of the Spine and Peripheral Nervous System 2023). Thirty-one pairs of spinal neurons from the spinal cord stretch out between the vertebrae. Two thin branches emerge from each nerve (roots). One side presents towards the front called motor or anterior roots which are responsible for carrying different commands from the spinal cord as well as the brain to the skeletal muscles to control movement, while the other side is the posterior roots present at the back, which are sensory roots carrying information such as temperature, position, vibration, pain, etc. from other bodily parts to the brain (Maiese 2023b). The spinal cord is responsible for reflex action that involves quick response to a particular threat or substance. This reflex action may be vestibulo-ocular reflex, knee jerk reflex or the reflex action of the hands.

1.3.3 Nerve

It constitutes the peripheral nervous system also called nerves, which consists of more than hundred billion of nerves or neurons. The neurons are like threads that traverse the body. They connect the brain and the other parts of the human body. Bundles of nerve fibres make up peripheral nerves. These fibres are covered in several layers of tissue made of myelin, a fatty material. The myelin sheath, which is formed by these layers, quickens the movement of electrical impulses along a nerve fibre. The nerve consists of two main parts: the somatic nervous system and the autonomic nervous system. The nerves which link/interconnect the brain and spinal cord as well as the voluntary muscles and skin sensory receptors make up the somatic nervous system. The brainstem and spinal cord are linked to the internal organs through the autonomic nervous system, which controls the body's automatic internal operations. The autonomic nervous system comprises two divisions, i.e. sympathetic (prepares the body for difficult or urgent circumstances) and parasympathetic (maintains normal body functions in times of normal situations) (mentioned below) (Maiese 2023b). Different functions of the autonomic nervous system include maintaining blood pressure and heart rate as well as breathing rates; digestion; temperature of the body; body metabolism; body fluid production such as saliva, tears and sweat; urination; defecation; sexual response; and the balance of electrolytes and water in the body (Low 2023).

Sympathetic	Parasympathetic
• Dilates pupil • Inhibits salivation • Relaxes airways • Constricts blood vessels • Accelerates heart beat • Stimulates sweat production • Celiac ganglion • Superior mesenteric ganglion • Inferior mesenteric ganglion • Stimulates secretion of epinephrine and norepinephrine • Inhibits digestion • Inhibits secretion • Inhibits voiding • Stimulates orgasm and ejaculation	• Constricts pupils • Stimulates tear production and salivation • Constricts airways • Slows heartbeat • Stimulates bile release • Stimulates digestion • Stimulates secretion • Promotes voiding (of urinary bladder) • Stimulates erection

1.4 Origin and Causes of Neurological Disorders

1.4.1 Risk Factors for Neurological Disorders

There are various risk factors that are related to neurological disorders. One of the most important risk factors is sleep disorder. Sleep disorder may constitute stroke, where the risk factor of sleep deprivation may or may not be directly involved. Indirect mechanism for sleep disorder affects the glucose metabolism in the body, and it has been linked to hypertension, type 2 diabetes mellitus and even obesity. All of which has been linked to increase the likeliness of stroke. The risk of hypertension, cardiovascular disease and stroke may also rise if periodic limb movement (PLM) and restless leg syndrome (RLS) coexist with other sleep-related movement disorders. Sleep disorder is also related to diseases like AD. Sleep deprivation is especially dangerous for some neurocognitive domains, such as executive attention and other higher cognitive processes. A study suggests that sleep deprivation is also related to loss of memory. In fact, imaging and pathological studies in the part of the brain linked to sleep have demonstrated some abnormalities, revealing the impact of sleep deprivation on AD. Patients with Parkinson's disease (PD) have also revealed such abnormalities related to the daytime sleepiness. Patients with multiple sclerosis and epilepsy revealed history of disrupted sleep patterns and disorders. Other factor involves smoking which phosphorylates the tau protein, which in turn creates neurofibrillary tangles resulting in synaptic dysfunction/degeneration leading to amyloid plaque. A study also reveals that smoking increases the risk of ischaemic stroke and neurotoxicity. Exposure to neurotoxic metals such as aluminium also affects neurological functions, leading to neurodegenerative disorders like AD. This is because higher immunoglobulin Al (IgA1) secretion increases amyloid aggregation. Al also promotes inflammatory signalization, one of the key characteristics of AD in the brain, through the pro-inflammatory transcription factor nuclear factor kappa B (NF-kB). Al which has the ability to activate monoamine oxidase B increases patients with PD. Even with multiple sclerosis, Al concentrations were higher in the urine samples of the patients.

1.4.2 Symptoms of Various Neurological Disorders

There are various neurological disorders and every such disorder may show unique signs and symptoms, while some of the most common symptoms may include memory loss, muscle cramps, weakness and numbness (Symptoms of Neurologic Disorders–Merck Manuals Professional Edition 2023). Different diseases such as multiple sclerosis exhibit different symptoms; these may include fatigue and loss of balance. In cerebral palsy, there is severe harm to the motor nerve system, which may even lead to paralysis (Da Cunha De Sá-Caputo et al. 2021). Some of the manifestations of PD, which is the most commonly known neurological disorder, are tremor and bradykinesia; however, it is also a rarely known fact that patients with Parkinson's disease experience the disease very differently and also the symptoms fluctuate from hour to hour and day to day. Indeed, some may not feel the tremor as onset of early symptoms, while some may not experience this on years to come (https://www.michaeljfox.org/news/seven-misconceptions-about-living-parkinsons-disease). Some of the most important neurological disorders we shall study are presented in the table below.

Disease name	Prevalence	Symptoms	Reference
Alzheimer's disease	Usually occurs above the age of 65 years	• Disorientation • Mood swings • Behavioural changes • Confusion about time, events/places • Memory loss • Difficulty in swallowing, speaking and walking • Suspicion about friends, family and professional caregivers	(Shea et al. 2021) (Fowler et al. 2021) (Mayeux and Stern 2012)
Multiple sclerosis	Usually affects the age group of 20–40 years old	• Loss of vision • Fatigue • Loss of balance • Loss of bladder control • Demyelinating lesions in the brain and the spinal cord • Progressive disability leading to negatively impact on cognition as well as on daily social and physical activities	(Momtazmanesh et al. 2021) (Naseri et al. 2022) (Marin et al. 2021)
Parkinson's disease	Male and female above the age of 80 have higher chances of suffering from the disease	• Resting tremor • Cogwheel rigidity • Bradykinesia • Postural instability • Autonomic dysfunction	(Hirsch et al. 2016) (Postuma and Berg 2016) (Muangpaisan et al. 2011)
Cerebral palsy	Occurs before, during or shortly after birth	• Diminution in the gestures' motor repertoire a drop in the frequency of regular motor patterns and a loss of movement quality • Both moving and being motionless positions are challenging for the young child	(McNamara et al. 2021) (McMorris et al. 2021)
Autism spectrum disorder	Age group of 8–11 has maximum occurrence	• Aggression • Self-mutilation • Crying • Lack of eye contact • Shouting • Hyperactivity • Involuntary imitation of	(Hodges et al. 2020) (Association 2013)

(continued)

Disease name	Prevalence	Symptoms	Reference
		other's movement • Impulsivity • Improper social interaction • Irritability • Repetition of meaningless words • Delay of speech in a child • Lack of cognition, attention • Depression • Intense interest in a few things • Anxiety • Lack of empathy • Lack of sensitivity to sound	
Amyotrophic lateral sclerosis	Age 55–75 is the most affected	• Muscle weakness usually affecting the arm, leg, neck and diaphragm, nasal and slurred speech, difficulty in chewing and swallowing • Muscle cramps • Stiff and tight muscles • Muscle twitches	(Amyotrophic Lateral Sclerosis (ALS) 2023)
Myasthenia gravis	Affects women more than men. Affects women under 40 and men above 60 years of age	• Compromise in working of muscles responsible for eyelid movement, facial expression, talking, swallowing and chewing • Ocular myasthenia. • Ptosis • Diplopia • Dysarthria • Weakness of hands, arms, fingers, neck, leg • Respiratory failure (severe cases)	(Nalbantoglu et al. 2021) (Hou et al. 2021) (Spillane et al. 2014) (Li et al. 2013) (Myasthenia Gravis 2023)
Strokes	Affects at any age More prevalent below 65 years of age	• Dizziness • Loss of balance • Lack of coordination • Severe headache • Sudden trouble in seeing • Sudden trouble in walking • Sudden trouble in speaking • Weakness or numbness in one part/side of the body or the face, leg, arm	(Mozaffarian et al. 2016) (Esmael et al. 2021) (stroke facts | cdc.gov 2022)

1.5 Different Neurological Disorders

1.5.1 Autonomic Nervous System Disorders

The autonomic nervous system (ANS) controls involuntary physiological activities including heart rate, blood pressure, digestion and breathing as well as sexual arousal, which is a part of the peripheral nervous system. It has three physically separate divisions: parasympathetic, sympathetic and enteric. Furthermore, it also plays a part in maintaining homeostasis (McCorry 2007). The rapid, immediate effects of the ANS on the body's most diverse physiological functions, such as gland secretion and the activation of visceral smooth muscle in arterial beds. On the other hand, tubular organs, balance the slow-acting, long-lasting effects of the endocrine system (Waxenbaum et al. 2021). In both healthy and diseased individuals, the ANS dynamically regulates the heart. Autonomic nerve problems may be the primary cause of ANS dysfunction or other systemic diseases may be the secondary cause. Changes in the cardiac autonomic function that are both anatomical and functional may be facilitated by heart problems. These modifications may, in turn, aid in the development of arrhythmias or contribute to the disease's progression (Vaseghi et al. 2017). The most well-known autonomic intervention connected to better results is a β-adrenergic inhibitor. Other methods, such as the cardiac sympathetic denervation, have shown promising results in treating persistent ventricular arrhythmias (Vaseghi et al. 2017). The two causes of autonomic dysfunction are intrinsic and extrinsic. Intrinsic autonomic dysfunction is impacted by diseases like diabetes mellitus and other primary autonomic failures that directly affect the autonomic nerves. The onset of both peripheral and autonomic neuropathies is linked to diabetes mellitus. Due to its significant occurrence, diabetes is the main reason of primary autonomic dysfunction. In the pathological research, myelinated vagus nerve axons have been found to be lost or damaged (Guo et al. 1987; Kristensson et al. 1971).

The presence of large nerve cells, vacuolization and neuroaxonal degeneration in the sympathetic nerves and ganglia has been questioned (Appenzeller and Richardson 1966; Schmidt 2002). In the study by Appenzeller and Richardson, anomalies were not found in the four patients without neuropathy and in four out of five patients having diabetes as well as neuropathy. In this research, people having diabetes mellitus and those who had a disease as well as peripheral neuropathy were found to have significant pathological alterations in their autonomic nerves (Appenzeller and Richardson 1966). It is probable that heterogeneous results are due to interplay or interaction amongst several pathways in the pathogenesis of the diabetic cardiac autonomic neuropathy which is done in vivo (Schmidt 2002; Schönauer et al. 2008; Vinik et al. 2003). The possible aberrant parasympathetic innervation of the heart is frequently the first indication of the cardiac autonomic neuropathy since the vagus nerve is the longest autonomic nerve in the body (Pop-Busui 2010).

Various conditions, such as idiopathic orthostatic hypotension, pure autonomic failure, multiple system atrophy and PD with autonomic failure, are amongst the other possible primary disorders of the autonomic dysfunction. Orthostatic hypotension and other symptoms combine to make up these diseases (Schatz et al. 1996). The complicated syndrome known as postural orthostatic tachycardia syndrome has a significant autonomic component. Some of these disorders may have autoimmune origin which possibly targets the ganglionic receptors (specially the nicotinic acetylcholine receptors) (Vernino et al. 2000).

1.6 Brain Dysfunction and Infection

1.6.1 Brain Dysfunction

Numerous dysfunctions can result from brain damage. Such dysfunctions can affect one or more of the several distinct processes that contribute to conscious experience, ranging from impairment of one or more of these processes to total absence/loss of consciousness (like the one occurring in coma) to the disorientation as well as failure to pay attention (state of delirium) (Tindall 1990). Coma is characterized as a state of profound unconsciousness and unresponsiveness with the eyes closed. While it is possible for a coma to last a lifetime or even for a long time, it is often a temporary condition. The brain's alerting and arousal mechanisms, awareness and conscious content are all impacted (Huff et al. 2012). Coma indicates brain dysfunction and may be caused by a metabolic process across the body or by a central nervous system process. The causes of coma can range from correctable metabolic irregularities to serious life-threatening mass lesions (Kadapatti and Iyer 2022).

Coma is a consequence of bilateral supratentorial or diencephalic brainstem dysfunction. This classification, however, usually is not in practice. To take an example, unilateral hemisphere lesions have a mass impact that might result in diencephalic-mesencephalic dysfunction and contralateral hemispheric dysfunction. The classification of underlying causes is according to pragmatic use. Major categories under this are global anoxia-ischaemia, cerebrovascular disease (haemorrhagic/ischaemic), etc. (Sakusic and Rabinstein 2021). The recognized pathophysiology of a coma is neuronal dysfunction carried on by a decreased amount of glucose or oxygen reaching the brain. The most serious clinical conditions that might arise from a range of aetiologies that alter the underlying substrate are diffuse CNS dysfunction and coma. For instance, any clinical intervention that results in substantial hypoxia or circulatory collapse could be the cause of a coma (Kadapatti and Iyer 2022).

In clinical practice, hypoglycaemia is a common occurrence, most usually in connection with the treatment of diabetes mellitus or as a side effect of alcoholism. Hyponatraemia and hypercalcaemia are two electrolyte disorders that might interfere with normal brain metabolism. Although the biology of various metabolic coma causes is unclear, it may involve erroneous neurotransmitters as in hepatic encephalopathy (Kadapatti and Iyer 2022). Coma may result from structural CNS lesions, such as intracerebral haemorrhage, that directly harm the brain's arousal centres or from secondary damage caused by the shifting of intracranial structures, vascular compression or elevated intracranial pressure. Clinically observable physical screening characteristics known as herniation syndromes may be used to identify the anatomic site of a CNS lesion (Maramattom and Wijdicks 2005).

Even though many physicians rush to order tests when they encounter a patient in a coma, the most important information regarding the cause of the coma is frequently discovered through a thorough history and an expert medical examination (Edlow et al. 2014). The Glasgow Coma Scale, the FOUR score and other rating systems can be used to roughly quantify the recording of the neurologic examination. While effective for sequential patient monitoring, short rating scales do have some drawbacks. Therefore, these systems record the eye response, motor response and verbal response which ranges from the different scores (Stead et al. 2009; Sternbach 2000). Treatment for all patients with altered mental status, point-of-care testing or empiric glucose administration is advised in order to determine the serum glucose level. Patients who exhibit symptoms of the narcotic toxidrome, such as reduced breathing rate, narrow pupils or changed mental status, should be given naloxone (Jordan and Morrisonponce 2022). Further testing is advised if a quickly reversible coma is not found. The main goal of the management strategy is to maintain the mean arterial pressure while maintaining the cerebral perfusion pressure. It is vital to continue providing supportive care while maintaining blood pressure and protecting the airways (Stevens et al. 2012).

Another disease of brain dysfunction is delirium which is an extreme neuropsychiatric syndrome described by acute occurrence of attentional as well as cognitive impairments. Frequently altered arousal is also experienced by a patient, ranging from hypervigilance and acute agitation to decreased responsiveness (at a near coma level). Additionally, one may have extremely distressing psychotic symptoms such as hallucinations as well as delusions and even mood swings (Wilson et al. 2020). Although delirium can have a single origin, it usually has several causes due to the interaction of predisposing and precipitating events. Lower magnitudes of precipitating factors are essential reasons for delirium the more heavily the predisposing elements are burdened (Inouye et al. 1990). There are four predisposing factors for delirium and five precipitating causes. These latter ones consist of malnutrition or poor nutritional status, the use of restrictions of any kind, the recent prescriptions of three or even more drugs and urinary bladder catheterization as well as other iatrogenic causes. Dementia or cognitive impairment, sensory deprivation and dehydration as well as the intensity of an acute illness are examples of the former (Inouye and Charpentier 1996).

The epidemiology of delirium has been studied in a variety of treatment settings throughout the world. The incidence and the prevalence of delirium are affected by the population studied (elders, children, adults and mixed age groups), the estimation method (screening instrument, diagnostic instrument) and the treatment setting (ICU or intensive care units, medical/surgical wards and also postoperative patients as well as consultation-liaison psychiatry services) (Grover and Avasthi 2018). Delirium is a psychiatric situation which necessitates precise detection and identification of the underlying cause as well as treatment of the symptoms. Delirium is a mental emergency that has to be precisely detected and its underlying cause found as well as its symptoms treated. This first and foremost requires a precise diagnosis. Since suitable rating scales must be employed to assess the severity of the symptoms and subtypes of delirium, the diagnosis should be made using any accepted nosological system (Grover and Kate 2012). Identifying potential causes, addressing or eliminating the etiological variables and managing delirium symptoms by both pharmaceutical and non-pharmacological treatment are all necessary once a patient has been diagnosed with delirium (Grover and Avasthi 2018).

1.6.2 Brain Infection

Any abnormalities, disorders or disease conditions in the brain have an impact on the entire body. The brain is prone to infections of the neurons or tissues as well as neurological diseases. Meningitis probably is the commonly found infectious disease that affects the CNS which results in inflammation of the brain's meningeal membranes. The dura, arachnoid and pia mater are the three membranes, also called meninges, which cover the vertebral canal of the skull, which encloses the brain as well as the spinal cord. On the other hand, encephalitis is an infection of the brain (Flægstad 2020). The disease causation is due to many pathogens such as fungi, bacteria and viruses; however, bacterial meningitis (common bacteria like *Haemophilus influenzae*, *Streptococcus pneumoniae* and *Neisseria meningitidis*) is seen highly worldwide. In the year 2015, there were around 8.7 million meningitis cases observed to be reported worldwide, having 379,000 fatalities as a consequence, despite advances in detection, treatment and vaccine (Flægstad 2020). Prior to the development of antibiotics, the condition was always fatal.

However, despite enormous advancements in healthcare, the disease still has a death rate that is very close to 25% (Hersi et al. 2022). Meningitis risk factors include chronic medical problems (such as renal or kidney failure, adrenal insufficiency and diabetes mellitus as well as cystic fibrosis), old age and lack of immunization, as well as immuno-suppressed states (such as transplant recipients, iatrogenic, AIDS, congenital deficiencies and so on). Meningitis can manifest clinically in a number of ways, depending on the host's age as well as immunological status. Symptoms frequently include fever and stiff/painful neck, as well as photophobia. Vomiting, headache, disorientation, dizziness, delirium and nausea as well as agitation are amongst the more general symptoms. Seizures, altered mental status and neurologic impairments are warning signs of elevated intracranial pressure and indicate a poor prognosis. In the neonates as well as infants, symptoms as well as signs are way less noticeable. Indeed, they may have a bulging fontanelle, reduced oral intake, altered state of mind or mental status, hypothermia or even fever and/or any combination of these. It is important to obtain information on vaccinations and the whole parental history. Meningitis can be prevented by vaccination against some of the most common causes, including measles, pneumococcus and varicella as well as *Haemophilus influenzae* type B (Hersi et al. 2022).

In bacterial meningitis, antibiotics and supportive care play an important role. Additionally, managing the airways, maintaining oxygenation and providing fever control while giving adequate intravenous fluids are factors of the foundation of meningitis management. The probable organism responsible for the infection determines the type of antibiotic to be used. The patient's demographics and prior medical history must be taken into consideration by the clinician in order to offer the most effective antibacterial coverage (Hersi et al. 2022).

1.7 Cranial Nerve Disorders and Peripheral Nerve Disorders

Our whole body is powered by nerves, yet those nerves may be harmed by trauma or diseases like diabetes. Our capacity to feel and move is impacted by the illness known as neuropathy, which damages the nerves. Depending on where in the body the injured nerves reside, different things might happen to our body and how we move. Cranial neuropathy is the term used to describe the condition of nerves in the brain or brainstem. The cranial nerves, which originate in your brain or brainstem, are those nerves that frequently have an impact on the face and eyes. Microvascular cranial nerve palsy, Bell's palsy and third, fourth or sixth nerve palsies are a few of the several cranial neuropathies. When the face nerve (seventh cranial nerve) is damaged, Bell's palsy develops.

People with diabetes and those who have high blood pressure are more likely to experience it. The third cranial nerve is impacted by this disorder. This nerve aids in controlling the eye movement muscle. Fourth nerve paralysis is sometimes termed superior oblique palsy. It has an impact on the superior oblique muscle, which aids in eye convergence (to look at the tip of your nose). Sixth nerve paralysis is also known as abducens palsy or cranial nerve VI palsy. The sixth cranial nerve, which also aids in regulating eye movement, is impacted. A virus that results in swelling frequently causes Bell's palsy. The facial nerve is under strain as a result. People with excessive blood pressure may develop microvascular cranial nerve palsy. Third nerve palsy can occasionally be present at birth. But an illness or a brain injury is also a potential cause. Third nerve palsy can also be caused by a condition that affects the brain, such as a brain tumour or aneurysm. Other factors include migraines and diabetes. A newborn is often born with fourth nerve palsy as a congenital birth condition. Fourth nerve palsy, however, can also result from a brain injury, stroke or malignancy. Infections, tumours, strokes, elevated brain pressure and even headaches can harm the sixth cranial nerve. (Multiple Cranial Neuropathies 2020).

Now when we talk about peripheral nerve disorders, weakness or numbness and pain are usually the most common symptoms belonging to peripheral neuropathy that develops when the peripheral nerves, which are situated outside the brain or spinal cord, are destroyed or damaged. It can affect several physiological functions like digestion, urination and circulation. The remainder of your body receives information from our brain as well as spinal cord (central nervous system) through our PNS. Additionally, the CNS receives sensory data from the peripheral nerves. Infections, traumatic injuries, metabolic issues and genetic causes, as well as exposure to various toxins, might be the potential reasons for peripheral neuropathy. Diabetes is one of the most typical causes. Patients with peripheral neuropathy typically describe their pain as stabbing, tingling or even burning. Symptoms frequently improve, especially if they were caused by a treatable disease. Medications can be used to treat the pain of a peripheral neuropathy. The symptoms vary depending on the type of nerves injured since each nerve in your peripheral nervous system has a distinct purpose. According to their classification, sensory nerves that receive sensations from the skin, such as heat, pain, vibration or touch, muscular movement is controlled by motor neurons.

Autonomic nerves regulate bodily processes including sweating, heat rate, blood pressure and urination as well as digestion. Peripheral neuropathy symptoms and signs may include progressive start of tingling, numbness or even prickling in the hands/feet that may eventually move up legs and arms; throbbing, stabbing, sharp or scorching pain and a high threshold for touch. Signs and symptoms of autonomic nerve dysfunction include inability to control intestinal or excessive perspiration, heat sensitivity, bladder or bowel issues and drops in blood pressure that make people feel lightheaded or dizzy. Nerve damage known as peripheral neuropathy is caused by a variety of disorders. The medical disorders can result in peripheral neuropathy: autoimmune conditions, and these include lupus, rheumatoid arthritis, Guillain-Barre syndrome, chronic inflammatory demyelinating polyneuropathy and vasculitis as well as Sjogren's disease. More than half of diabetics have been observed with some form of neuropathy. These include specific viral or bacterial illnesses, such as HIV, leprosy, hepatitis C or B, diphtheria, shingles, Epstein-Barr virus and Lyme disease, as well as hepatitis B and C inherited illnesses. Neuropathies that run in families include Charcot-Marie-Tooth disease and other conditions. Tumours on the nerves or near the nerves can form both cancerous (malignant) and non-cancerous (benign) growths. Additionally, some malignancies linked to the body's immunological response can lead to polyneuropathy. These are a subtype of paraneoplastic syndrome, which is a degenerative bone marrow condition. These include monoclonal gammopathies, a rare illness called amyloidosis, lymphoma and a kind of bone cancer called myeloma.

Additional reasons for neuropathies include alcoholism. Vitamin deficits can result from poor food choices made by alcoholics and toxins being exposed. Industrial chemicals, heavy metals like lead and mercury and medications are examples of toxic compounds. Peripheral neuropathy can be caused by some drugs, notably those used to treat cancer (chemotherapy) and a nerve injury or pressure. Peripheral nerves can be severed or damaged by injuries, such as those caused by car accidents, slips and falls or sports injuries. Nerve pressure can be caused by wearing a cast, using crutches or repeatedly performing an action like typing and deficits in vitamins. Niacin, vitamin E and B vitamins such as B1, B6 and B12, are essential for maintaining healthy nerves. Several risk factors for peripheral neuropathy include diabetes mellitus, particularly even if the blood sugar levels are inadequately managed, abuse of alcohol and vitamin deficits, especially those of the B vitamins. Infections such as human immunodeficiency virus, hepatitis B or C, Lyme disease and shingles as well as Epstein-Barr virus are all autoimmune illnesses where the body's immune system assaults its own tissues (Peripheral Neuropathy–Symptoms and Causes–Mayo Clinic 2022).

1.8 Cranio-Cervical Junction Disorders

The cranio-cervical junction, a complicated region where the skull and upper cervical spine converge, is the site of several spinal diseases. The area of the cranio-cervical junction, near the base of the brainstem, is where the brain and spinal cord are connected. Disorders of the cranio-cervical junction can weaken the spine and harm the nervous system. An issue with the cranio-cervical junction may cause neck discomfort, headaches, balance issues, voice changes, difficulty swallowing, respiratory troubles or sleep apnoea, articulation difficulties due to motor speech impairments, spinal cord compression, spasticity, twisted or rotated neck and other symptoms. The tops of the vertebrae shift upward, which causes the skull's hole where the spinal cord and brain meet to close and perhaps put pressure on the brainstem. Segmentation faults are a condition when the vertebrae are out of place. Hypoplasia syndrome entails congenital fusions and the lack of some spinal segments. Anomalies in growth and development, achondroplastic stenosis, osteogenesis imperfecta and renal rickets are some of the diseases that affect development. Platybasia, i.e. flattening of the occipital bone at the base of the skull, causes this disorder. When the brain tissue collapses into a wide hole, it results in a Chiari malformation. Spina bifida and encephalocele disorders are caused by the spinal cord's or its coverings' inadequate development (Craniocervical Junction Disorders | Children's Healthcare of Atlanta 2022).

1.9 Headaches, Meningitis, Delirium and Dementia

1.9.1 Headache

Pain in any part of the head, which includes the face, scalp and upper neck and inside the head, is referred to as a headache. One of the most frequent causes of doctor visits is headaches. Some people suffers with headaches oftenly. These are two categories of headaches: primary headaches, and secondary headaches. Primary headache conditions include migraine, cluster headache, and various trigeminal autonomic cephalalgia such as hemicrania continua, chronic paroxysmal hemicrania and short-lasting unilateral neuralgiform headache along with conjunctival injection and tears. Secondary headaches may be caused by conditions affecting the eyes, brain, throat, nose, teeth, ears, jaws, throat or neck, as well as by conditions that affect the entire body (systemic). Primary headaches have two main reasons that are frequently encountered: tension-type (most common overall) and migraine. Less frequent reasons include cluster headaches, and a less frequent main headache condition or one of the several secondary headache disorders is less frequently the cause of headaches. Few of the secondary headache conditions, especially those affecting the human brain, such as meningitis, bleeding or a brain tumour inside the brain, are dangerous (intracerebral haemorrhage). The reason of the headache determines how to treat it. Acetaminophen or a nonsteroidal anti-inflammatory drug (NSAID) can be utilized to treat tension headaches or headaches that come with a small viral illness (Silberstein 2023).

1.9.2 Delirium and Dementia

The common frequent reason for mental (cognitive) dysfunction (the failure or the inability to learn, remember and apply information normally) are dementia as well as delirium. Both disorders are quite harmful and may or may not coexists. Although, both disorders are quite different. Usually Dilirium starts off abruptly, alters mental state/function and are usually reversible, whereas dementia develops incrementally, advances slowly and are irreversible. Additionally, there are two conditions that have various effects on mental function: Mainly, attention is impacted by delirium. Memory is significantly impacted by dementia. Although delirium and dementia can strike anybody at any age, they are significantly more prevalent in older individuals due to changes in the brain caused by aging. One such example of dementia is AD; deterioration of the brain tissue is the primary cause of the AD-related gradual loss of mental function, which includes the death or loss of several nerve cells, usual build-up of an aberrant protein of beta-amyloid and the emergence of neurofibrillary tangles. Another example is chronic traumatic encephalopathy, which generally affects athletes but can also affect troops who have been exposed to explosions and is a gradual degradation of brain cells caused by various head injuries. Lewy bodies are formed in the nerve cells because of the progressive loss of the mental function which characterizes dementia with Lewy bodies. Dementia caused by Parkinson's disease is characterized by Lewy bodies developing in the brains of those who have the condition. Though some of these disorders cannot be treated completely but controlled upto some extent. Therefore, much emphasis is required to research about more forms of dementia as well as delirium related disorders (Huang 2023a, b, c, d).

1.9.3 Meningitis

Meningitis is an inflammation of the fluid-filled area in-between the meninges as well as the layers of tissues which surround the human brain as well as the spinal cord (subarachnoid space). Meningitis can be caused by bacteria, viruses, or fungi, as well as by illnesses other than infections and medications. Fever, headaches, and a stiff neck which makes dropping the chin towards the chest difficult or nearly impossible are all signs of the meningitis. It is notable that infants may not develop a stiff neck, but the end symptoms may usually vary amongst the extremely elderly as well as in those who use immunosuppressive medications. To collect a sample of the cerebrospinal fluid for conducting an experiment, a technique is performed known as, spinal tap. Drugs to alleviate symptoms are also used in the treatment of meningitis, which is dependent on the aetiology (for instance, antibiotics for bacterial meningitis). Different kinds of meningitis may result in various symptoms.

Furthermore, the severity and rate of onset of the symptoms vary. However, they all lead to the following: a very tight neck that makes it difficult or impossible to drop the chin towards the chest and headache/fever. Meningitis treatment is based on the underlying cause. Antimicrobial medications (like antiviral, antibiotic or even antifungal medications) are given as necessary if meningitis is caused by an infection. Since bacterial meningitis advances quickly and is life threatening. Therefore, antibiotices and corticosteroids may also be administered to patients to minimize brain swelling (Greenlee 2023).

1.9.4 Prion Disease

Prion disease is a very uncommon neurological condition. It is lethal to both humans and animals, in humans various kinds of prion disease occur like Creutzfeldt-Jakob disease (CJD), variant Creutzfeldt-Jakob disease (vCJD), Gerstmann-Straussler-Scheinker syndrome, fatal familial insomnia and kuru. In animals bovine spongiform encephalopathy (BSE), chronic wasting disease (CWD), scrapie, transmissible mink encephalopathy, feline spongiform encephalopathy and ungulate spongiform encephalopathy (CDC Works 24/7 2023) are common prion diseases. It is a series of illnesses caused by an irregularly shaped host protein known as a prion, specifically the misfolding of the PrP protein of the prion family. The condition can be:

- Sporadic—misfolded PrP appears without a known reason. Patients are sporadic in 80% to 95% of cases like Creutzfeldt-Jakob disease.
- Genetic—usually caused by a mutation in PrP-coding gene. About 10% to 15% are inherited cases (typically from a family member) like Creutzfeldt-Jakob disease, Gerstmann-Straussler-Scheinker syndrome and fatal familial insomnia.
- Acquired—less than 1% that are acquired due to contaminated blood and medical equipment like kuru, variant Jakob-Creutzfeldt disease and iatrogenic Creutzfeldt-Jakob disease (Geschwind 2015; Seladi-Schulman 2022).

In the 1920s, the first case of human prion disease was discovered. Human prion disorders affect most of the developed world, with 1 to 1.5 cases per million people per year (Geschwind 2015). Misfolded PrP accumulates in the brain tissue, forming a sponge-like structure under a microscope. It is also known as spongiform encephalopathies or TSEs (Prion Diseases 2019). This build-up causes memory loss, personality changes, hallucinations, immobility, weariness and other symptoms. A biopsy of the brain tissue can be used to diagnose it. Prion disease cannot be cured; however, it can be slowed down by medicine (Prion Diseases 2021).

1.9.5 Sleep Disorder

Sleep disorders, usually referred to as sleep-wake disorders, are issues with sleep amount, quality and timing that can cause daytime sleepiness and other symptoms. The National Sleep Foundation recommends that adults get between 7 and 9 h of sleep per night. Sleepiness disrupts brain functions, which can lead to personality changes, sadness and memory loss in people of all ages as well as learning disabilities in children. Physical, medical, neurological and environmental factors, working the night shift, drugs, heredity and ageing are the main reasons for sleep disturbances. About 80 different types of sleep problems exist. The following stands out:

(a) Insomnia: People with insomnia have trouble getting to sleep or staying asleep. Both chronic and acute insomnia exist. Stress can induce acute insomnia, which can last anywhere from one night to a few weeks, but chronic insomnia, which affects at least three nights per week for at least 3 months or more, is caused by depression and ongoing stress. Amongst individuals who reported having sleep issues, 97% indicated that they were depressed and 59% said it significantly affected their quality of life (Nutt et al. 2008).
(b) Chronic Sleep Apnoea: When a person's breathing is disrupted while they are asleep, they are at risk of developing the potentially deadly sleep disorder known as sleep apnoea. It might be both obstructive and central. Obstructive apnoea occurs when the throat's soft tissue at the back tightens or constricts, when a person is sleeping which frequently blocks the airway. Central does not have a blockage in the airway, however the brain is unable to signal the body for breath.
(c) Restless Leg Syndrome: Restless leg syndrome is a neurological disorder that causes inability to fall and remain asleep due to an irrepressible impulse to twitch the legs. Issues with daytime tiredness, irritability and attention may be related to it.
(d) Narcolepsy: Narcolepsy is a kind of neurological disorder which potentially affects how well the body regulates sleep as well as wakefulness. There are uncontrollable episodes, excessive daytime sleepiness and sporadic episodes of daytime sleepiness. (Common Sleep Disorders: Symptoms, Causes, and Treatment 2020).
(e) Additional Sleep Disorders: Slow eye movement disorders of sleep arousal, nightmares, hypersomnolence, circadian rhythm sleep-wake disturbances and rapid eye movement sleep behaviour (What Are Sleep Disorders? 2020).

1.9.6 Spinal Cord Disorder

Ailments that damage and erode the spinal cord or adjacent tissues and bones are referred to as spinal cord diseases. One could lose mobility or function in various bodily areas depending on the degree of the injury. Spinal cord disorder symptoms are limb weakness or paralysis, loss of touch, alterations in reflexes, loss of bowel or bladder control, spontaneous muscle contractions and backache. Surgery, medicine and physical therapy are all available forms of treatment. Disorders of the spinal cord frequently result from causes outside the cord, including the following:

(a) Compression because of spinal stenosis.
(b) Problem of herniated or rubbery disc.
(c) Tumour.
(d) Abscess (collection of pus with great pain).
(e) Haematoma (blood collection but outside blood vessel).

Less often than other types of disorders, cord-specific disorders occur. Infarction of the spinal cord, haemorrhage, transverse myelitis, poliovirus infection, HIV infection, infection caused by West Nile, syphilis (that can result in tabes dorsalis), vitamin B12 insufficiency and trauma are only a few examples of intrinsic illnesses.

1.9.7 Stroke and Tumour

Stroke usually happens when the artery leading to our brain narrows or bursts, killing a portion of the brain tissue (cerebral infarction) and causing symptoms to appear quickly. Most strokes are ischaemic (often caused by arterial blockage), but others are usually haemorrhagic (results because of rupture of an artery). Like ischaemic strokes, transient ischaemic episodes have symptoms which usually go away in an hour, but there is no permanent brain damage. Symptoms that arise suddenly include muscle weakness paralysis or altered or lost feelings on one side of the body, trouble in speaking, disorientation, vision problems, dizziness, loss of balance as well as coordination and even in certain cases haemorrhagic strokes and a sudden, severe headache. Imaging and blood tests are also performed; however, symptoms are the major basis for diagnosis. The extent and location of brain injury, the patient's age and the existence of other illnesses are only a few of the variables that influence recovery after a stroke. Strokes can be avoided by controlling higher blood pressure or cholesterol and excessive blood sugar as well as quitting smoking. Treatment options include surgery to remove a clot, different treatments to address blocked or restricted arteries and medications to reduce the likelihood of blood clotting or break up existing clots such as angioplasty (Chong 2023a, b, c).

Any abnormal development, whether cancerous or benign, is referred to as a tumour. The effects of a non-cancerous tumour are minimal or non-existent in many areas of the body. However, because the structures that shelter the brain as well as the spinal cord, i.e. the skull as well as the spine, cannot enlarge to accommodate any development in their contents, any growth or mass in the brain or the spinal cord. Probably a brain or a spinal cord tumour can be extremely harmful. Nerve tissue in the brain or spinal cord can give rise to tumours, whether they are malignant or benign. From other parts of the body, cancerous tumours can spread (metastasize) to the brain or spinal cord. Even when there is no proof that the nerve tissue has been invaded, some tumours that are in other parts of the body can nonetheless generate symptoms of nervous system failure.

Paraneoplastic syndromes are the name given to these conditions. The most prevalent paraneoplastic disorders include muscular weakness, numbness and tingling because of peripheral nerve damage (polyneuropathy). Seizures, incoordination, double vision, aberrant eye movements and psychosis (which may include delusions, hallucinations and unusual behaviours) are all symptoms of more severe paraneoplastic disorders. Even when the tumour is stable, these symptoms might be lethal. In these situations, therapy entails eliminating blood antibodies that might result in the paraneoplastic syndrome (plasmapheresis). But removing the tumour is the best course of action. Surgery, radiation treatment, chemotherapy or—most frequently—a combination of these approaches can be used to treat nervous system tumours. Despite all attempts to avoid it, radiation treatment occasionally causes harm to the neurological system. Because chemotherapy might impact brain function, clinicians carefully choose chemotherapy medications to prevent unintended side effects (Goldman 2023).

1.10 Diagnosis of the Brain, Spinal Cord and Nerve Disorders

A diagnosis that is inferred from the medical history and neurologic examination may require diagnostic tests to be confirmed.

Electroencephalography: The electrical activity of the brain is captured as wave patterns through an easy, painless process called electroencephalography (EEG), which is then written on a paper or can be stored in a computer. EEG can assist in determining the following, such as:

(a) Various disorders of seizures.
(b) Disruptions in sleep.
(c) Certain brain structural or metabolic conditions.

EEG, for instance, can assist in locating the source of a seizure and revealing abnormalities in electrical activity linked to disorientation, which may be caused by diseases like liver failure (liver encephalopathy) or medications. An examiner applies tiny, circular adhesive sensors (electrodes) to the subject's scalp as part of the operation. The electrodes are wired up to a device that creates a record or tracing of the minute voltage alterations that each electrode detects. The electroencephalogram consists of these tracings (the EEG).

Even when the first EEG is normal and a seizure disorder is suspected, a second EEG is performed after employing a technique that increases the likelihood of seizure activity. For example, the patient may be made to experience sleep deprivation, may be encouraged to breathe rapidly and deeply (hyperventilate) or may exposed to a flashing light (stroboscope). Sometimes when patients are being observed in a hospital via a video camera, the brain's electrical activity is recorded for 24 h or even more (for instance, when a behaviour that mimics a seizure is challenging to separate from a mental disease). This technique is known as EEG. When a seizure-like behaviour is captured by the camera, the physicians can assess if it is a seizure or whether the brain activity is normal and then suggests a psychiatric condition by looking or observing at the EEG at the same instant.

1.10.1 Electromyography

A tiny needle is placed into a muscle during electromyography (EMG) to capture the muscle's electrical activity both at rest and during contraction. Normally, electrical activity in resting muscle is absent. Electrical activity is generated by even a little contraction, and it grows as the contraction gets stronger. The electromyogram is the name given to the EMG record. It is abnormal if muscle weakness results from a problem with a spinal nerve root, a peripheral nerve, a muscle or a neuromuscular junction. Based on the patient's symptoms, the examination and electromyography findings, each type of issue has a specific pattern of anomalies that may be recognized. In contrast to CT and EEG, which may be performed often by technicians, EMG calls for a neurologist's competence since they must choose the proper nerves and muscles to test and interpret the data.

1.10.2 Nerve Conduction Studies

The rate upon which the motor or the sensory nerves transmit impulses is measured by nerve conduction experiments. An impulse is stimulated along the nerve being tested by a little electrical current. Several electrodes positioned on the skin's surface or numerous needles put along the nerve's course can both give electricity. The muscle contracts as the impulse travels along the nerve and ultimately reaches it. Physicians can determine the speed or responsiveness of nerve transmission by monitoring the time it takes for the nerve impulse to reach the target muscle as well as the distance from the impulse to reach the muscle and the distance from the stimulating electrode or needle to the muscle. A single or several stimulations of the nerve are possible (to determine how well the neuromuscular junction is functioning). For instance, a single nerve condition like carpal tunnel syndrome may be the root of delayed nerve conduction (painful compression of a nerve in the wrist). Alternately, the issue might be caused by a polyneuropathy, a condition that destroys nerves all over the body, starting with those in the feet. After repeated stimulation, if the muscle's reaction becomes gradually weaker, there may be an issue at the neuromuscular junction (as in myasthenia gravis).

1.10.3 Evoked Responses and Imaging Tests

For this exam, medical professionals utilize stimuli for sight, hearing and touch to trigger brain regions and elicit reactions. The reaction that the stimuli trigger is discovered via EEG. Doctors can determine how effectively those brain regions are functioning based on these answers. For instance, a flashing light can excite the retina of the eye and the optic nerve as well as the nerve pathway to the back of the brain, where vision is received and processed or interpreted. Evoked reactions are particularly helpful for assessing how effectively a child's or infant's senses are working. For instance, by making a clicking sound at each ear and listening for a reaction, clinicians can assess a baby's hearing. Evoked responses are also helpful in determining how the optic nerve, brainstem and spinal cord are affected by multiple sclerosis and other diseases. MRI may or may not be able to find these effects. The imaging tests are often used to identify neurological (nervous system) disorders; they are magnetic resonance imaging (MRI), positron emission tomography, angiography, Doppler ultrasonography and computed tomography (CT).

1.10.4 Myelography

In myelography, a radiopaque contrast agent is given via a spinal tap into the subarachnoid space before x-ray images of the spinal cord are obtained. MRI has essentially supplanted myelography since it is quicker and safer and typically generates more detailed pictures. When MRI cannot give the necessary level of information of the spinal cord and surrounding bone, myelography using CT is employed. When MRI is unavailable or cannot be performed safely, myelography with CT is also employed (if someone has a cardiac pacemaker, for instance).

1.10.5 Spinal Tap

Between the tissue layers (meninges) which surround the brain as well as the spinal cord, cerebrospinal fluid travels via a passageway called the subarachnoid space. The fluid that surrounds the brain as well as the spinal cord guards them against slight trauma and rapid shock. A spinal tap also called as lumbar punctures involves the removal of a cerebrospinal fluid along with a needle and sending it to a lab for analysis. Tumours, infections and even bleeding in the brain as well as the spinal cord are looked for in the cerebrospinal fluid. The cerebrospinal fluid, which typically includes few red and white blood cells and is clear and colourless, may vary in composition and appearance because of certain illnesses. For instance, the subsequent data points to specific disorders are as follows:

- The presence of more white blood cells in cerebrospinal fluid is basically a sign that the brain or spinal cord may be infected or inflamed.
- Cloudy fluid, which is caused by the presence of many white blood cells, may indicate encephalitis or meningitis and an inflammation as well as infection of the tissues surrounding the spinal cord as well as the brain (infection and inflammation of the brain).

- Any lesion to the spinal nerve root, the brain or the spinal cord might cause rise of protein levels in the fluid (the part of a spinal nerve next to the spinal cord).
- Multiple sclerosis or an infection is suggested by abnormal antibodies found in the fluid.
- Fall of sugar or glucose levels may signal malignancy or even meningitis.
- Blood present in the fluid may be a sign of a brain haemorrhage, such as when an aneurysm, a round swelling or a bulge present in a weak artery in the brain bursts or ruptures.
- Many diseases, including meningitis and brain tumours, can cause a rise in the fluid's pressure.

Other tests include biopsy of the skin, muscle and nerves and echoencephalography which utilizes ultrasound images to generate image of the brain as well as genetic testing (Neurological Diagnostic Tests and Procedures 2020; Levin 2023).

1.11 Treatment of Neurological Disorders

The field of neuroscience medicine includes neuroradiology. Problems with the neurological system are the primary focus of the treatment. In interventional neuroradiology, tiny, flexible tubes also known as catheters are inserted into the blood arteries supplying the brain. Hence, this makes it possible for the physician to treat blood vessel conditions like stroke which can have an impact on the neurological system of the body. The following are examples of interventions used in interventional neuroradiology:

(a) Balloon angioplasty as well as stenting of carotid or vertebral arteries.
(b) Endovascular embolization and coiling to treat cerebral aneurysms.
(c) Intra-arterial therapy (for stroke).
(d) Myelography, a technique for photographing the spinal structures.
(e) Brain and spine radiation oncology.
(f) Soft tissue needle biopsies.
(g) Kyphoplasty and vertebroplasty for treating vertebral fractures.

For the treatment of issues with the brain and its supporting structures, open or conventional neurosurgery may be required. This type of invasive surgery calls for the creation of a craniotomy or opening in the skull. Using a microscope and tiny, precise equipment, the surgeon may perform microsurgery on very minute brain regions. Some forms of nervous system problems may need stereotactic radiosurgery. Other methods of treating diseases or abnormalities of the nervous system, perhaps administered via a drug pump, and medications (as those prescribed to persons who experience acute muscular spasms) include:

(a) Spinal cord stimulation.
(b) Deep brain stimulation.
(c) Physical treatment as well as rehabilitation after a stroke or brain damage.
(d) Spinal surgery (Neurosciences: MedlinePlus Medical Encyclopedia 2023).

1.12 Conclusion

Neurological disorders pose a great challenge towards the society as it plays with the functional activity of brain. Now a days, significant part of populations are affected with these neurological complications. These complications are the result of impairements in the nervous system including the brain, spinal cord and the nerves. Some complications are easily curable while some are controlled under proper medications only. Furthermore, some complications are getting worst with the time. The research is needed to maintain a better condition with minimum side effects. In order to have promising treatments, we need to understand the nature of such complications and disorders. It's also remarkable to see that though many developments have been made, biggest achievements like the deep brain stimulation are still needed. Possible therapeutic agents can be searched for neurodegenerative disorders in interdisciplinary science such as stem cells technology and their derived factors like exosomes, virology, immunology, biotechnology and nanotechnology. Further research for personalized therapies and testing regimens can also solve the issue of such disorders.

Acknowledgement The author would like to thank Deanship of Scientific Research at Majmaah University for supporting this work.

References

Amyotrophic Lateral Sclerosis (ALS) (2023) National institute of neurological disorders and stroke. https://www.ninds.nih.gov/health-information/disorders/amyotrophic-lateral-sclerosis-als. Retrieved 15 Feb 2023

Anatomy of the Brain (2023). http://johnshopkinshealthcare.staywellsolutionsonline.com/Conditions/Neuroscience/About/85,P00773. Retrieved 1 Feb 2023

A Neurosurgeon's Overview of the Anatomy of the Spine and Peripheral Nervous System (2023). https://www.aans.org/en/Patients/Neurosurgical-Conditions-and-Treatments/Anatomy-of-the-Spine-and-Peripheral-Nervous-System#:~:text=The%20spinal%20cord%20is%20an,cone%20called%20the%20conus%20medullaris. Retrieved 1 Feb 2023

Appenzeller O, Richardson EP (1966) The sympathetic chain in patients with diabetic and alcoholic polyneuropathy. Neurology 16(12):1205–1209. https://doi.org/10.1212/WNL.16.12.1205

Association NBM (2013) Diagnostic and statistical manual of mental disorders. https://doi.org/10.1176/appi.books.9780890425596

Brown RJC (2019) Why study the history of neuroscience? Front Behav Neurosci 13. https://doi.org/10.3389/fnbeh.2019.00082

CDC Works (2023) Ctr Dis Control Prev 24/7. https://www.cdc.gov/

Chong JY (2023a) Ischemic stroke. MSD Manual Consumer Version. https://www.msdmanuals.com/en-in/home/brain,-spinal-cord,-and-nerve-disorders/stroke-cva/ischemic-stroke

Chong JY (2023b) Overview of hemorrhagic stroke. MSD manual consumer version. https://www.msdmanuals.com/en-in/home/brain,-spinal-cord,-and-nerve-disorders/stroke-cva/overview-of-hemorrhagic-stroke

Chong JY (2023c) Overview of stroke. MSD manual consumer version. https://www.msdmanuals.com/en-in/home/brain,-spinal-cord,-and-nerve-disorders/stroke-cva/overview-of-stroke

Common Sleep Disorders: Symptoms, Causes & Treatment. (2020) Cleveland clinic. https://my.clevelandclinic.org/health/articles/11429-common-sleep-disorders. Retrieved 30 Oct 2022

Craniocervical Junction Disorders | Children's Healthcare of Atlanta (2022) Children's healthcare of Atlanta. https://www.choa.org/medical-services/neurosciences/craniocervical-junction-disorders. Retrieved 3 Nov 2022

Da Cunha De Sá-Caputo D, Bernardo-Filho M, Seixas A, Taiar R (2021) Introductory chapter: neurological disorders–therapy approaches. IntechOpen EBooks. https://doi.org/10.5772/intechopen.97178

Edlow JA, Rabinstein A, Traub SJ, Wijdicks EFM (2014) Diagnosis of reversible causes of coma. Lancet (London, England) 384(9959):2064–2076. https://doi.org/10.1016/S0140-6736(13)62184-4

Esmael A, Elsherif M, Eltoukhy K (2021) Predictive value of the Alberta stroke program early CT score (ASPECTS) in the outcome of the acute ischemic stroke and its correlation with stroke subtypes, NIHSS, and cognitive impairment. Stroke Res Treat 2021:1–10. https://doi.org/10.1155/2021/5935170

Flægstad T (2020) Meningococcal meningitis. Disorders of consciousness–a review of important issues. https://doi.org/10.5772/INTECHOPEN.90687

Fowler NR, Judge KS, Lucas K, Gowan T, Stutz PV, Shan M, Wilhelm L, Parry T, Johns SA (2021) Feasibility and acceptability of an acceptance and commitment therapy intervention for caregivers of adults with Alzheimer's disease and related dementias. DOAJ 21:127. https://doi.org/10.1186/s12877-021-02078-0

Gardner-Thorpe C (2000) A short history of neurology. Brain 123:2573. https://doi.org/10.1093/brain/123.12.2573

Geschwind MD (2015) Prion diseases. Continuum 21:1612–1638. https://doi.org/10.1212/con.0000000000000251

Goldman SA (2023) Overview of tumors of the nervous system. MSD manual consumer version. https://www.msdmanuals.com/en-in/home/brain,-spinal-cord,-and-nerve-disorders/tumors-of-the-nervous-system/overview-of-tumors-of-the-nervous-system

Greenlee JE (2023) Introduction to Meningitis. MSD manual consumer version. https://www.msdmanuals.com/en-in/home/brain,-spinal-cord,-and-nerve-disorders/meningitis/introduction-to-meningitis

Grover S, Avasthi A (2018) Clinical practice guidelines for management of delirium in elderly. Indian J Psychiatry 60(Suppl 3):S329. https://doi.org/10.4103/0019-5545.224473

Grover S, Kate N (2012) Assessment scales for delirium: a review. World J Psyc 2(4):58. https://doi.org/10.5498/WJP.V2.I4.58

Guo YP, Mcleod JG, Baverstock J (1987) Pathological changes in the vagus nerve in diabetes and chronic alcoholism. J Neurol Neurosurg Psychiatry 50(11):1449–1453. https://doi.org/10.1136/JNNP.50.11.1449

Hersi K, Gonzalez FJ, Kondamudi NP (2022) Meningitis. StatPearls

Hirsch L, Jette N, Frolkis AD, Steeves T, Pringsheim T (2016) The incidence of Parkinson's disease: a systematic review and meta-analysis. Neuroepidemiology 46(4):292–300. https://doi.org/10.1159/000445751

Hodges H, Fealko C, Soares N (2020) Autism spectrum disorder: definition, epidemiology, causes, and clinical evaluation. Transl Pediatr 9(S1):S55–S65. https://doi.org/10.21037/tp.2019.09.09

Hou J, Cao J, Tan P, Yu Y (2021) Pneumocystis jirovecii pneumonia, Nocardia brasiliensis, and mycobacterium tuberculosis co-infection in a myasthenia gravis patient. Medicine 100(1): e24245. https://doi.org/10.1097/md.0000000000024245

Huang J (2023a) Alzheimer disease. MSD manual consumer version. https://www.msdmanuals.com/en-in/home/brain,-spinal-cord,-and-nerve-disorders/delirium-and-dementia/alzheimer-disease
Huang J (2023b) Chronic traumatic encephalopathy (CTE). MSD Manual Consumer Version. https://www.msdmanuals.com/en-in/home/brain,-spinal-cord,-and-nerve-disorders/delirium-and-dementia/chronic-traumatic-encephalopathy-cte
Huang J (2023c) Dementia with lewy bodies and Parkinson disease dementia. MSD manual consumer version. https://www.msdmanuals.com/en-in/home/brain,-spinal-cord,-and-nerve-disorders/delirium-and-dementia/dementia-with-lewy-bodies-and-parkinson-disease-dementia
Huang J (2023d) Overview of delirium and dementia. MSD Manual Consumer Version. https://www.msdmanuals.com/en-in/home/brain,-spinal-cord,-and-nerve-disorders/delirium-and-dementia/overview-of-delirium-and-dementia
Huff JS, Stevens RD, Weingart SD, Smith WS (2012) Emergency neurological life support: approach to the patient with coma. Neurocrit Care 17 Suppl 1(SUPPL:1. https://doi.org/10.1007/S12028-012-9755-4
Inouye SK, Charpentier PA (1996) Precipitating factors for delirium in hospitalized elderly persons: predictive model and interrelationship with baseline vulnerability. JAMA 275(11):852–857. https://doi.org/10.1001/JAMA.1996.03530350034031
Inouye SK, Van Dyck CH, Alessi CA, Balkin S, Siegal AP, Horwitz RI (1990) Clarifying confusion: the confusion assessment method. A new method for detection of delirium. Ann Intern Med 113(12):941–948. https://doi.org/10.7326/0003-4819-113-12-941
Jordan MR, Morrisonponce D (2022) Naloxone. XPharm: The Comprehensive Pharmacology Reference 1–6. https://doi.org/10.1016/B978-008055232-3.62239-1
Kadapatti K, Iyer SK (2022) Coma. ICU protocols: a step-wise approach, 1:319–325. https://doi.org/10.1007/978-981-15-0898-1_32
Kristensson K, Nordborg C, Olsson Y, Sourander P (1971) Changes in the vagus nerve in diabetes mellitus. Acta Pathol Microbiol Scand A 79(6):684–685. https://doi.org/10.1111/J.1699-0463.1971.TB01872.X
Kumar DD, Aslinia F, Yale SH, Mazza JJ (2011) Jean-Martin Charcot: the father of neurology. Clin Med Res 9(1):46–49. https://doi.org/10.3121/cmr.2009.883
Levin MC (2023) Tests for brain, spinal cord, and nerve disorders. MSD manual consumer version. https://www.msdmanuals.com/en-in/home/brain,-spinal-cord,-and-nerve-disorders/diagnosis-of-brain,-spinal-cord,-and-nerve-disorders/tests-for-brain-spinal-cord-and-nerve-disorders
Li Y, Arora Y, Levin KH (2013) Myasthenia gravis: newer therapies offer sustained improvement. Cleve Clin J Med 80(11):711–721. https://doi.org/10.3949/ccjm.80a.13044
Low P (2023) Overview of the autonomic nervous system. Merck manuals consumer version. https://www.merckmanuals.com/home/brain,-spinal-cord,-and-nerve-disorders/autonomic-nervous-system-disorders/overview-of-the-autonomic-nervous-system
Maiese K (2023a) Brain. Merck manuals consumer version. https://www.merckmanuals.com/home/brain,-spinal-cord,-and-nerve-disorders/biology-of-the-nervous-system/brain
Maiese K (2023b) Nerves. Merck manuals consumer version. https://www.merckmanuals.com/home/brain,-spinal-cord,-and-nerve-disorders/biology-of-the-nervous-system/nerves
Maiese K (2023c) Spinal cord. Merck manuals consumer version. https://www.merckmanuals.com/home/brain,-spinal-cord,-and-nerve-disorders/biology-of-the-nervous-system/spinal-cord
Maramattom BV, Wijdicks EFM (2005) Uncal herniation. Arch Neurol 62(12):1932–1935. https://doi.org/10.1001/ARCHNEUR.62.12.1932
Marin CE, Kfouri PBP, Callegaro D, Lana-Peixoto MA, Neto AHC, Vasconcelos CCF, D'Almeida JRM, Goncalves M, Mendes MA, Parolin MB, Nascimento OJM, Da Gama PD, Dias-Carneiro RPC, Dias R, Damasceno A, Martin GS, Becker J (2021) Patients and neurologists have different perceptions of multiple sclerosis symptoms, care and challenges. Mult Scler Relat Disord 50:102806. https://doi.org/10.1016/j.msard.2021.102806
Mayeux R, Stern Y (2012) Epidemiology of Alzheimer disease. Cold Spring Harb Perspect Med 2(8):a006239. https://doi.org/10.1101/cshperspect.a006239

Mayfield Brain & Spine (2018) MAYFIELD. https://mayfieldclinic.com/pe-anatbrain.htm#:~:text=The%20brain%20has%20three%20main,and%20fine%20control%20of%20movement. Retrieved 1 Nov 2022

McCorry LK (2007) Physiology of the autonomic nervous system. Am J Pharm Educ 71(4):78. https://doi.org/10.5688/AJ710478

McMorris CA, Lake JK, Dobranowski K, McGarry C, Lin EHB, Wilton D, Lunsky Y, Balogh R (2021) Psychiatric disorders in adults with cerebral palsy. Res Dev Disabil 111:103859. https://doi.org/10.1016/j.ridd.2021.103859

McNamara L, Scott KP, Boyd RN, Novak I (2021) Consensus of physician behaviours to target for early diagnosis of cerebral palsy: a Delphi study. J Paediatr Child Health 57:1009. https://doi.org/10.1111/jpc.15369

Medical, A. C. for G (2013) ACGME program requirements for graduate medical education in anesthesiology. 2015(August 3), pp. 1–3. Available at: http://www.acgme.org/acgmeweb/portals/0/pfassets/2013-pr-faq-pif/110_emergency_medicine_07012013.pdf

Mental health: neurological disorders (2016). https://www.who.int/news-room/questions-and-answers/item/mental-health-neurological-disorders. Retrieved 3 Oct 2022

Mishra S, Trikamji B, Singh S, Singh P, Nair RS (2013) Historical perspective of Indian neurology. Ann Indian Acad Neurol 16(4):467. https://doi.org/10.4103/0972-2327.120422

Momtazmanesh S, Shobeiri P, Saghazadeh A, Teunissen CE, Burman J, Szalárdy L, Klivényi P, Bartos A, Fernandes A, Rezaei N (2021) Neuronal and glial CSF biomarkers in multiple sclerosis: a systematic review and meta-analysis. Rev Neurosci 32(6):573–595. https://doi.org/10.1515/revneuro-2020-0145

Mozaffarian D, Benjamin EJ, Go AS, Arnett DK, Blaha MJ, Cushman M, Das SR, De Ferranti SD, Després J, Fullerton HJ, Howard VJ, Huffman MD, Isasi CR, Jiménez MC, Judd SE, Kissela BM, Lichtman JH, Lisabeth LD, Liu S, Turner M et al (2016) Heart disease and stroke statistics—2016 update. Circulation 133(4):e38. https://doi.org/10.1161/cir.0000000000000350

Muangpaisan W, Mathews A, Hori H, Seidel D (2011) A systematic review of the worldwide prevalence and incidence of Parkinson's disease. J Med Assoc Thailand 94(6):749–755

Multiple Cranial Neuropathies (2020) Johns Hopkins Medicine. https://www.hopkinsmedicine.org/health/conditions-and-diseases/multiple-cranial-neuropathies

Myasthenia Gravis (2023) National institute of neurological disorders and stroke. https://www.ninds.nih.gov/myasthenia-gravis-fact-sheet#:~:text=Myasthenia%20gravis%20affects%20both%20men,at%20any%20age%2C%20including%20childhood. Retrieved 1 Feb 2023

Nalbantoglu M, Akalin M, Gunduz A, Kiziltan ME (2021) Electrophysiological investigation for autonomic dysfunction in patients with myasthenia gravis: a prospective study. Ideggyogyaszati Szemle-Clin Neurosci 74:33. https://doi.org/10.18071/isz.74.0033

Naseri A, Nasiri E, Sahraian MA, Daneshvar S, Talebi M (2022) MRI findings in late-onset multiple Sclerosis; a systematic review and meta-analysis. Mult Scler Relat Disord 59: 103594. https://doi.org/10.1016/j.msard.2022.103594

Neurological Diagnostic Tests and Procedures (2020) NINDS catalog. https://catalog.ninds.nih.gov/publications/neurological-diagnostic-tests-and-procedures

Neurological Disorders (2022). https://dphhs.mt.gov/schoolhealth/chronichealth/neurologicaldisorders#:~:text=Neurological%20disorders%20are%20medically%20defined,in%20a%20range%20of%20symptoms. Retrieved 3 Oct 2022

Neurosciences: MedlinePlus Medical Encyclopedia. (2023). https://medlineplus.gov/ency/article/007456.htm. Retrieved 1 Nov 2022

Nutt DJ, Wilson S, Paterson LM (2008) Sleep disorders as core symptoms of depression. Dialogues Clin Neurosci 10(3):329–336. https://doi.org/10.31887/dcns.2008.10.3/dnutt

Peripheral neuropathy–Symptoms and causes–Mayo Clinic (2022) Mayo clinic. https://www.mayoclinic.org/diseases-conditions/peripheral-neuropathy/symptoms-causes/syc-20352061#:~:text=Peripheral%20neuropathy%20is%20nerve%20damage,inflammatory%20demyelinating%20polyneuropathy%20and%20vasculitis

Pop-Busui R (2010) Cardiac autonomic neuropathy in diabetes: a clinical perspective. Diabetes Care 33(2):434–441. https://doi.org/10.2337/DC09-1294
Postuma RB, Berg D (2016) Advances in markers of prodromal Parkinson disease. Nat Rev Neurol 12(11):622–634. https://doi.org/10.1038/nrneurol.2016.152
Prion Diseases (2019) NIH: national institute of allergy and infectious diseases. https://www.niaid.nih.gov/diseases-conditions/prion-diseases
Prion Diseases (2021) Johns Hopkins Medicine. https://www.hopkinsmedicine.org/health/conditions-and-diseases/prion-diseases
Sakusic A, Rabinstein AA (2021) Acute Coma. Neurol Clin 39(2):257–272. https://doi.org/10.1016/j.ncl.2021.01.001
Schatz IJ, Bannister R, Freeman RL, Goetz CG, Jankovic J, Kaufmann HC, Koller WC, Low PA, Mathias CJ, Polinsky RJ, Quinn NP, Robertson D, Streeten DHP (1996) Consensus statement on the definition of orthostatic hypotension, pure autonomic failure, and multiple system atrophy. The consensus Committee of the American Autonomic Society and the American Academy of Neurology. Neurology 46(5):1470. https://doi.org/10.1212/WNL.46.5.1470
Schmidt RE (2002) Neuropathology and pathogenesis of diabetic autonomic neuropathy. Int Rev Neurobiol 50:257–292. https://doi.org/10.1016/S0074-7742(02)50080-5
Schönauer M, Thomas A, Morbach S, Niebauer J, Schönauer U, Thiele H (2008) Cardiac autonomic diabetic neuropathy. Diab Vasc Dis Res 5(4):336–344. https://doi.org/10.3132/DVDR.2008.047
Seladi-Schulman J (2022) What is prion disease? Healthline. https://www.healthline.com/health/prion-disease
Shea YF, Pan Y, Mak HK, Bao Y, Lee S, Chiu P, Chan H (2021) A systematic review of atypical Alzheimer's disease including behavioural and psychological symptoms. Psychogeriatrics 21(3):396–406. https://doi.org/10.1111/psyg.12665
Silberstein SD (2023) Overview of headache. MSD manual consumer version. https://www.msdmanuals.com/en-in/home/brain,-spinal-cord,-and-nerve-disorders/headaches/overview-of-headache
Spillane J, Hirsch N, Kullmann DM, Taylor CB, Howard RA (2014) Myasthenia gravis–treatment of acute severe exacerbations in the intensive care unit results in a favourable long-term prognosis. Eur J Neurol 21(1):171–173. https://doi.org/10.1111/ene.12115
Stead LG, Wijdicks EFM, Bhagra A, Kashyap R, Bellolio MF, Nash DL, Enduri S, Schears R, William B (2009) Validation of a new coma scale, the FOUR score, in the emergency department. Neurocrit Care 10(1):50–54. https://doi.org/10.1007/S12028-008-9145-0
Sternbach GL (2000) The Glasgow coma scale. J Emerg Med 19(1):67–71. https://doi.org/10.1016/S0736-4679(00)00182-7
Stevens RD, Huff JS, Duckworth J, Papangelou A, Weingart SD, Smith WS (2012) Emergency neurological life support: intracranial hypertension and herniation. Neurocrit Care 17(Suppl 1). https://doi.org/10.1007/S12028-012-9754-5
Stroke Facts | cdc.gov (2022) Centers for disease control and prevention. https://www.cdc.gov/stroke/facts.htm#:~:text=Stroke%20risk%20increases%20with%20age,less%20than%2065%20years%20old
Symptoms of Neurologic Disorders–Merck Manuals Professional Edition. (2023). Merck manuals professional edition. https://www.merckmanuals.com/en-pr/professional/neurologic-disorders/symptoms-of-neurologic-disorders. Retrieved 23 Feb 2023
Tindall SC (1990) Level of consciousness. Definitions. https://doi.org/10.32388/ajo2fj
Tyler KL, York GWP, Steinberg DM, Okun MS, Steinbach M, Satran R, Fine EJ, Manteghi T, Bleck TP, Swanson JW, Mishra S, Meador KJ, Clifford DB, Toole JF, Melson L (2003) Part 2: history of 20th century neurology: decade by decade. Ann Neurol 53:S27. https://doi.org/10.1002/ana.1346
Vaseghi M, Barwad P, Malavassi Corrales FJ, Tandri H, Mathuria N, Shah R, Sorg JM, Gima J, Mandal K, Sàenz Morales LC, Lokhandwala Y, Shivkumar K (2017) Cardiac sympathetic

denervation for refractory ventricular arrhythmias. J Am Coll Cardiol 69(25):3070. https://doi.org/10.1016/J.JACC.2017.04.035
Vernino S, Low PA, Fealey RD, Stewart JD, Farrugia G, Lennon VA (2000) Autoantibodies to ganglionic acetylcholine receptors in autoimmune autonomic neuropathies. N Engl J Med 343(12):847–855. https://doi.org/10.1056/NEJM200009213431204
Vinik AI, Maser RE, Mitchell BD, Freeman R (2003) Diabetic autonomic neuropathy. Diabetes Care 26(5):1553–1579. https://doi.org/10.2337/DIACARE.26.5.1553
Waxenbaum JA, Reddy V, Varacallo M (2021) Anatomy, autonomic nervous system in StatPearls. StatPearls Publishing
What are Sleep Disorders? (2020). https://www.psychiatry.org/patients-families/sleep-disorders/what-are-sleep-disorders. Retrieved 22 Dec 2022
Wilson JE, Mart MF, Cunningham C, Shehabi Y, Girard TD, MacLullich AMJ, Slooter AJC, Ely EW (2020) Delirium. Nature Reviews Disease Primers 2020 6:1 6(1):1–26. https://doi.org/10.1038/s41572-020-00223-4

The Mechanistic Approach Involved in the Progression of Neurodegenerative Disorders

2

Mashoque Ahmed Rather, Andleeb Khan, Saeed Alsharani, Sadaf Jahan, Arif Jamal Siddiqui, Sidharth Mehan, and Athanasios Alexiou

Abstract

Neurodegenerative disorders are rare hereditary disorders, which lead to the gradual deterioration of neural function in specific regions of the central nervous system. Neurodegenerative disorders are commonly depicted by the aggregation of extracellular and intracellular protein misfolds and neuronal degeneration in the selective areas of the nervous system. The misfolded proteins in neurodegenerative diseases possess unusual conformational effects, and the interaction between toxic proteins may escalate in different regions in the central nervous system, thereby causing degeneration of neurons. There are multiple pieces of evidence that specify substantial interaction between amyloidosis, tauopathies,

M. A. Rather
Department of Molecular Pharmacology & Physiology, Morsani College of Medicine, University of South Florida, Tampa, USA

A. Khan (✉) · S. Alsharani
Department of Pharmacology and Toxicology, College of Pharmacy, Jazan University, Jazan, Saudi Arabia

S. Jahan
Department of Medical Laboratory Sciences, College of Applied Medical Sciences, Majmaah University, Al-Majmaah, Saudi Arabia

A. J. Siddiqui
Department of Biology, College of Science, University of Hail, Hail City, Saudi Arabia

S. Mehan
Neuropharmacology Division, Department of Pharmacology, ISF College of Pharmacy, Moga, Punjab, India

A. Alexiou
Novel Global Community Educational Foundation, Hebersham, NSW, Australia

AFNP Med Austria, Wien, Austria

S. Jahan, A. J. Siddiqui (eds.), *Applications of Stem Cells and Derived Exosomes in Neurodegenerative Disorders*, https://doi.org/10.1007/978-981-99-3848-3_2

synucleinopathies, and unusual protein aggregation of other disorders. The escalation of protein aggregates worsens over time due to abnormal functioning of intracellular mechanisms, leading to the inappropriate and unusual supplies of the essential factors and the mislaid interaction between the neurons. The molecular mechanism involves the dispersion and the transmission between the misfolded pathogenic proteins and the impact of the neuronal cells on the configuration and the effects of abnormal inclusions. In this regard, this chapter aims to highlight the mechanisms involved in neurodegenerative diseases, misfolding of proteins, the role of proteostasis, involvement of chaperones, and the association of ubiquitin-proteasome system and to emphasize the oxidative stress involvement and mitochondrial deficit in the progression of the diseases and the alternative splicing events in the neurodegenerative disorders.

Keywords

Alzheimer's disease · Parkinson's disease · Proteostasis · Protein aggregation · Protein misfolding · Mitochondrial dysfunction · UPS · Splicing

2.1 Introduction

Neurodegenerative disorders (NDs) can be extensively characterized by their clinical manifestations, with cognitive and movement disorders, which affect millions of populations globally. These diseases are linked with the progressive neural disintegration in the different regions of the brain. Their pathogenesis includes intermittent aggregation and deposition of specific protein moieties in different areas of the brain as extracellular accumulations or intracellular inclusions. However, the NDs allocate several other cellular processes linked with the progressive neuronal damage and eventually the death of neurons, which include oxidative stress, mitochondrial impairment, apoptosis, proteotoxic stress, neuroinflammation, and irregularities in the ubiquitin-proteasomal system. The most common NDs are Alzheimer's disease (AD) and Parkinson's disease (PD), which widely affect the elderly population of the world. The unusual protein configuration and their dispersal in the neuronal cells comprise the key events of pathological structures in these disorders such as α-synuclein (α-syn) in Lewy bodies, Tau in neurofibrillary tangles, and Aβ (amyloid-β) in senile plaques.

The abnormal aggregations of α-syn, Tau, and amyloid-β are comprised of cellular components and intrinsic neuronal proteins and vary in different diseases like Aβ and Tau aggregates in AD and abnormal α-syn protein moiety in PD. Multiple pieces of evidence have documented the toxicity produced by these abnormal protein aggregates in the different regions of the brain and examined the molecular mechanism intricated in the neuronal degeneration and impairment (Clavaguera et al. 2014; Laferriere et al. 2019). Also, the critical function of misfolded protein has been emphasized in the progression and pathogenesis of NDs. The dispersal of misfolded proteins in unhealthy individuals observes stereotypical configurations (Braak and Braak 1991; Braak et al. 2003), which describes the alterations in the susceptibility of neurons in the CNS (Surmeier et al. 2017). Over the years, several studies have presented that the protein intricated in the pathogenesis of different NDs experiences communication between the cells by using various cellular and animal models. Following communication, the misfolded proteins perform as templates to prompt typical correlative proteins to dysregulate in the recipient cell, triggering the augmentation of the misfolded protein moiety configuration (Guo et al. 2016; Luk et al. 2012; Volpicelli-Daley et al. 2011). Therefore, the intracellular communication and augmentation of these misfolded proteins could play an instrumental role in the distribution of abnormal proteins in the CNS. Besides, several studies have observed the complexity of the abnormal protein lesions and the mechanisms involved that could inflect the communication between the neurons. The aim of this chapter is to abridge the mechanistic approach associated with the progression of NDs.

2.2 Mechanisms Involved in ND Progression

The mechanisms associated with the progression of NDs are multiple factors including external, genetic, and endogenous influences and are categorized based on their known genetic markers or their protein accumulation. Irregular protein interactions and conformations leading to the toxic extracellular and intracellular accumulations of misformed proteins including the generation of insoluble fibrils are pathological features of several NDs. Irregular communications between the proteins and the afflictions generated from them cause neuronal impairment and eventually the death of neurons (Palop et al. 2006; Seeley et al. 2009). The unusual communication progressively translates the normal proteins into insoluble fibrils along with the aggregates of irregular forms (Ecroyd and Carver 2008).

The toxicity of the elements generated during early events is considered to be reliable for the production of protein aggregates and their association with the NDs (Israeli and Sharon 2009; Selkoe 2008; Soto and Estrada 2008). In toxic oligomerization, domain antibody alone could identify a conformational antigenic detriment that is presented by proteins linked with several dysfunctions such as α-syn, β-amyloid, Tau protein, etc. (Aguzzi et al. 2008; Lafaye et al. 2009; Moore et al. 2009; Williams and Paulson 2008). In AD, soluble Aβ peptides are upregulated in plasma and brain tissues, and their varied concentrations are associated with cognitive impairment (Xia et al. 2009; Tomic et al. 2009). Several pieces of observation support the existence of diverse forms of Aβ42, in which some strains may generate toxicity and some aggregate into oligomers without conforming to toxic configurations (Harmeier et al. 2009). The amyloid precursor protein (APP)-like protein-1 (APLP1) has been anticipated as a viable arbitrator of Aβ-incited synaptic impairment (Lauren et al. 2009). The Aβ peptides function as an indispensable component in the synchronization of the liberation of synapse vesicle, which indicates its principal pathological role in leading to the loss of synapse formation in AD (Abramov et al. 2009). The APLP1 is a dominant marker receptor for Aβ toxic oligomer, which is mostly present on the cell surface, making up heterodimers with APP protein (Nygaard and Strittmatter 2009). These proteins conceivably affect the presynaptic and postsynaptic area, the transmission of synapses, and the agitation of APP synaptic binding action which may eventually lead to synaptic impairment and AD pathology (Nimmrich and Ebert 2009; Hoe et al. 2009).

The Tau proteins are responsible for providing microtubule network stability, assembly, and support for the axonal transport as well as synaptic activity. The oxidation, ubiquitination, and glycosylation processes modify the Tau protein posttranslationally (Goedert et al. 2006). The Tau protein is unable to bind the microtubules when it is hyperphosphorylated, which breaks the consistency of microtubules. The pathological elements of AD trigger multiple protein kinases that cleave the Tau protein in distinct regions, which in turn initiates the generation of Tau-specific fragments (Arnaud et al. 2006). The Tau protein fragments could present an enhanced inclination for self-association prior to the production of exposed aggregates. Thus, the cells become deficient to clear the oligomers effectively, which diminishes the neuronal function. Consequently, the stability between the clearance and processing of Tau protein is important to maintain the appropriate Tau levels. It is documented that Tau oligomeric forms are toxic, and its aggregates disperse via cellular transmission of abnormal proteins (Frost and Jacks 2009; Ren et al. 2009; Clavaguera et al. 2009).

The α-syn accretion is central to several disorders including PD. The assemblage of α-syn into soluble oligomers might lead to the generation of indecipherable aggregates that play a vital part in the degeneration of neurons in PD (Kazantsev and Kolchinsky 2008). The α-syn protein is much more susceptible to misfolding and contributes to the death of neurons (Uversky 2007). The protein aggregation, processing, and toxicity are potentially affected by cathepsin D, a lysosomal protease, and α-syn has been found to be intricated in the phosphorylation of Tau by glycogen synthase kinase 3β in PD models (Cullen et al. 2009; Duka et al. 2009). In relation to dementia brains with Lewy body subjects, the enhanced extents of α-syn oligomers were identified (Paleologou et al. 2009). Induction of α-syn oligomerization might cause the dispersion of its pathogenicity and contribute to the Lewy body generation (Danzer et al. 2009; Luk et al. 2009), and the transmission of α-syn aggregates between the cells through endocytosis triggers nuclear disintegration and the stimulation of caspase-3 (Desplats et al. 2009).

2.3 Protein Aggregation as a Primary Disease-Initiating Event

2.3.1 Alzheimer's Disease

AD is a neurodegenerative disorder manifested by the progressive loss in memory, diminished execution of tasks, low speech, and difficulty with naming objects. Several adverse internal and external factors lead to neural damage in the hippocampal region and in the basal forebrain of the CNS, which triggers inaccurate neuronal connections and synaptic impairment (Selkoe 2002; Hardy and Selkoe 2002). Two major protein aggregates are associated with AD pathogenesis, including extracellular amyloid deposits and intracellular tangles. The primary element of amyloid plaques is the Aβ peptide, which is produced through proteolytic APP processing, and the neurofibrillary tangles are produced by the Tau protein hyperphosphorylation. The proteolytic cleavage could play a crucial part in neurodegenerative conditions including AD, in which the formation of Aβ peptide is carried by the biochemical activity of β-secretase and γ-secretase (Esler and Wolfe 2001; Citron 2002). In contrast, the α-secretase and γ-secretase cleave the APP protein, which possesses less tendency to aggregate while in intact form.

2.3.2 Parkinson's Disease

PD is another neurodegenerative disease in which the dopaminergic neurons in the substantia nigra get degenerated, leading to resting tremors, postural instability, rigidity, and bradykinesia. There are multiple genetic markers involved in the progression of the disease, but the main causative agent includes α-syn expressed in the presynaptic region of neurons. The genetic mutations in several genes could trigger the early onset of the disease, like α-syn gene point mutation that led to autosomal dominant PD, whereas the mutations in PINK1 or DJ-1 genes cause recessive early-onset PD (Valente et al. 2004). Lewy bodies are the detrimental elements of adult-onset PD existing in the cytoplasm of neurons, which are densely occurring in the substantia nigra but are also seen in other areas of the brain. The principal component of Lewy bodies is accumulated α-syn protein.

2.4 Protein Misfolding

2.4.1 Alzheimer's Disease

Generally, in abnormal conditions, the toxicity of proteins occurs following their switch from α-helix state to β-sheet confirmation. The β-sheet configuration is existing in various innate proteins, and the shifting from one state of α-helix to β-sheet conformation is the distinctive hallmark of amyloid accumulation (Kirkitadze et al. 2001). The occurrence of this transition is mostly linked with the transformation of a normal role to a pathological state, and this irregularity in conformational changes displays hydrophobic amino acid residues and triggers the aggregation of proteins (Dobson 1999, 2002). Subsequently, the toxic proteins involve other innate proteins to change their normal conformation into toxic ones, and those freshly generated abnormal proteins can repeat this transition to introduce a self-sustaining twist, thus stimulating the amplification of the generated toxicity with more impairment. The proteins perform their activity smoothly when their amino acids are folded appropriately, whereas they produce adverse effects when in misfolded conditions, following their deposition into insoluble forms that trigger toxicity to the cells (Dobson 1999, 2002).

The two factors of the unfolded protein response (UPR) including inositol-requiring enzyme 1 (IRE1) and activating transcription factor 6 (ATF6) are involved in the proteolytic function of presenilins (PS). The UPR system stimulates the proteolytic dissolution of IRE1, which initiates the transferring of cytosolic domains into the nucleus to take part in RNA splicing. However, the UPR mechanism gets reduced in the cells without PS1, which confirms that PS1 is a crucial element of UPR. It is mentioned that PS1 alterations damage the UPR system which leads to the deposition of unfolded proteins following endoplasmic stress (ER) (Katayama et al. 2001). The enhanced Aβ1–42 levels in these cells could be the effect of the preservation of unfolded APP in ER due to the defective protein folding mechanism.

2.4.2 Parkinson's Disease

In the ubiquitin/proteasome system, the aggregation and impairment of proteins are considered to be the essential step in the progression of PD. In protein folding mechanism, syn is an active constituent that has modified structures in different environments, such as a coil structure in aqueous solution, but displays significant upregulation in its helical state while having communication with phospholipids (Conway et al. 1998; Jo et al. 2000; Perrin et al. 2000). The confirmation switching of syn with other proteins could be vital for the normal physiological protein function. However, the relaxation in conformational transitions might trigger the exhibition of hydrophobic residues, which enhances the formation of insoluble aggregates of synuclein protein. It has been evidenced that syn protein possesses a tendency to generate fibrils with enhanced β-sheet confirmation which come across with the filaments present in LB (Lewy body) (Rochet et al. 2000; Serpell et al. 2000). The deposits of α-syn protein in the cell could interrupt the cellular mechanism and aggravate mitochondrial impairment, ER stress, and defective protein quality regulator structures (Giasson et al. 2000; Rosenberg 2000). Several studies have postulated that α-syn in the oligomeric and protofibrillarly forms also triggers toxicity to damage neuronal cells. Moreover, the occurrence of mutations in the Parkin was believed to be because of the impairment in enzymatic action, but subsequently, it has been manifested that most of the missense mutants of Parkin recollect their enzymatic ability (Kahle et al. 2000; Sherrington et al. 1995). Rather, the inactivation of the Parkin is believed to be due to the instigation of misfolding by these mutations. These missense mutations affect its solubility and consequently initiate its aggregation (Abeliovich et al. 2000; Shimura et al. 2001) (Fig. 2.1).

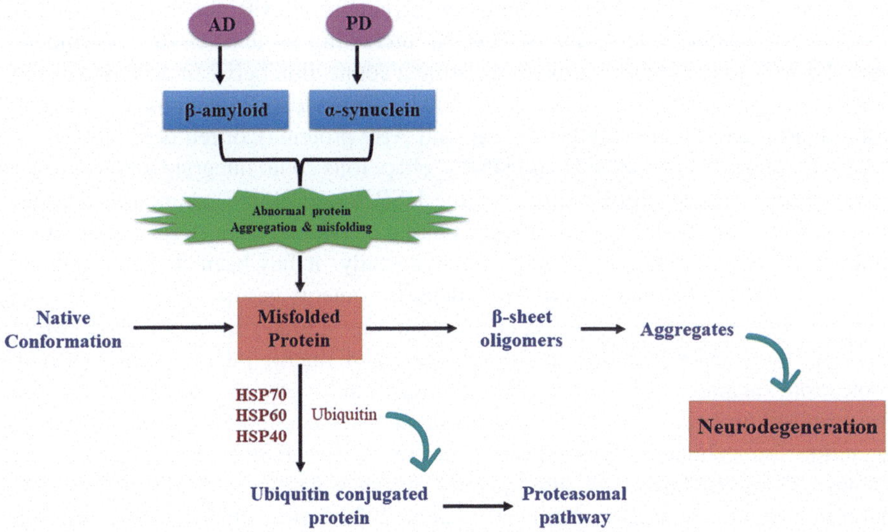

Fig. 2.1 Protein aggregation, misfolding, and protein conformation changes in AD and PD leading to the accumulation of aggregated proteins via the activity of ubiquitin proteasome system, leading to the degeneration of neurons

2.5 Proteostasis and Molecular Chaperons

2.5.1 Alzheimer's Disease

The impairment in proteome function and maintenance of consistency of cells occur in the living system, and to maintain the appropriate protein configuration, the protein misfolding must be rectified efficiently. In this regard, the molecular chaperones are considered to be the vital elements in either initiating the process of refolding or proceeding with the protein-degradation process. In AD, Aβ, Tau metabolism, and alteration in the proteostasis lead to neuronal damage and eventually the death of neurons (Ross and Poirier 2004; Ahmad et al. 2018a, b; Khan et al. 2012). The altered proteostasis initiates the deposition of aggregated and misfolded proteins, which are the major events in neurodegeneration as depicted in AD. The HSP70 and HSP90 are the two key chaperon scaffolds along with the co-chaperons that delineate the release and binding of substrate and are constrained via heat shock elements (HSE) through the HSF1 transcription factor. It has been proposed that the promoter region of the APP gene displayed HSEs which signify the role of HSPs on AD pathogenesis (Dewji and Do 1996; Salbaum et al. 1988). The APP protein moiety develops in Golgi bodies and the endoplasmic reticulum (ER), and the APP ectodomain is connected with the luminal ER chaperon (GRP78) glucose-related protein 78, the isotype of HSP70. The occurrence of this interplay deteriorates its development which leads to the production of Aβ40 and Aβ42 peptides (Hoshino et al. 2007; Yang et al. 1998). In the Golgi apparatus and ER, the cytosolic HSP70 co-chaperone carboxy terminus of HSC70-interacting protein (CHIP) coordinates with the APP intracellular region, facilitating a connection between APP processing and molecular chaperons (Kumar et al. 2007). The cytosolic chaperons are also in coordination with the metabolism of Aβ and APP protein (Cottrell et al. 2005).

It is documented that HSP22 and HSP27 chaperons could integrate with Aβ fibrils to regulate additional fibrilization, and the HSP70 and HSP90 chaperons are centrally linked with the inhibition of early phases of Aβ accretion (Evans et al. 2006; Wilhelmus et al. 2006). In an experimental study, it has been detected that the HSP70 and HSP90 enhanced the Aβ42 peptide contents in microglia after day one and subsequently diminished its concentration after a time period of 3 days, which suggests that the HSP action in the instigation of microglial cells could provide protection against Aβ accumulation (Kakimura et al. 2002). The Tau hyperphosphorylation affects its transmission with microtubules and triggers its accretion intracellularly. It is reported that the CHIP, HSP27, and HSP70 could identify Tau irregularity and subsequently diminish its extent by the processes of dephosphorylation and degradation (Dou et al. 2003; Petrucelli et al. 2004; Shimura et al. 2004).

2.5.2 Parkinson's Disease

The α-syn exists extensively in the brain tissues and the aggregation of its abnormal form led to the initiation of PD. The α-syn accumulation begins when its non-amyloid-β component (NAC) hydrophobic region encounters a configurational modification into β-sheets and incites polymerization which in turn give rises to oligomers and the formation of fibrils (Meade et al. 2019). There are several proteins responsible for the folding, deposition, and misfolding of α-syn protein. The two configurations of HSP90, i.e., HSP90α and HSP90β, are existing in mammals, and the HSP90 chaperon is extensively present in the cytosol but its isoforms have been reported in the ER and mitochondria (Hoter et al. 2018). It has N-terminal, central, and C-terminal domains and is the principal chaperon that triggers the α-syn fibrilization in PD (Bohush et al. 2019). Its involvement in the maturation of fibrils in an ATP-dependent manner has been found in a study by interacting with the NAC region of the α-syn (Falsone et al. 2009). The toxicity caused by α-syn in HSP90 cells was found to be in the enhanced state compared to the normal cells in the PD model of yeast (Liang et al. 2008). However, its inhibition supports that diminishing results of oligomerization and α-syn fibril formation in the H4 neuroglioma cells (Putcha et al. 2010). The HSP90 isoform overexpression in mitochondria includes TNF receptor-associated protein 1 (TRAP1), attenuating toxicity caused by α-syn in primary neurons of rat, HEK293 cells, and fruit fly (Butler et al. 2012).

Thus, the chaperons imperatively participate in the regulation of proteostasis by rectifying the abnormal proteins and in the folding of novel proteins (Hartl and Hayer-Hartl 2002). The HSP70 chaperon is intricated in the attenuation of α-syn accumulation by participating in the misfolded protein degradation or by fixing

Table 2.1 Proteins affecting proteostasis in Alzheimer's and Parkinson's disease

	Proteins	Function	Reference
AD	Aβ40/42	Diminishes proteasome activity and regulates autophagy via mTOR signaling	(Spilman et al. 2010)
	Tau	Tau accumulates induced modulatory effect on proteasome activity	(Keck et al. 2003)
	HSPs (HSP22, 27, 70, 90)	HSP22 and 27 tend to fix with fibrillar amyloid deposits to regulate their fibrilization, whereas HSP70 and 90 play a crucial role in regulating early phases of amyloid accumulation	(Wilhelmus et al. 2006)
	PS1	PS1 augments Aβ42 generation and is vital for the lysosomal autophagy and proteolysis	(Citron et al. 1997)
	BAG1	Participates in the modulation of Tau proteasomal degradation in the presence of HSP70	(Elliott et al. 2007)
	HSF1	It plays an instrumental role in the induction of APP gene during stress	(Salbaum et al. 1988)
PD	PINK1 (HSP90)	Takes part in the balancing of cleaved patterns of PINK1 gene	(Lin and Kang 2008)
	PINK1 (HSP60)	It plays an essential role in protein folding in mitochondria	(Rakovic et al. 2011)
	LRRK2 (HSP90)	Maintains the stability of LRRK2	(Wang et al. 2008)
	LRRK2 (HSC70)	Participates in the LRRK2 proteasomal degradation	(Wang et al. 2008)

refolding through the ubiquitin proteasome pathway (UPS). In PD rodent model, the enhanced toxicity of α-syn was observed due to alterations in the ATPase domain of HSP70 chaperon, which signifies that the refolding action of HSP70 was crucial for its defensive activity (Klucken et al. 2004). The dopaminergic neurons were degenerated via the overexpression of human WT α-syn in flies, which was fixed by α-syn and HSP70 co-expression, highlighting the HSP70 activity in attenuating the toxicity produced by α-syn protein (Auluck et al. 2002) (Table 2.1).

2.6 The Role of the Ubiquitin-Proteasome System

2.6.1 Alzheimer's Disease

The ubiquitin-proteasome system (UPS) plays an indispensable function in the degradation pathway of the cell. The two oligomers Aβ40 and Aβ42 could obstruct the activity of proteasome, while the internal surface binding of Aβ40 oligomer to the proteasome prevents the 20S chymotrypsin-like activity (Lopez Salon et al. 2003; Oh et al. 2005). It is evidenced in a dose-dependent study that these Aβ oligomers diminished the proteolytic function of proteasome (Tseng et al. 2008). Moreover, the 20S proteasome was detected in the degradation of Aβ oligomers in vitro, whereas the proteasome inhibition in cultured cells augmented the intracellular levels of these oligomers (Tseng et al. 2008). The Aβ42 oligomer has been manifested to deposit extensively in the healthy and AD brain in multivesicular bodies, which suggests that these patterns are the potential communication region of Aβ and the proteasome (Takahashi et al. 2002). It has also been manifested that the proteasome degrades the PS1 and PS2 proteins, which indicates the impact of UPS on the metabolism of Aβ oligomers. Therefore, the attenuation of the activity of proteasome could enhance the γ-secretase function and supports the generation of Aβ (Checler et al. 2000), which defines the Aβ and proteostasis association while emphasizing the crucial role of UPS in Aβ processing.

The 26S and 20S proteasomes have been found in the deterioration of Tau protein, and the proteasome inhibitors employed in the cellular and animal models enhanced the possibility of Tau protein deposition (Cardozo and Michaud 2002; David et al. 2002; Liu et al. 2009; Ren et al. 2006). A study has manifested that co-precipitation of Tau aggregates and 20S proteasome exhibited the attenuated levels of Tau aggregates in samples with least action of proteasome, highlighting the obstructive interaction between proteasome and Tau aggregates (Keck et al. 2003). Moreover, Tau accumulation from the human brain with AD has been found in the inhibition of proteasome directly in an in vitro study, whereas normal Tau isolates had no impact on the proteasome activity (Keck et al. 2003). The deposition and the maintenance of hyperphosphorylated Tau protein in AD might limit the interaction between Tau protein and CHIP (Dickey et al. 2006). Also, the Tau protein is acetylated posttranslationally, leads to the dysfunction of proteasomal degradation activity, and facilitates the enhanced Tau accumulation, which specifies the critical function of proteasome activity in the degradation of Tau aggregates (Min et al. 2010).

2.6.2 Parkinson's Disease

The internal occurrence of ubiquitinated proteins in neural insertions is one of the factors of NDs and holds several UPS structures which are involved in the intracellular exoneration of proteins. The impairment in the UPS system led to the initiation of neurodegenerative processes. In PD, proteasome and autophagy are critically allied with the degradation of α-syn that also depicts the association between UPS and autophagy lysosome pathway (Cuervo et al. 2004). The approximal occurrence of PARKIN and α-syn in LBs highlights that the PARKIN is centrally involved in the posttranslational alteration and α-syn ubiquitination (Schlossmacher et al. 2002). Another protein, i.e., PINK1, which coordinates with Beclin1 triggers the process of autophagy (Michiorri et al. 2010), whereas PARKIN protein via proteasome-dependent pathway conveniently augments the clearance of cellular Aβ (Rosen et al. 2010). An efficient Ub-3-ligase complex encompassing mutations in PINK1, PARKIN, and DJ1 proteins triggers misfolded protein degradation and might create a toxic process for recessive inherited types of PD (Eliezer 2009; Xiong et al. 2009). The inhibition of UPS could be responsible for the severe worsening of NDs, and the transportation of integral elements to the nervous system that augments the toxic protein degradation might get challenges in the abnormal functioning of UPS.

2.7 Aberrant Alternative Splicing Events in Neurodegenerative Disease

2.7.1 Alzheimer's Disease

The aberrant alternative splicing performs an instrumental function in the organism's complexity, and approximately 95% genes are spliced, alternatively stimulating the production of 100,000 proteins in the human genome (Mills et al. 2013). The alternative splicing events appears co-transcriptionally or generates posttranscriptionally various variants of mRNA from a single gene. The spliceosomes bring about the splicing which is made of U1 to U6 small RNA molecules and various other proteins (Qian and Liu 2014), whereas the specific nucleotide sequences present inside mRNA are involved in the regulation of splicing events. These nucleotide sequences or cis-elements are comprised of intronic splicing silencers, exonic splicing enhancers, and silencers (Qian and Liu 2014). In addition to the cis-elements, trans-elements are also involved in the splicing events in association with the sis-elements to regulate the process of alternative splicing (Stilling et al. 2014), and the mutations in this splicing system could trigger the onset of the disease.

The alternative splicing events and the differential gene expression could affect several signaling pathways in AD. The overexpressed genes in AD participate in the processes of transmission, synaptogenesis, long-term potentiation, and postsynaptic density, which collectively promote AD pathogenesis (Stilling et al. 2014). The splicing events could initiate the enhanced gene expression in AD associated with immune pathways, stimulating the upregulated responses of neuroinflammation in the hippocampal region (Stilling et al. 2014). Consequently, it has been anticipated that the diminishing specific AD genes might trigger chromosomal instability due to improper DNA repair mechanism (Penna et al. 2013). The Aβ oligomer aggregation gets affected by the mutation incidences in AD-associated factors such as APP, PSEN1, and PSEN2 (Schellenberg et al. 1992; Levy-Lahad et al. 1995). The PSEN1 and PSEN2 are part of the γ-secretase complex, and the abnormal cleaving of APP by this complex triggers the deposition of β-amyloid (Baulac et al. 2003). The alterations in PSEN1 gene occur in intron-4, leading to the defective splicing and elimination of exon-4 (De Jonghe et al. 1999), with the incidence of other alteration events that could promote the β-amyloid expression (Cruts and Van Broeckhoven 1998).

The alternative splicing of exons 2, 3, and 10 of the Tau protein marks the formation of six homologs expressed in the brain (Andreadis et al. 1992). The second microtubule binding repeat encoded by exon 10, and its insertion, produces isoforms of Tau gene with three or four binding sites such as 3R-tau and 4R-tau (Goedert et al. 1989). The proteins rich in serine/arginine (SR) residues are the family of splicing agents which promote Tau protein alternative splicing events, in which an SR protein, i.e., SC35, participates in the inclusion of exon 10 while acting as an enhancer on SC35 protein at 5' end of the Tau protein RNA transcript (Qian et al. 2011). In SC35 phosphorylation event by PKA (protein kinase A) regulates the inclusion of exon 10 that triggers the enhanced 3R-tau isotype expression (Chen et al. 2014). The elevation of exon 10 inclusion is directly associated with the 3R-tau to 4R-tau imbalance which occurs due to attenuation of the PKA phosphorylation pathway (Chen et al. 2014). Additionally, the exon 10 inclusion has been reported to produce 4R-tau isoform that promotes the amelioration in Tau aggregation and neurofibrillary tangles (Goedert et al. 1989). Several other SR proteins also execute the splicing of Tau gene regulation such as SRSR6, SRSF9, and SRSF1, which are also involved in the advancement of exon 10 inclusion, whereas SRSF3-SRSF4, SRSF7, and SRSF11 proteins are critically involved in suppressing exon 10 inclusion (Qian and Liu 2014). Thus, the role of PKA is believed to be a common mechanistic approach to regulate splicing events by acting on several SR proteins.

2.7.2 Parkinson's Disease

In PD, multiple related factors are involved in the unusual alternatively splicing events. The PD-associated gene PARK2 contains 12 exons which encodes 465 amino acids, and due to alternative splicing events, directs the production of seven isoforms including transcript variants (D'Agata and Cavallaro 2004). Their altered expression patterns have been associated with LB disorder. It has been demonstrated that in frame deletions, the transcript variant 3 expression lacks three to five exons and the transcript variant 12 lacks two to seven exons, which were ameliorated in PD conditions compared to normal conditions (Beyer et al. 2008; Humbert et al. 2007). Another gene, i.e., SNCAIP, which a presynaptic protein, is present at synaptic terminals which encodes synphilin-1 (Ribeiro et al. 2002). The SNAIP gene alternative splicing results in the generation of eight splicing variants, and among them four variants are condensed proteins of the C-terminus. However, the small length isoforms of the synphilin-1 might present robust interactions compared to the normal-sized proteins, and their altered expression patterns have been detected in LB disorder conditions (Humbert et al. 2007). Several studies have presented that synphilin-1A contains exon 9A but lacks exons 3 and 4 in PD. Hence, its overexpression enhances aggregation and triggers the inclusion generation and proteasome saturation, and protein toxicity, which specifies that synphilin-1A could play an indispensable role in neuronal degeneration (Eyal et al. 2006; Humbert et al. 2007).

In addition, LRRK2 is a PD-associated gene containing 51 exons which encode almost 2527 amino acids, and its genetic alteration is the common cause of PD (Farrer 2006). To date, 51 exons of LRRK2 have been sequenced and two missense alterations such as T23561 and G2019s have been reported which are disease-specific (Johnson et al. 2007). Moreover, at the splice donor site of exon 19, a

4-bp intronic deletion has been found to obliterate normal splicing events, and the mutation events in LRRK2 gene contribute approximately to 5% of the familial PD cases (Johnson et al. 2007). The first gene detected in PD pathophysiology is SNCA gene which encodes a presynaptic α-syn protein (Polymeropoulos et al. 1997). The SNCA gene consists of seven exons, and its alternative splicing isoforms have been found to be allied with abnormal expression patterns in disease conditions (Beyer et al. 2008). Four major isoforms have been distinguished in SNCA gene in which SNCA-140 is the full-length form, and SNCA-98 lacking exons 3 and 5, SNCA-112 lacking exon 5, and SNCA-126 lacking exon 3 are short-length transcripts (Campion et al. 1995; Ueda et al. 1994). The exon 5 excision, that is, SNCA-112, occurs at the 3' end of the SNCA gene, which could produce functional impairment pertinent to LB pathogenesis. The excision leads to the C-terminus condensation that promotes the improved aggregation of α-syn, which in turn directs the production of LB (Beyer 2006; Hoyer et al. 2004; Lee et al. 2002; Spillantini et al. 1997).

2.8 Mitochondrial Dysfunction in NDs

2.8.1 Alzheimer's Disease

The cytoplasmic organelle mitochondria are crucial for functions and survival of cells and neurons, which are critically intricated in the cell propagation, protein folding, cellular motility, and intermediary metabolism (Galluzzi et al. 2012; Ahmed et al. 2013, 2018a, b). The main purpose of mitochondria is its role in oxidative metabolism, and incidences of impairment in the production of ATP, which triggers loss of energy and cellular impairment and eventually causes cell death, as evidenced from several NDs (Beal 1996; Ahmad et al. 2019). In case of AD, impairment in the mitochondrial activity is considered to be the initial sign, prior to other involved fatal lesions, and any oxidative injury in mitochondria lead to the advancement of the disease (Hirai et al. 2001; Readnower et al. 2011; Rather et al. 2021). The incidences of alternations in the activity of mitochondria play a central role in the generation of oxidative stress, which triggers impairment to lipids, DNA, and proteins of mitochondria. It is documented that deficits in the energy production and upregulation of oxidative damage in mitochondria are the initial pathological signs that trigger neuronal degeneration (Johri and Beal 2012; Javed et al. 2015). The damage caused by the reactive oxygen species (ROS) initiates glial cells to produce pro-inflammatory markers and NO, following reactive nitrogen species (RNS) formation (Jekabsone et al. 2006). Subsequently, activated glial cells release H2O2 that is converted into a more toxic prooxidant called hypochlorous acid (Halliwell 1992). Therefore, the oxidative stress, mitochondrial deficit, and the instigation of inflammatory mediators leads to the pathogenesis of AD.

2.8.2 Parkinson's Disease

Mitochondrial impairment following oxidative stress is linked with the PD pathogenesis. Cardiolipin, a specific mitochondrial lipid, gets peroxidized by oxidative stress, results in cytochrome-c release in the cytosol, and subsequently triggers apoptosis. The cells and neurons require ATP for energy, and superoxide radicals and H2O2 are generated during oxidative phosphorylation in the mitochondria. In the pathological state, mitochondria get affected, leading to its impairment, which in turn enhances ROS production and damages the antioxidant mechanism. The dopaminergic neurons are more susceptible to the ROS products, and occurrence of additional oxidative damage could be detrimental to the cells and neurons. It is believed that electron transport chain complex I is crucial for the survival of neurons, and the intraperitoneal induction of MPTP and rotenone produced toxicity to the dopaminergic neurons (Betarbet 2002). Moreover, the downregulated activity of complex I has been observed in the PD subjects (Benecke 1993). Also, it was reported that in the somatodendritic area of nigral neurons, the mitochondrial density was remarkably reduced (Liang 2007). Mutations in the PARKIN, PINK1, and DJ-1 proteins are linked with the mitochondrial dysfunction and are associated with familial PD forms. The PARKIN is usually linked with the external membrane of mitochondria, and the cells obtained from PARKIN gene-mutated patients exhibited attenuated ATP generation and complex I activity (Darios 2003; Grunewald 2010; Mortiboys 2008). The incidence of mutations in another protein kinase, i.e., PINK1, present in mitochondria causes decline in ATP generation, diminished mitochondrial DNA, and formation of free radicals (Gegg 2009; Grunewald 2009). DJ-1 is a redox-sensitive protein existing in the mitochondria, which is involved in the elimination of H2O2 (Andres-Mateos 2007). Furthermore, α-syn interacts with the mitochondrial membranes, though it is present in cytosol, but has been found to obstruct complex I system (Liu et al. 2009). Also, it is reported that enhanced expression of α-syn in mice model manifested structural and functional irregularity in mitochondria (Martin 2006).

2.9 Role of Oxidative Stress and ROS in NDs

2.9.1 Alzheimer's Disease

The imbalance between the antioxidant defensive mechanism and the free radical formation is referred to as oxidative stress, and the capability of the intracellular system to counter and remove the adverse effects via antioxidants. A free radical immensely reacts with other molecules by carrying unpaired electrons, which is capable of interacting chemically with lipids, proteins, and DNA components to achieve stability among themselves. The stability of free radicals poses a great threat to destabilization of cellular components, following their extensive chain reactions. Due to the extra lipid contents in the brain makes it more susceptible target to oxidative stress, and its production is greatly involved in the progression of NDs such as AD (Pratico et al. 2002). Incidences of oxidative products of lipids, DNA, and proteins in postmortem tissues of AD subjects have been evident, which confirms its role in NDs (Giasson et al. 2002).

The extensive generation of ROS products has been observed in Aβ aggregates to advance amyloidogenesis and eventually causes death of neurons (Frank and Gupta 2005). The inference of ROS progression is reinforced by several clinical confirmations in AD including DNA and protein oxidation, lipid peroxidation, enhanced concentrations of neurotoxic trace metal ions, and diminished energy metabolism (Markesbery 1997). Thus, antioxidant defensive mechanism is an essential cellular defense system to eliminate the generation of ROS products and protects ROS-induced neuronal damage.

2.9.2 Parkinson's Disease

In the progression of PD, oxidative stress is considered to be the initial event that halts cellular defense mechanism and destabilize the normal function of the cells and tissues (Andersen 2004; Jenner 2003). The enhanced oxidized levels of DNA, proteins, and lipids were found in the substantia nigra of PD subjects (Bosco 2006; Nakabeppu 2007). Because of the presence of ROS-producing kinases which include tyrosinase, monoamine oxidase, and tyrosine hydroxylase, the dopaminergic neurons are predominantly susceptible to ROS. A temperate level of oxidative stress could also promote a cascade of oxidative events that directs the death of cells, because of the extra intracellular susceptibility to ROS. The impairment in mitochondria, dopamine metabolism, and inflammation are involved in the generation of ROS products and are considered to be the main sources of oxidative affliction for dopaminergic neurons. Environmental and genetic factors such as dopaminergic toxins and genetic mutations in PARKIN, PINK1, and DJ-1, whose gene products are essential for the normal functioning of mitochondria, are responsible for bringing dysfunctionality in mitochondria. Thus, mitochondrial impairment causes ROS deposition and cytochrome-c release and a cascade of apoptotic events (Fig. 2.2).

2.10 Conclusion

There are multiple components and mutations involved in the pathogenesis of NDs, and in the past several years, research had focused on exploring the mechanistic approach involved in the progression of NDs. In this chapter, we have summarized the evidences procured by mechanistic approach to the pathogenesis of NDs. The molecular mechanisms involved in the progression of NDs are the transportation of pathological proteins, conformational changes, formation of fibrils, and pathological protein aggregates in the cells and neurons. The impairment in proteostasis is documented as a predominant mechanism which involves protein misfolding and aggregation, following deficient clearance approach, which results in dysfunction of

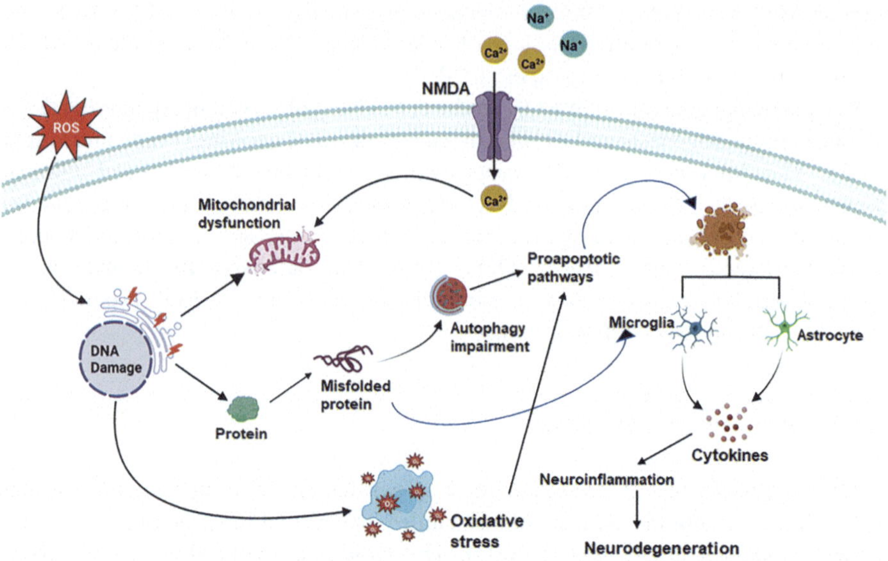

Fig. 2.2 Mechanisms involved in the progression of neurodegenerative disorders: Mitochondrial function, inflammatory cytokines, autophagy, protein homeostasis, and nuclear structure are damaged in NDs, which triggers aggregation of misfolded proteins, mitochondrial impairment, altered calcium homeostasis, damage to the antioxidant defense mechanism, and cell death, which eventually leads to neurodegeneration

various cellular processes and ultimately leads to cell death. The active functional configuration is absent in subsequent protein aggregates, which produces local stress and toxicity to the cellular environment. The formation of protein aggregates is the crucial step in the pathogenesis of NDs, and the contribution of UPS protein degradation is possibly a subsidiary event that advances the production of fibrils. The accumulated evidences supported that UPS is intricated in the constructive response caused by the protein deposits that transforms its action, which results in generation of enhanced protein aggregates. However, various pathological proteins have been detected, and the occurrence of conformational changes in them has emphasized the molecular pathogenesis in NDs. The advancement of novel techniques to detect, isolate, and scrutinize the pathological proteins involved in the progression of NDs would further provide the novel insights of molecular pathogenesis in NDs.

References

Abeliovich A, Schmitz Y, Farinas I, Choi-Lundberg D, Ho WH, Castillo PE et al (2000) Mice lacking alpha-synuclein display functional deficits in the nigrostriatal dopamine system. Neuron 25:239–252

Abramov E, Dolev I, Fogel H et al (2009) Amyloid-beta as a positive endogenous regulator of release probability at hippocampal synapses. Nat Neurosci 12:1567–1576

Aguzzi A, Sigurdson C, Heikenwaelder M (2008) Molecular mechanisms of prion pathogenesis. Annu Rev Pathol 3:11–40

Ahmad RM, Thenmozhi AJ, Manivasagam T, Bharathi DM, Essa MM, Guillemin GJ (2018a) Neuroprotective role of Asiatic acid in aluminium chloride induced rat model of Alzheimer's disease. Front Biosci (Schol Ed) 10(2):262–275

Ahmad RM, Thenmozhi AJ, Manivasagam T, Nataraj J, Essa MM, Chidambaram SB (2018b) Asiatic acid nullified aluminium toxicity in in vitro model of Alzheimer's disease. Front Biosci (Elite Ed) 10(2):287–299

Ahmad RM, Thenmozhi AJ, Manivasagam T, Saravanababu C, Guillemin GJ, Essa MM (2019) Asiatic acid attenuated aluminum chloride-induced tau pathology, oxidative stress and apoptosis via AKT/GSK-3β signaling pathway in Wistar rats. Neurotox Res 35(4):955–968

Ahmed ME, Javed H, Khan MM, Vaibhav K, Ahmad A, Khan A, Tabassum R, Islam F, Safhi MM, Islam F (2013) Attenuation of oxidative damage-associated cognitive decline by Withania somnifera in rat model of streptozotocin-induced cognitive impairment. Protoplasma 250(5): 1067–1078

Andersen JK (2004) Oxidative stress in neurodegeneration: cause or consequence? Nat Med 10: S18–S25

Andreadis A, Brown WM, Kosik KS (1992) Structure and novel exons of the human tau gene. Biochemistry 31:10626–10633

Andres-Mateos E (2007) DJ-1 gene deletion reveals that DJ-1 is an atypical peroxiredoxin like peroxidase. Proc Natl Acad Sci U S A 104:14807–14812

Arnaud L, Robakis NK, Figueiredo-Pereia ME (2006) It may take inflammation, phosphorylation and ubiquitination to 'tangle' in Alzheimer's disease. Neurodegener Dis 3:313–319

Auluck PK, Chan HE, Trojanowski JQ, Lee VMY, Bonini NM (2002) Chaperone suppression of α-synuclein toxicity in a drosophila model for Parkinson's disease. Science 295:865–868

Baulac S, LaVoie MJ, Kimberly WT, Strahle J, Wolfe MS et al (2003) Functional gamma-secretase complex assembly in Golgi/trans-Golgi network: interactions among presenilin, nicastrin, Aph1, Pen-2, and gamma-secretase substrates. Neurobiol Dis 14:194–204

Beal MF (1996) Mitochondria, free radicals, and neurodegeneration. Curr Opin Neurobiol 6:661–666

Benecke R (1993) Electron transfer complexes I and IV of platelets are abnormal in Parkinson's disease but normal in Parkinson-plus syndromes. Brain 116(6):1451–1463

Betarbet R (2002) Animal models of Parkinson's disease. BioEssays 24:308–318

Beyer K (2006) α-Synuclein structure, posttranslational modification and alternative splicing as aggregation enhancers. Acta Neuropathol 112:237–251

Beyer K, Domingo-Sabat M, Humbert J, Carrato C, Ferrer I, Ariza A (2008) Differential expression of a-synuclein, parkin, and synphilin-1 isoforms in Lewy body disease. Neurogenetics 9:163–172

Bohush A, Bieganowski P, Filipek A (2019) Hsp90 and its co-chaperones in neurodegenerative diseases. Int J Mol Sci 20:4976

Bosco DA (2006) Elevated levels of oxidized cholesterol metabolites in Lewy body disease brains accelerate alpha-synuclein fibrilization. Nat Chem Biol 2:249–253

Braak H et al (2003) Staging of brain pathology related to sporadic Parkinson's disease. Neurobiol Aging 24:197–211

Braak H, Braak E (1991) Neuropathological staging of Alzheimer-related changes. Acta Neuropathol 82:239–259

Butler EK, Voigt A, Lutz AK, Toegel JP, Gerhardt E, Karsten P, Falkenburger B, Reinartz A, Winklhofer KF, Schulz JB (2012) The mitochondrial chaperone protein TRAP1 mitigates α-Synuclein toxicity. PLoS Genet 8:e1002488

Campion D, Martin C, Heilig R, Charbonnier F, Moreau V, Flaman JM, Petit JL, Hannequin D, Brice A, Frebourg T (1995) The NACP/synuclein gene: chromosomal assignment and screening for alterations in Alzheimer disease. Genomics 26:254–257

Cardozo C, Michaud C (2002) Proteasome-mediated degradation of tau proteins occurs independently of the chymotrypsin-like activity by a nonprocessive pathway. Arch Biochem Biophys 408:103–110

Checler F, da Costa CA, Ancolio K, Chevallier N, Lopez-Perez E, Marambaud P (2000) Role of the proteasome in Alzheimer's disease. Biochim Biophys Acta 1502:133–138

Chen C, Jin N, Qian W, Liu W, Tan X et al (2014) Cyclic AMP-dependent protein kinase enhances SC35-promoted tau exon 10 inclusion. Mol Neurobiol 49:615–624

Citron M (2002) Alzheimer's disease: treatments in discovery and development. Nat Neurosci 5: 1055–1057

Citron M, Westaway D, Xia W, Carlson G, Diehl T, Levesque G et al (1997) Mutant presenilins of Alzheimer's disease increase production of 42-residue amyloid beta-protein in both transfected cells and transgenic mice. Nat Med 3:67–72

Clavaguera F et al (2014) Peripheral administration of tau aggregates triggers intracerebral tauopathy in transgenic mice. Acta Neuropathol 127:299–301

Clavaguera F, Bolmont T, Crowther RA et al (2009) Transmission and spreading of tauopathy in transgenic mouse brain. Nat Cell Biol 11:909–913

Conway KA, Harper JD, Lansbury PT (1998) Accelerated in vitro fibril formation by a mutant alpha-synuclein linked to early-onset Parkinson disease. Nat Med 4:1318–1320

Cottrell BA, Galvan V, Banwait S, Gorostiza O, Lombardo CR, Williams T, Schilling B, Peel A, Gibson B, Koo EH, Link CD, Bredesen DE (2005) A pilot proteomic study of amyloid precursor interactors in Alzheimer's disease. Ann Neurol 58:277–289

Cruts M, Van Broeckhoven C (1998) Presenilin mutations in Alzheimer's disease. Hum Mutat 11: 183–190

Cuervo AM, Stefanis L, Fredenburg R et al (2004) Impaired degradation of mutant alpha-synuclein by chaperone-mediated autophagy. Science 305:1292–1295

Cullen V, Lindfors M, Ng J et al (2009) Cathepsin D expression level affects alpha-synuclein processing, aggregation, and toxicity in vivo. Mol Brain 2. https://doi.org/10.1186/1756-6606-2-5

D'Agata V, Cavallaro S (2004) Parkin transcript variants in rat and human brain. Neurochem Res 29:1715–1724

Danzer KM, Krebs SK, Wolff M et al (2009) Seeding induced by alpha-synuclein oligomers provides evidence for spreading of alpha-synuclein pathology. J Neurochem 111:192–203

Darios F (2003) Parkin prevents mitochondrial swelling and cytochrome c release in mitochondria-dependent cell death. Hum Mol Genet 12:517–526

David DC, Layfield R, Serpell L, Narain Y, Goedert M, Spillantini MG (2002) Proteasomal degradation of tau protein. J Neurochem 83:176–185

De Jonghe C, Cruts M, Rogaeva EA, Tysoe C, Singleton A et al (1999) Aberrant splicing in the presenilin-1 intron 4 mutation causes presenile Alzheimer's disease by increased Abeta42 secretion. Hum Mol Genet 8:1529–1540

Desplats P, Lee HJ, Bae EJ et al (2009) Inclusion formation and neuronal cell death through neuron-to-neuron transmission of alpha-synuclein. Proc Natl Acad Sci U S A 106:13010–13015

Dewji NN, Do C (1996) Heat shock factor-1 mediates the transcriptional activation of Alzheimer's beta-amyloid precursor protein gene in response to stress. Brain Res Mol Brain Res 35:325–328

Dickey CA, Yue M, Lin WL, Dickson DW, Dunmore JH, Lee WC, Zehr C, West G, Cao S, Clark AM, Caldwell GA, Caldwell KA, Eckman C, Patterson C, Hutton M, Petrucelli L (2006) Deletion of the ubiquitin ligase CHIP leads to the accumulation, but not the aggregation, of both endogenous phospho- and caspase-3-cleaved tau species. J Neurosci 26:6985–6996

Dobson CM (1999) Protein misfolding, evolution and disease. Trends Biochem Sci 24(9):329–332

Dobson CM (2002) Protein-misfolding diseases: getting out of shape. Nature 418(6899):729–730

Dou F, Netzer WJ, Tanemura K, Li F, Hartl FU, Takashima A, Gouras GK, Greengard P, Xu H (2003) Chaperones increase association of tau protein with microtubules. Proc Natl Acad Sci U S A 100:721–726

Duka T, Duka V, Joyce JN et al (2009) Alpha-Synuclein contributes to GSK-3beta-catalyzed tau phosphorylation in Parkinson's disease models. FASEB J 23:2820–2830

Ecroyd H, Carver JA (2008) Unraveling the mysteries of protein folding and misfolding. IUBMB Life 60:769–774

Eliezer D (2009) Biophysical characterization of intrinsically disordered proteins. Curr Opin Struct Biol 19:23–30

Elliott E, Tsvetkov P, Ginzburg I (2007) BAG-1 associates with Hsc70.Tau complex and regulates the proteasomal degradation of tau protein. J Biol Chem 282:37276–37284

Esler WP, Wolfe MS (2001) A portrait of Alzheimer's secretases-new features and familiar faces. Science 293:1449–1454

Evans CG, Wisen S, Gestwicki JE (2006) Heat shock proteins 70 and 90 inhibit early stages of amyloid beta-(1-42) aggregation in vitro. J Biol Chem 281:33182–33191

Eyal A, Szargel R, Avraham E, Liani E, Haskin J, Rott R, Engelender S (2006) Synphilin-1A: an aggregation-prone isoform of synphilin-1 that causes neuronal death and is present in aggregates from a-synucleinopathy patients. Proc Natl Acad Sci U S A 103:5917–5922

Falsone SF, Kungl AJ, Rek A, Cappai R, Zangger K (2009) The molecular chaperone Hsp90 modulates intermediate steps of amyloid assembly of the Parkinson-related protein α-synuclein. J Biol Chem 284:31190–31199

Farrer MJ (2006) Genetics of Parkinson disease: paradigm shifts and future prospects. Nat Rev Genet 7:306–318

Frank B, Gupta S (2005) A review of antioxidants and Alzheimer's disease. Ann Clin Psychiatry 17: 269–286

Frost B, Jacks RL (2009) Diamond MI. Propagation of tau misfolding from the outside to the inside of a cell. J Biol Chem 284:12845–12852

Galluzzi L, Keppo O, Trojel-Hansen C, Kroemer G (2012) Mitochondrial control of cellular life, stress, and death. Circ Res 111:1198–1207

Gegg ME (2009) Silencing of PINK1 expression affects mitochondrial DNA and oxidative phosphorylation in dopaminergic cells. PLoS One 4:e4756

Giasson BI, Duda JE, Murray IV, Chen Q, Souza JM, Hurtig HI et al (2000) Oxidative damage linked to neurodegeneration by selective alpha-synuclein nitration in synucleinopathy lesions. Science 290:985–989

Giasson BI, Ischiropoulos H, Lee VM, Trojanowski JO (2002) The relationship between oxidative/ nitrative stress and pathological inclusions in Alzheimer's and Parkinson's diseases. Free Radic Biol Med 32:1264–1275

Goedert M, Klug A, Crowther RA (2006) Tau protein, the paired helical filament and Alzheimer's disease. J Alzheimers Dis 9:195–207

Goedert M, Spillantini MG, Jakes R, Rutherford D, Crowther RA (1989) Multiple isoforms of human microtubule-associated protein tau: sequences and localization in neurofibrillary tangles of Alzheimer's disease. Neuron 3:519–526

Grunewald A (2009) Differential effects of PINK1 nonsense and missense mutations on mitochondrial function and morphology. Exp Neurol 219:266–273

Grunewald A (2010) Mutant Parkin impairs mitochondrial function and morphology in human fibroblasts. PLoS One 5:e12962

Guo JL et al (2016) Unique pathological tau conformers from Alzheimer's brains transmit tau pathology in nontransgenic mice. J Exp Med 213:2635–2654

Halliwell B (1992) Reactive oxygen species and the central nervous system. J Neurochem 59:1609–1623

Hardy J, Selkoe DJ (2002) The amyloid hypothesis of Alzheimer's disease: progress and problems on the road to therapeutics. Science 297:353–356

Harmeier A, Wozny C, Rost BR et al (2009) Role of amyloid-beta glycine 33 in oligomerization, toxicity, and neuronal plasticity. J Neurosci 29:7582–7590

Hartl FU, Hayer-Hartl M (2002) Molecular chaperones in the cytosol: from nascent chain to folded protein. Science 295:1852–1858

Hirai K, Aliev G, Nunomura A, Fujioka H, Russell RL, Atwood CS, Johnson AB, Kress Y, Vinters HV, Tabaton M, SHimohama S, Cash AD, Siedlak SL, Harris LR, Jones PK, Petersen RB, Perry G, Smith M (2001) Mitochondrial abnormalities in Alzheimer's disease. J Neurosci 21: 3017–3023

Hoe HS, Fu Z, Makarova A et al (2009) The effects of amyloid precursor protein on postsynaptic composition and activity. J Biol Chem 284:8495–8506

Hoshino T, Nakaya T, Araki W, Suzuki K, Suzuki T, Mizushima T (2007) Endoplasmic reticulum chaperones inhibit the production of amyloid-beta peptides. Biochem J 402:581–589

Hoter A, El-Sabban ME, Naim HY (2018) The HSP90 family: structure, regulation, function, and implications in health and disease. Int J Mol Sci 19:2560

Hoyer W, Cherny D, Subramaniam V, Jovin TM (2004) Impact of the acidic C-terminal region comprising amino acids 109–140 on a-synuclein aggregation in vitro. Biochemistry 43:16233–16242

Humbert J, Beyer K, Carrato C, Mate JL, Ferrer I, Ariza A (2007) Parkin and synphilin-1 isoform expression changes in Lewy body diseases. Neurobiol Dis 26:681–687

Israeli E, Sharon R (2009) Beta-synuclein occurs in vivo in lipid-associated oligomers and forms hetero-oligomers with alpha-synuclein. J Neurochem 108:465–474

Javed H, Vaibhav K, Ahmed ME, Khan A, Tabassum R, Islam F, Safhi MM, Islam F (2015) Effect of hesperidin on neurobehavioral, neuroinflammation, oxidative stress and lipid alteration in intracerebroventricular streptozotocin induced cognitive impairment in mice. J Neurol Sci 348(1–2):51–59

Jekabsone A, Mander PK, Tickler A, Sharpe M, Brown GC (2006) Fibrillar beta-amyloid peptide Abeta1-40 activates microglial proliferation via stimulating TNF-alpha release and H2O2 derived from NADPH oxidase: a cell culture study. J Neuroinflammation 3:24

Jenner P (2003) Oxidative stress in Parkinson's disease. Ann Neurol 53(3):S26–S36

Jo E, McLaurin J, Yip CM, St. George-Hyslop P, Fraser PE (2000) Alpha-synuclein membrane interactions and lipid specificity. J Biol Chem 275:34328–34334

Johnson J, Paisan-Ruiz C, Lopez G, Crews C, Britton A, Malkani R, Evans EW, McInerney-Leo A et al (2007) Comprehensive screening of a north American Parkinson's disease cohort for LRRK2 mutation. Neurodegener Dis 4:386–391

Johri A, Beal MF (2012) Mitochondrial dysfunction in neurodegenerative diseases. J Pharmacol Exp Ther 342:619–630

Kahle PJ, Leimer U, Haass C (2000) Does failure of Parkin-mediated ubiquitination cause juvenile parkinsonism? Trends Biochem Sci 25:524–527

Kakimura J, Kitamura Y, Takata K, Umeki M, Suzuki S, Shibagaki K, Taniguchi T, Nomura Y, Gebicke-Haerter PJ, Smith MA, Perry G, Shimohama S (2002) Microglial activation and amyloid-beta clearance induced by exogenous heat-shock proteins. FASEB J 16:601–603

Katayama T, Imaizumi K, Honda A, Yoneda T, Kudo T, Takeda M et al (2001) Disturbed activation of endoplasmic reticulum stress transducers by familial Alzheimer's disease-linked presenilin 1 mutation. J Biol Chem 16(276):43446–43454

Kazantsev AG, Kolchinsky AM (2008) Central role of alpha-synuclein oligomers in neurodegeneration in Parkinson disease. Arch Neurol 65:1577–1581

Keck S, Nitsch R, Grune T, Ullrich O (2003) Proteasome inhibition by paired helical filament-tau in brains of patients with Alzheimer's disease. J Neurochem 85:115–122

Khan A, Vaibhav K, Javed H, Khan MM, Tabassum R, Ahmed ME, Srivastava P, Khuwaja G, Islam F, Siddiqui MS, Safhi MM, Islam F (2012) Attenuation of Aβ-induced neurotoxicity by thymoquinone via inhibition of mitochondrial dysfunction and oxidative stress. Mol Cell Biochem 369(1–2):55–65

Kirkitadze MD, Condron MM, Teplow DB (2001) Identification and characterization of key kinetic intermediates in amyloid beta-protein fibrillogenesis. J Mol Biol 312(5):1103–1119

Klucken J, Shin Y, Masliah E, Hyman BT, McLean PJ (2004) Hsp70 reduces α-synuclein aggregation and toxicity. J Biol Chem 279:25497–25502

Kumar P, Ambasta RK, Veereshwarayya V, Rosen KM, Kosik KS, Band H, Mestril R, Patterson C, Querfurth HW (2007) CHIP and HSPs interact with beta-APP in a proteasome-dependent manner and influence Abeta metabolism. Hum Mol Genet 16:848–864

Lafaye P, Achour I, England P et al (2009) Single-domain antibodies recognize selectively small oligomeric forms of amyloid beta, prevent Abeta-induced neurotoxicity and inhibit fibril formation. Mol Immunol 46:695–704

Laferriere F et al (2019) TDP-43 extracted from frontotemporal lobar degeneration subject brains displays distinct aggregate assemblies and neurotoxic effects reflecting disease progression rates. Nat Neurosci 22:65–77

Lauren J, Gimbel DA, Nygaard HB et al (2009) Cellular prion protein mediates impairment of synaptic plasticity by amyloid-beta oligomers. Nature 457:1128–1132

Lee HJ, Choi C, Lee SJ (2002) Membrane-bound a-synuclein has a high aggregation propensity and the ability to seed the aggregation of the cytosolic form. J Biol Chem 277:671–678

Levy-Lahad E, Wijsman EM, Nemens E, Anderson L, Goddard KA et al (1995) A familial Alzheimer's disease locus on chromosome 1. Science 269:970–973

Liang CL (2007) Mitochondria mass is low in mouse substantia nigra dopamine neurons: implications for Parkinson's disease. Exp Neurol 203:370–380

Liang J, Clark-Dixon C, Wang S, Flower TR, Williams-Hart T, Zweig R, Robinson LC, Tatchell K, Witt SN (2008) Novel suppressors of α-synuclein toxicity identified using yeast. Hum Mol Genet 17:3784–3795

Lin W, Kang UJ (2008) Characterization of PINK1 processing, stability, and subcellular localization. J Neurochem 106:464–474

Liu YH, Wei W, Yin J, Liu GP, Wang Q, Cao FY, Wang JZ (2009) Proteasome inhibition increases tau accumulation independent of phosphorylation. Neurobiol Aging 30:1949–1961

Lopez Salon M, Pasquini L, Besio Moreno M, Pasquini JM, Soto E (2003) Relationship between beta-amyloid degradation and the 26S proteasome in neural cells. Exp Neurol 180:131–143

Luk KC et al (2012) Pathological alpha- synuclein transmission initiates Parkinson-like neurodegeneration in nontransgenic mice. Science 338:949–953

Luk KC, Song C, O'Brien P et al (2009) Exogenous alpha-synuclein fibrils seed the formation of Lewy body-like intracellular inclusions in cultured cells. Proc Natl Acad Sci U S A 106:20051–20056

Markesbery WR (1997) Oxidative stress hypothesis in Alzheimer's disease. Free Radic Biol Med 23:134–147

Martin LJ (2006) Parkinson's disease alpha-synuclein transgenic mice develop neuronal mitochondrial degeneration and cell death. J Neurosci 26:41–50

Meade RM, Fairlie DP, Mason JM (2019) Alpha-synuclein structure and Parkinson's disease–lessons and emerging principles. Mol Neurodegener 14:29

Michiorri S, Gelmetti V, Giarda E et al (2010) The Parkinson-associated protein PINK1 interacts with Beclin 1 and promotes autophagy. Cell Death Differ 17:962. https://doi.org/10.1038/cdd.2009.200

Mills JD, Nalpathamkalam T, Jacobs HI, Janitz C, Merico D et al (2013) RNA-Seq analysis of the parietal cortex in Alzheimer's disease reveals alternatively spliced isoforms related to lipid metabolism. Neurosci Lett 536:90–95

Min SW, Cho SH, Zhou Y, Schroeder S, Haroutunian V, Seeley WW, Huang EJ, Shen Y, Masliah E, Mukherjee C, Meyers D, Cole PA, Ott M, Gan L (2010) Acetylation of tau inhibits its degradation and contributes to tauopathy. Neuron 67:953–966

Moore RA, Taubner LM, Priola SA (2009) Prion protein misfolding and disease. Curr Opin Struct Biol 19:14–22

Mortiboys H (2008) Mitochondrial function and morphology are impaired in parkin-mutant fibroblasts. Ann Neurol 64:555–565

Nakabeppu Y (2007) Oxidative damage in nucleic acids and Parkinson's disease. J Neurosci Res 85:919–934

Nimmrich V, Ebert U (2009) Is Alzheimer's disease a result of presynaptic failure? Synaptic dysfunctions induced by oligomeric beta-amyloid. Rev Neurosci 20:1–12

Nygaard HB, Strittmatter SM (2009) Cellular prion protein mediates the toxicity of beta-amyloid oligomers: implications for Alzheimer disease. Arch Neurol 66:1325–1328

Oh S, Hong HS, Hwang E, Sim HJ, Lee W, Shin SJ, Mook-Jung I (2005) Amyloid peptide attenuates the proteasome activity in neuronal cells. Mech Ageing Dev 126:1292–1299

Paleologou KE, Kragh CL, Mann DM et al (2009) Detection of elevated levels of soluble alpha-synuclein oligomers in post-mortem brain extracts from patients with dementia with Lewy bodies. Brain 132:1093–1101

Palop JJ, Chin J, Mucke L (2006) A network dysfunction perspective on neurodegenerative diseases. Nature 443:768–773

Penna I, Vassallo I, Nizzari M, Russo D, Costa D et al (2013) A novel snRNA-like transcript affects amyloidogenesis and cell cycle progression through perturbation of Fe65L1 (APBB2) alternative splicing. Biochim Biophys Acta 1833:1511–1526

Perrin RJ, Woods WS, Clayton DF, George JM (2000) Interaction of human alpha-synuclein and Parkinson's disease variants with phospholipids. Structural analysis using site-directed mutagenesis. J Biol Chem 275:34393–34398

Petrucelli L, Dickson D, Kehoe K, Taylor J, Snyder H, Grover A, De Lucia M, McGowan E, Lewis J, Prihar G, Kim J, Dillmann WH, Browne SE, Hall A, Voellmy R, Tsuboi Y, Dawson TM, Wolozin B, Hardy J, Hutton M (2004) CHIP and Hsp70 regulate tau ubiquitination, degradation and aggregation. Hum Mol Genet 13:703–714

Polymeropoulos MH, Lavedan C, Leroy E, Ide SE, Dehejia A, Dutra A, Pike B, Root H, Rubenstein J, Boyer R et al (1997) Mutation in the a-synuclein gene identified in families with Parkinson's disease. Science 276:2045–2047

Pratico D, Uryu K, Sung S, Tang S, Trojanowski JQ, Lee VM (2002) Aluminum modulates brain amyloidosis through oxidative stress in APP transgenic mice. FASEB J 16:1138–1140

Putcha P, Danzer KM, Kranich LR, Scott A, Silinski M, Mabbett S, Hicks CD, Veal JM, Steed PM, Hyman BT (2010) Brain-permeable small-molecule inhibitors of Hsp90 prevent α-synuclein oligomer formation and rescue α-synuclein-induced toxicity. J Pharmacol Exp Ther 332:849–857

Qian W, Liang H, Shi J, Jin N, Grundke-Iqbal I et al (2011) Regulation of the alternative splicing of tau exon 10 by SC35 and Dyrk1A. Nucleic Acids Res 39:6161–6171

Qian W, Liu F (2014) Regulation of alternative splicing of tau exon 10. Neurosci Bull 30:367–377

Rather MA, Khan A, Alshahrani S, Rashid H, Qadri M, Rashid S, Alsaffar RM, Kamal MA, Rehman MU (2021) Inflammation and Alzheimer's disease: mechanisms and therapeutic implications by natural products. Mediat Inflamm 31:9982954

Rakovic A, Grünewald A, Voges L, Hofmann S, Orolicki S, Lohmann K, Klein C (2011) PINK1-interacting proteins: proteomic analysis of overexpressed PINK1. Parkinsons Dis 6:153979

Readnower RD, Sauerbeck AD, Sullivan PG (2011) Mitochondria, amyloid β, and Alzheimer's disease. Int J Alzheimers Dis 10:1042–1045

Ren PH, Lauckner JE, Kachirskaia I et al (2009) Cytoplasmic penetration and persistent infection of mammalian cells by polyglutamine aggregates. Nat Cell Biol 11:219–225

Ren QG, Liao XM, Wang ZF, Qu ZS, Wang JZ (2006) The involvement of glycogen synthase kinase-3 and protein phosphatase-2A in lactacystin-induced tau accumulation. FEBS Lett 580: 2503–2511

Ribeiro CS, Carneiro K, Ross CA, Menezes JR, Engelender S (2002) Synphilin-1 is developmentally localized to synaptic terminals, and its association with synaptic vesicles is modulated by a-synuclein. J Biol Chem 277:23927–23933

Rochet JC, Conway KA, Lansbury PT (2000) Inhibition of fibrillization and accumulation of prefibrillar oligomers in mixtures of human and mouse alpha-synuclein. Biochemistry 39: 10619–10626

Rosen KM, Moussa CE, Lee HK et al (2010) Parkin reverses intracellular beta-amyloid accumulation and its negative effects on proteasome function. J Neurosci Res 88:167–178

Rosenberg RN (2000) The molecular and genetic basis of AD: the end of the beginning. Neurology 54:2045–2054
Ross CA, Poirier MA (2004) Protein aggregation and neurodegenerative disease. Nat Med 10:S10–S17
Salbaum JM, Weidemann A, Lemaire HG, Masters CL, Beyreuther K (1988) The promoter of Alzheimer's disease amyloid A4 precursor gene. EMBO J 7:2807–2813
Schellenberg GD, Bird TD, Wijsman EM, Orr HT, Anderson L et al (1992) Genetic linkage evidence for a familial Alzheimer's disease locus on chromosome 14. Science 258:668–671
Schlossmacher MG, Frosch MP, Gai WP et al (2002) Parkin localizes to the Lewy bodies of Parkinson disease and dementia with Lewy bodies. Am J Pathol 160:1655–1667
Seeley WW, Crawford RK, Zhou J et al (2009) Neurodegenerative diseases target large-scale human brain networks. Neuron 62:42–52
Selkoe DJ (2002) Alzheimer's disease is a synaptic failure. Science 298:789–791
Selkoe DJ (2008) Soluble oligomers of the amyloid beta-protein impair synaptic plasticity and behavior. Behav Brain Res 192:106–113
Serpell LC, Berrimann J, Jakes R, Goedert M, Crowther RA (2000) Fiber diffraction of synthetic alpha-synuclein filaments shows amyloid-like cross-beta conformation. Proc Natl Acad Sci U S A 97:4897–4902
Sherrington R, Rogaev EI, Liang Y, Rogaeva EA, Levesque G, Ikeda M et al (1995) Cloning of a gene bearing missense mutations in early-onset familial Alzheimer's disease. Nature 375:754–760
Shimura H, Miura-Shimura Y, Kosik KS (2004) Binding of tau to heat shock protein 27 leads to decreased concentration of hyperphosphorylated tau and enhanced cell survival. J Biol Chem 279:17957–17962
Shimura H, Schlossmacher MG, Hattori N, Frosch MP, Trockenbacher A, Schneider R et al (2001) Ubiquitination of a new form of alpha-synuclein by Parkin from human brain: implications for Parkinson's disease. Science 293:263–269
Soto C, Estrada LD (2008) Protein misfolding and neurodegeneration. Arch Neurol 65:184–189
Spillantini MG, Schmidt ML, Lee VM, Trojanowski JQ, Jakes R, Goedert M (1997) α-Synuclein in Lewy bodies. Nature 388:839–840
Spilman P, Podlutskaya N, Hart MJ, Debnath J, Gorostiza O, Bredesen D, Richardson A, Strong R, Galvan V (2010) Inhibition of mTOR by rapamycin abolishes cognitive deficits and reduces amyloid-beta levels in a mouse model of Alzheimer's disease. PLoS One 5:e9979
Stilling RM, Benito E, Gertig M, Barth J, Capece V et al (2014) De-regulation of gene expression and alternative splicing affects distinct cellular pathways in the aging hippocampus. Front Cell Neurosci 8:373
Surmeier DJ, Obeso JA, Halliday GM (2017) Selective neuronal vulnerability in Parkinson disease. Nat Rev Neurosci 18:101–113
Takahashi RH, Milner TA, Li F, Nam EE, Edgar MA, YamaguchiH BMF, Xu H, Greengard P, Gouras GK (2002) Intraneuronal Alzheimer abeta42 accumulates in multivesicular bodies and is associated with synaptic pathology. Am J Pathol 161:1869–1879
Tomic JL, Pensalfini A, Head E et al (2009) Soluble fibrillar oligomer levels are elevated in Alzheimer's disease brain and correlate with cognitive dysfunction. Neurobiol Dis 35:352–358
Tseng BP, Green KN, Chan JL, Blurton-Jones M, LaFerla FM (2008) Abeta inhibits the proteasome and enhances amyloid and tau accumulation. Neurobiol Aging 29:1607–1618
Ueda K, Saitoh T, Mori H (1994) Tissue-dependent alternative splicing of mRNA for NACP, the precursor of non-ab component of Alzheimer's disease amyloid. Biochem Biophys Res Commun 205:1366–1372
Uversky VN (2007) Neuropathology, biochemistry, and biophysics of alpha-synuclein aggregation. J Neurochem 103:17–37
Valente EM et al (2004) Hereditary early-onset Parkinson's disease caused by mutations in PINK1. Science 304:1158–1160

Volpicelli-Daley LA et al (2011) Exogenous α-synuclein fibrils induce Lewy body pathology leading to synaptic dysfunction and neuron death. Neuron 72:57–71

Wang L, Xie C, Greggio E, Parisiadou L, Shim H, Sun L, Chandran J, Lin X, Lai C, Yang WJ (2008) The chaperone activity of heat shock protein 90 is critical for maintaining the stability of leucine-rich repeat kinase 2. J Neurosci 28:3384–3391

Wilhelmus MM, Otte-Holler I, Wesseling P, de Waal RM, Boelens WC, Verbeek MM (2006) Specific association of small heat shock proteins with the pathological hallmarks of Alzheimer's disease brains. Neuropathol Appl Neurobiol 32:119–130

Williams AJ, Paulson HL (2008) Polyglutamine neurodegeneration: protein misfolding revisited. Trends Neurosci 31:521–528

Xia W, Yang T, Shankar G et al (2009) A specific enzyme-linked immunosorbent assay for measuring beta-amyloid protein oligomers in human plasma and brain tissue of patients with Alzheimer disease. Arch Neurol 66:190–199

Xiong H, Wang D, Chen L et al (2009) Parkin, PINK1, and DJ-1 form a ubiquitin E3 ligase complex promoting unfolded protein degradation. J Clin Invest 119:650–660

Yang Y, Turner RS, Gaut JR (1998) The chaperone BiP/GRP78 binds to amyloid precursor protein and decreases Abeta40 and Abeta42 secretion. J Biol Chem 273:25552–25555

3 Isolation, Characterization and Differentiation of Stem Cells

Hmed Ben-Nasr, Mongi Saoudi, and Riadh Badraoui

Abstract

Throughout the development of metazoan, a pool of genes and factors, the black box cassette (BBC), regulates all processes deciding cells' proliferation and differentiation, in a fascinating and synchronized hierarchical pathway. At maturity, only a tiny fraction of cell subsets, the totipotent cells, retain in memory an active BBC in order to achieve regeneration and repair of damaged tissues. Totipotency, mainly the memorial BBC, forged a new era of scientific, agronomic and medical researches, among which the cellular therapy is intended to cut off with several diseases of degenerative nature or needing tissue replacement. However, the application of SCs is still refrained at its different steps. This chapter discusses recent status of totipotent cell identification, selection and conditioning for stem cell (SC)-based application, particularly for clinical use.

H. Ben-Nasr (✉)
Laboratory of Pharmacology. Medicine faculty of Sfax, Sfax, Tunisia

Department of Life Sciences, Faculty of Sciences of Gafsa, Gafsa, Tunisia

M. Saoudi
Laboratory of Animal Ecophysiology, Faculty of Sciences of Sfax, University of Sfax, Sfax, Tunisia

R. Badraoui
Laboratory of General Biology, Department of Biology, University of Ha'il, Ha'il City, Saudi Arabia

Laboratory of induced and developed diseases LR19ES12, Medicine Faculty of Sfax. University of Sfax, Sfax, Tunisia

Section of Histology–Cytology, Medicine Faculty of Tunis El Manar University, Tunis, Tunisia

S. Jahan, A. J. Siddiqui (eds.), *Applications of Stem Cells and Derived Exosomes in Neurodegenerative Disorders*, https://doi.org/10.1007/978-981-99-3848-3_3

Keywords

Totipotent cells · Origin · Selection · Regulation · Therapy

3.1 Introduction

The story of stem cells begun in the mid-seventeenth century when Robert Hooke first discovered that all living organisms are made of a unified structure, the cell (Hyllner et al. 2015). Details of how a unique cell arises into metazoan body were progressively disclosed based on the great advances in embryology and oncology. In fact, the fertilized egg undergoes a sequential series of cellular division and differentiation to build specialized tissues and organs of a new organism. Hundreds of cells with different shapes and functions are consequently formed. The term totipotency was coined to describe this peculiar phenomenon that is progressively lost throughout the developmental course, i.e. from zygote to maturation. In adulthood, totipotency is substantially retained in some tiny subpopulations, namely, stem cells (SCs) that are featured to repair some damaged tissues, to ensure immune defence or to reproduce (germinal cells: GCs). The unique cells that keep their total totipotency belong to the germinal line. Other cellular types, the somatic ones, are generally classified according to their ability to regenerate tissues or to differentiate into other cell types, e.g., multi-, pluri-, or uni-potent stem cells. Those totally losing this capability are termed somatic cells.

SCs are a keystone basis in several domains including assisted reproduction and the production of hybrids, transgenic organisms and clones. Perhaps, the genesis of genetically modified organisms intended to ameliorate food production is the most commonly known application based on stem cells (Heller 2003; Slack 2018). Recognition of the cellular stemness and its potential to regenerate damaged parts of the body instigates a new genera of pharmacological treatment, i.e. 'the cellular therapy' (CT). It is conceived that bone marrow (BM) replacement in haematological cancer represents the first successful application of such therapy (Thomas et al. 1975a, b). Early assays of BM replacement were limited by the immune self-recognition sentinel, leading to graft rejection. The immune system of patients is then profoundly compromised by total irradiation for grafting success. Great knowledge in molecular biology, and in particular oncology, divulgated the elemental sequences involved in this process and allowed engineering of the function, proliferation and differentiation of cells, to give a novel type of stem cells, the induced pluripotent stem cells (iPSCs). iPSCs are basically crucial in almost cellular and molecular biotechnologies, and eventually CT (Gage 1998). This chapter addresses the major problems that limit application of SCs with special focus on selection, isolation and control techniques of stem cells.

3.2 The Concept of a Stem Cell

Stem cells have been defined as those cells capable of self-renewing and differentiating into other cell types and regenerating tissues or organs. Throughout their development, they can be isolated from the embryonic (ESC), extraembryonic (exESC) or foetus and adult tissues (ASC). Germinal stem cells (GSC), which are specifically intended for reproduction, are found in both embryonic and fetal-adult stages. ESC isolated from early embryos (2 to 16 cell blastocysts) and GSC have a versatile totipotency and can develop into a complete new body. Once the blastocyst compacts, an outer cell layer, forming a 'shell' for an inner mass cell (IMC), appears in its surface. The 'shell' layer develops into extraembryonic tissues that comprise the trophoblast, which is an important source of human pluripotent cells (hPSCs). IMC segregates into two layers: the hypoblast and the epiblast. The former is ideal to form the endoderm which is committed to differentiate into digestive tract and yolk sac. The epiblastic layer duplicates under mechanical and chemical inductive cues provided by the endoderm to give the ectoderm and the mesoderm. These three germ layers constitute the first progenitors of the future tissues of the organism. Epithelial and neuronal cells derive from ectoderm and other tissues (muscle, skin, bone, etc.) originate from the mesoderm layer (Pera et al. 2000; Baker and Pera 2018; Rossant and Tam 2022). Cell-to-extracellular matrix (EMC) and cell-to-cell contacts are determinants in inducing and guiding differentiation and movements of a PSC to reach its final functional status. Certainly, this physico-spatial environment of PSC varies with its progressive advancement towards its fate, alongside embryogenesis and foetal/adult growth. For example, if a naive PSC from the dorsal root of the epiblast is patterned to become a neurone, it is firstly submitted to influences of mesodermal inductions, ectodermal surrounding cells and organizer centres, during neurulation and ontogenesis. A naive PSC will never transform into the sought neurone if it is directly engrafted into the final tissue, since it will miss all intermediate commitments and differentiation events. Otherwise, there are time- and space-dependent restrictions and conditions for PSC to go throughout its destiny (Rao 1999). These step-wise differentiation and cells' fate decision are governed by two 'soft programmes'. The first one, we named herein the 'black box cassette' (BBC), is proper to the cell and involves genetic and cytoplasmic factors such as maternal mRNA. The second regulator involves the environmental/epigenetic factors (EF) imposed by surrounding conditions. BBC and EF are firmly coordinated and hierarchized to ensure the appropriate development of the future newborn. Disruption in the BBC programme and/or EF, at any developmental stage of the metazoan body, might lead to the appearance of particular organoids: tumors and cancers (Nasr et al. 2013). During morphogenesis and ontogenesis, cellular changes (moving from place to place, shape-shaping and connecting to other cells or EMC, achieving energetic demands, proliferation and apoptosis) occurs chronologically in an orchestrated and hierarchized fashion with surrounding environment, towards the fate of the committed cell in adulthood. Any alteration in these regulated variations of the cell leads to abnormal development and tumorigenesis (Fong et al. 2010).

A decade after the successful transfer of an epithelial into an enucleated embryo cell to clone the sheep Dolly (Tian et al. 2003), researchers were able to transform somatic cell into a new type of stem cell, the induced pluripotent stem cell (iPSC). The first iPSCs were obtained by modifying the expression of Oct4, Sox2, Klf4 and c-Myc genes in fibroblastic cells that acquired an ESC-like cell features (Bilic and Belmonte 2012; Shi et al. 2017). iPSC engineering guides researches to regulate and control the 'fate' of SC and subsequent application in scientific researches, biotechnologies and in particular, the sought CT.

3.3 Challenges to Characterize and Use SCs

Totipotent stem cells are isolated from developing embryos and their annexed tissues (trophoblast). Most of progenitors of adult PSC originate from IMC-derived tissues. During the development of organisms, cells' totipotency is regulated by a specific pool of genes (BBC) and surrounding conditions. In the earliest embryogenesis, the BBC is fully active and becomes progressively locked down. Only pluri- or uni-potent adult stem cells will retain a part of the active BBC devoted for tissue repair and regeneration (Fig. 3.1). Through cell reprogramming by retroviral introduction of genes, nuclear transfer, mutations, methylation/demethylation, or exposure to specific factors like the leukaemia inhibitory factor (LIF), it was possible to make changes in BBC to get iPSC of specific features.

In assisted reproduction, the fully totipotent stem cells, the fertilized egg and its earliest daughter cells require mere morphological screening through microscopic observation and in vitro conditioning to attain their ultimate goal. During transgenesis, techniques appear more difficult and involve nuclei handling of ESCs that should retain their full capacity to build a structured embryo. Up to the blastocyst compaction, further colossal interventions to isolate, identify and reprogramme SCs are needed to achieve the objective like CT.

Envisioned applications of SCs in scientific researches and therapy encompass tissue replacement, drug delivery, endocrine secretion and many other medical domains. Despite the accelerated research towards human pluripotent stem cells (hPSCs) and their characterization and utilization in CT, and generally in other

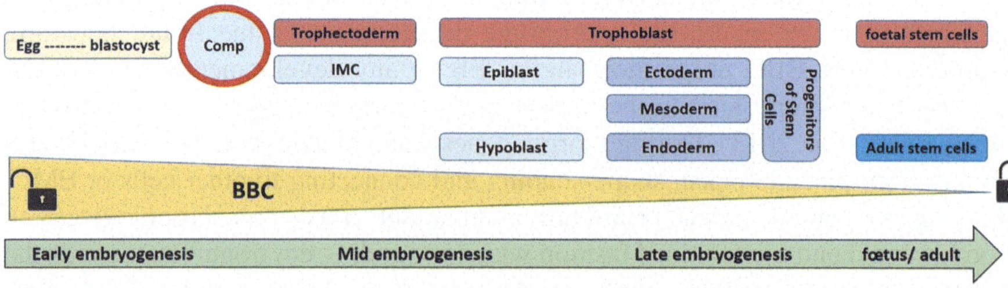

Fig. 3.1 Schematic representation of origins of SCs, adapted from (Pera et al. 2000; Nasr et al. 2013; Baker and Pera 2018; Rossant and Tam 2022)

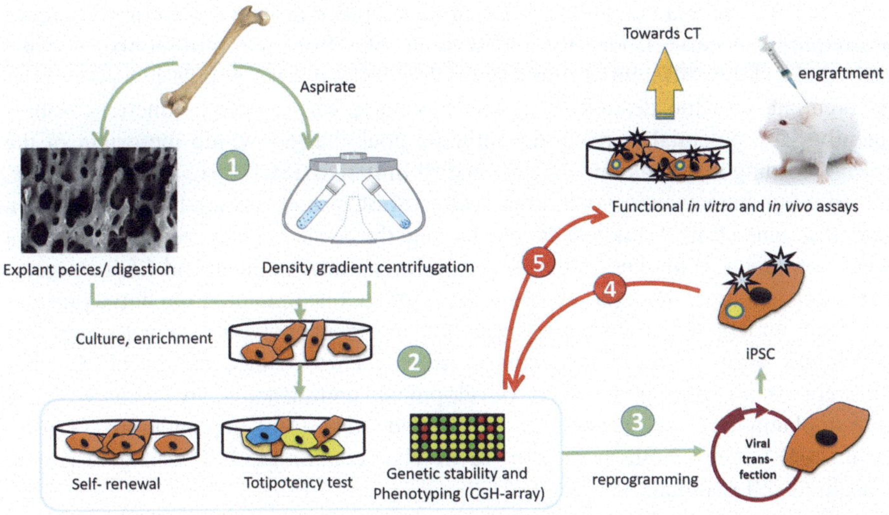

Fig. 3.2 Step-wise laboratory assays towards cellular therapy application (photomicrograph of biopsy from (Badraoui et al. 2007))

disciplines, many challenges remain unsolved. These constraints can be summarized in three major questions: Which stem cells fit the best target, how to isolate and prepare stem cells to attain their endpoint and what is the way to control the fate decision of SC? (Fig. 3.2).

To use pluripotent cells in CT, PSC need firstly to be isolated from a tissue of choice and identified through successive laboratory tests. In the third step, characterized PSC are reprogrammed in order to fit the ultimate goal of CT (drug delivery, secretome therapy or repair and replacement transplantation). Produced iPSC are then submitted to the same serial laboratory assays, before in vitro and in vivo experimental trials of transplantation. Because of the important conditional variations between the culture media, and the targeted tissue and which of the origin, accurate mathematical models might predict the fate of iPSC and CT to avoid tumorigenesis and wrong pathway of CT.

3.3.1 Which Stem Cells Fit Best the Ultimate Endpoint?

To produce transgenic and genetically modified organisms, fertilized oocyte or SCs isolated from early embryo are used by enucleation and pronuclei injection or gene transfection. Although it requires meticulous in vitro conditioning and nucleic handling, the final goal is easily attained by in utero embryo transfer to obtain the transgenic organism. SCs used for such endpoint are either in vitro-fertilized eggs or their first blastocyst-derived cells of known properties (Council 1989; Gama Sosa et al. 2010). For CT and molecular and genetic researches, SCs can be obtained from a large variety of tissues, including the bone marrow, brain, muscle, adipose tissue, lung, hair follicle and gut, and at different developmental stages. Several techniques are required to harvest, identify and separate SCs from non-proliferative somatic cells that exhibit different features, totipotencies and gene expression patterns to differentiate into the desired tissue or to accomplish a specific function. Subsequently SC's potential to reach the ultimate goal changes within the origin of the cell. According to the International Stem Cell Initiative (ISCI) there are, for example, 59 different human embryonic stem cells (hESCs) that are ready to be used in scientific and clinical studies. While having the same origin, these cells exhibit some common CD markers, such as glycolipid, keratan sulphate and CD9 antigens, but differ in many others (collection et al. 2007). Concomitantly, the naive pluripotent cell undergoes progressive variations in its phenotyping. Trusler and her colleagues reviewed these variations in hPSCs. They pointed that an hPSC takes different states during in vivo development commencing by a naive state (pre-implantation) and primed or committed state (post-implantation) (Trusler et al. 2018). Thus, both in in vitro or in vivo conditions, PSC can give rise to undesired cell populations.

SC totipotency declines from multipotency to unipotency alongside with ageing as a consequence of the progressive locking down of the BBC that can be inhibited by telemore shortening (Blasco 2007), thus enabling cells to have more specificity and stability. Therefore, it is recognized that uni-potent stem cells isolated from the CT-targeted tissue are the best SCs to ensure the success of the therapy. Such choice allows low time- and cost-consuming techniques and yields SC with adequate cell determinants (CD) that fit to the target. It also permits avoiding the matter of immune histocompatibility. Therefore, a neural stem cell, for example, which is isolated from the brain, more suitable in CT for than a mesenchymal stem cell (MSC) Alzheimer's disease. However, it is not always the case because of the morbidity of invasive techniques of isolation and problems of donor graft's host compatibility of the stem cell to be used in CT. Determining the origin and the timing for SC isolation is crucial for a successful CT.

3.3.2 How to Isolate, Identify and Prepare Stem Cells for their Endpoint?

The International Stem Cell Banking Initiative postulates that SC is best characterized through three laboratory assays: (i) the pluripotency tests, (ii) in vitro and in vivo differentiation tests and (iii) karyotype analysis to outline their genetic stability (Crook et al. 2010). SCs can be obtained from tissue's explant or in vitro engineering of cells. Because of their motility and ability to adhere to plastic surfaces, they are easily separated from other cells of the native tissue. Explants are excised into small pieces or submitted to digestion by proteolytic enzymes and cultured in adequate medium, for sufficient time to liberate SC that mostly adhere to plastic surfaces of culture dishes or flasks (Zhu et al. 2010). Successive cultures of the obtained SC yield a relatively pure population of SC. Condition of isolating SC greatly varies within the tissue of origin and the expected SC to be enriched. During processes of SC harvest, proliferation and differentiation assays and cryopreservation, bioreactor systems should be supplemented with prerequisite nutrients, growth factors, cytokines, antibiotics, foetal bovine serum (FBS), EMC substrate (e.g. Matrigel matrix) and many other factors, in order to maintain their survival and totipotency (Ulloa-Montoya et al. 2005). Many investigators outlined the requirement of coculture with feeder cells, like irradiated embryonic fibroblast, in addition to standard soluble factors in culture medium, in order to maintain hPSC survival and pluripotency. However, for mouse PSC, adequate soluble factors related to BMP-4, Stat 3 and LIF are sufficient (Villa-Diaz et al. 2013; Chen et al. 2014). The adjustment of culture media and conditions is therefore of great concern for successful SC isolation and monitoring. Pragmatically, the culture system might reproduce a similar condition close to these of original tissue. On peripheral blood or aspirates of the bone marrow, PSC are isolated using density gradient centrifugation, magnetic cell separation and flow cytometry coupled with immunophenotyping by fluorescent antigens. More accurate and sophisticated techniques, such as array comparative genomic hybridization (aCGH), are needed which permit the detection of many chromosomal abnormalities appearing in SC at the same time (Elliott et al. 2010). Isolated PSC might spontaneously change their phenotypes, because of their instability out of their original tissue. In effect, it was shown that many SCs differentiate after ten passages in culture medium and arise into heterogeneous-derived cell population (Blum and Benvenisty 2008). This fact breaks the concept of SC stability. Recourse for identifying and selecting the desired SC is, de novo, required, and several cellular markers including cluster differentiation (CD), metabolic enzymes and gene transcriptome are ready to be used for identification of PSC. Wang and his colleagues found that hiPSC and hESC could be rapidly identified by their capability to differentially bind to 13 lectins (extracted from different plant species) at the same time. This property is ensured by lectins' differential binding to hydrophobic glycoproteins moving in the outer surface of the cytoplasmic membrane. This glycomic array-based technique fairly separate PSC from non-totipotent cells (Wang et al. 2011).

Each of the 200 cellular types that are present in the human body expresses a unique set of genes. Analysing these transcriptomes might inform about the origin, developmental stage and eventually the pluripotency of the SC (Yilmaz and Benvenisty 2019). Because of the great biotechnological progress, scientists are now able to depict the transcriptome of these genes and introduce or retrieve some of them to produce a large variety of iPSC with specific patterns.

According to the ISCI (collection et al. 2007), all hESCs share in common the same surface markers and developmental gene transcriptomes. This signature includes membrane surface markers (glycolipids SSEA3 and SSEA4, keratan sulphate TRA-1-60 and TRA-1-81 (CD34-related transmembrane glycoproteins), GCTM2, GCT343, CD9, CD90, HLA-1, tissue-non-specific alkaline phosphatase enzyme and the transcriptome of a variety of development- regulating genes (NANOG, OCT4 or POU5F1, TDGF1, DNMT38, GABRB3 and GDF3) (collection et al. 2007). Interestingly, expression of most of these genes is also shared with cancer stem cells that develop into tumours (Hanahan and Weinberg 2011). We

Table 3.1 Examples of human pluripotent stem cells and their identifiers

Original tissue	Type of SC	Cellular identifiers	Reference
In vitro-cultured embryo	hESC	SSEA-3+, SSEA-4+, TRA-1-60+, TRA-1-81+, GDTM-2+, alkaline phosphatase+, Oct-4+, germ cell nuclear factor+, GDF-3+, TDGF-1+	(Pera et al. 2000)
Peripheral blood, bone marrow, umbilical cord blood	hHSC	$CD34^+$, $CD59^+$, $Thy1^+$, $CD38^{low/-}$, c-$kit^{-/\ low}$, Lin−	(Geraerts and Verfaillie 2009)
Bone marrow All tissues	hMSC	$CD10^+$, $CD13^+$, $CD73^+$, $CD90^+$, $CD105^+$, $CD140b^+$, $CD146^+$, $CD271^+$, $CD340\ ^+$, $CD349^+$, $W8B2^+$, $SSEA4^+$, Stro-1^+	
Bone marrow	hBMSC	$CD45^-$, MHC I/II^-, c-Kit^{low},$CD90^{low/-}$ / $Oct4^-$	
Amniotic fluid	hAFS	$cKit^+$, MHC-I^+, $MHCII^{low}$, $CD45^-$, $CD34^-$, $CD133^-$, $CD29^+$, $CD44^+$, $CD73^+$, $CD90^+$, $CD105^+$ and $SSEA4^+$ / $Oct4^+$	

coined the term the 'black box cassette (BBC)' to this pool of genes which are involved in both embryogenesis and tumorigenesis (Nasr et al. 2013). Because of this close similarity between PSC and cancer stem cell (CSC), tumours are produced from PSC, a fact that refrains the application of CT (Ghosh et al. 2011; Chen et al. 2012). Other genes are also differentially expressed during embryogenesis such as SOX and BMP. Some of them have a temporary expression or repression in relation to the developmental stage and the targeted organ. The distal enhancer gene OCT4 regulates PSC entrance in differentiation and maintains its pluripotency. In contrast to the Sox2 gene, the Oct4 gene is overexpressed in naive human mesenchymal stem cell from different origins such as the bone marrow, muscle and brain (Theunissen et al. 2014). Obviously, none of these gene modifications is sufficient to change totipotency. Thus, mathematical models are now developed to outline correlations between these hallmarks of SC's identification, differentiation and ontogenesis. These models might solve the problem of SC markers changes within development and predict the time and localization of the sought PSC to be selected and used in CT. They might also estimate the accuracy and safety of CT (Borsani et al. 2019). Nonetheless, standard methods and PSC identifiers exist and permit isolation and selection of various types with encouraging findings towards CT.

Table 3.1 summarizes the common markers of some human stem cells. An SC subtype is characterized by a combination of this marker expression (presence/absence or high/low expression). Some markers could be shared by different SCs such as the stage-specific embryonic antigen (SSEA4) which is commonly expressed by human embryonic stem cell (hESC), human mesenchymal stem cell (hMSC) and human amniotic fluid cell (hAFC). Interestingly, human hematopoietic stem cell (hHSC), human mesenchymal stem cell (hMSC) and human bone marrow-derived stem cell (hBMSC) can be isolated from the same origin, with the bone marrow while having enormous differences in its hallmarks. Such diversification reflects the ambiguity of the accurate characterization of SCs.

3.3.3 How to Control the Fate Decision of SC?

During normal embryogenesis, the regulation of SC proliferation and its differentiation is imposed by a cascade of multiple genes controlling the polarization, segmentation and ontogenesis of the embryo. Each pool of genes responds to several soluble growth factors (FGF, TGF, LIF etc.) and instructive induction that comprises physically established intercellular and extracellular matrix interactions and gradient-diffuse signals. Understanding these BBC and EF elements is a basic keystone in CT.

CT is suggested as a good solution to cut off with matters of organ transplantation, targeted- drug delivery and many endocrine dysfunctions, and using this approach in clinical trials is continuously reported (Trounson and McDonald 2015). But the possibility that transplanted PSC turns in a wrong way and leads to cancer development remains embarrassing (Cunningham et al. 2012). According to the ultimate goal of cellular therapy, there is a need to reprogramme the PSC to obtain the sought features. Because of their ability to home the injured tissues or tumours, PSC, are sought as good drug carriers to target injured tissues and tumours. In general, they handle the expression of exosomes carrying the active molecules and the release of exosomes at the disease's 'niche'. They can be also used to counteract cancer development through their proteic transcriptome that includes chemokines, growth factors and angiogenic factors. In particular, MSC overexpress the stromal cell-derived factor 1 that orients it to the injured tissue, thus facilitating the drug delivery to the 'cible' (Lai et al. 2013; Tran and Damaser 2015; Duan et al. 2021). These MSC are characterized by several markers including CD119, CD121, CD25, CD123, CD124, CD54, CD56, CD58, CD62, CD166 and many others that rely on chemotaxis and cell adhesion (Saeedi et al. 2019). While encouraging for repair and regeneration of many damaged tissues, transplantation of PSC for such issue remains inconclusive.

Techniques for producing iPSCs are multiple and involve genetic modifications through viral transduction and non-viral transfection. Associated adenovirus (AAV) and lentivirus (LV) vectors are the most used vectors for gene transfection in PSC. This technique inserts specific genes to be amplified or to produce transgenic cells (Shakhbazau et al. 2008; Park et al. 2015; Zubkova et al. 2021). Through such technique, Palma and his collaborators (2013) induced the overexpression of OCT4, a prominent gene in pluripotency, in both MSC and fibroblast. As a result, there was activation of the expression of some development related genes such SOX, NANOG and c-MYC (Palma et al. 2013). However, while promoting results and advances in clinical trials for some diseases, accurate genome editing for a successful and safe CT remains far to be established and needs further hard work (Hockemeyer and Jaenisch 2016).

3.4 Conclusion

Since their discovery, SCs gained much attention in biology and medicine. In addition to ethical considerations, their existence in numerous tissues and at very low amounts, particularly in adulthood, restrains their isolation and identification. Multipotent ESCs are already used in many domains such as assisted reproduction and the production of transgenic and genetically modified organisms that needs ease and 'simple' in vitro handling to attain the ultimate goal. However, in other applications like CT, it sounds technically more difficult because of the need to control the in vivo microenvironment and the fate of SCs that is yet far to be recognized. To endorse the applications of SCs, there should be predictive approaches modelling the choice of SC subtype and its origin and frameworks of technical methods of isolation, identification, genetic reprogramming and eventually in vitro and in vivo microenvironment conditioning. The choice of the SC type and its isolation technique from the original tissue must fit the best targeted endpoint. The more proximate the isolated SC to the featured targeted tissue/organ, the less handling will be needed. Up-to-date genetic reprogramming of SCs includes nanotechniques that present highly specific target at molecular and genetic levels and might strengthen their application accuracy.

References

Badraoui R, Abdelmoula NB, Sahnoun Z, Fakhfakh Z, Rebai T (2007) Effect of subchronic exposure to tetradifon on bone remodelling and metabolism in female rat. C R Biol 330(12): 897–904

Baker CL, Pera MF (2018) Capturing totipotent stem cells. Cell Stem Cell 22(1):25–34

Bilic J, Belmonte JCI (2012) Concise review: induced pluripotent stem cells versus embryonic stem cells: close enough or yet too far apart? Stem Cells 30(1):33–41

Blasco MA (2007) Telomere length, stem cells and aging. Nat Chem Biol 3(10):640–649

Blum B, Benvenisty N (2008) The tumorigenicity of human embryonic stem cells. Adv Cancer Res 100:133–158

Borsani E, Della Vedova AM, Rezzani R, Rodella LF, Cristini C (2019) Correlation between human nervous system development and acquisition of fetal skills: an overview. Brain Dev 41(3):225–233

Chen KG, Mallon BS, McKay RD, Robey PG (2014) Human pluripotent stem cell culture: considerations for maintenance, expansion, and therapeutics. Cell Stem Cell 14(1):13–26

Chen L, Kasai T, Li Y, Sugii Y, Jin G, Okada M, Vaidyanath A, Mizutani A, Satoh A, Kudoh T (2012) A model of cancer stem cells derived from mouse induced pluripotent stem cells. PLoS One 7(4):e33544

Collection, C. o. d., a. G. P. J. 5, C. o. m. distribution, s. c. H. L. E. 1, and s. m. o. t. I. p. w. p. b. a. s. c. c. A. P. W. B. N. M. B. S. M. R. D. P. M. F. 35 36 Rossant Janet 39 Semb Henrik 41 Stacey Glyn N 1 et al (2007) Characterization of human embryonic stem cell lines by the international stem cell initiative. Nat Biotechnol 25(7):803–816

Council NR (1989) Field testing genetically modified organisms: framework for decisions

Crook JM, Hei D, Stacey G (2010) The international stem cell banking initiative (ISCBI): raising standards to bank on. In Vitro Cell Dev Biol Anim 46(3):169–172

Cunningham JJ, Ulbright TM, Pera MF, Looijenga LH (2012) Lessons from human teratomas to guide development of safe stem cell therapies. Nat Biotechnol 30(9):849–857

Duan H, Liu Y, Gao Z, Huang W (2021) Recent advances in drug delivery systems for targeting cancer stem cells. Acta Pharm Sin B 11(1):55–70
Elliott AM, Hohenstein Elliott KA, Kammesheidt A (2010) High resolution array-CGH characterization of human stem cells using a stem cell focused microarray. Mol Biotechnol 46(3):234–242
Fong CY, Gauthaman K, Bongso A (2010) Teratomas from pluripotent stem cells: a clinical hurdle. J Cell Biochem 111(4):769–781
Gage FH (1998) Cell therapy. Nature 392(6679 Suppl):18–24
Gama Sosa MA, De Gasperi R, Elder GA (2010) Animal transgenesis: an overview. Brain Struct Funct 214(2):91–109
Geraerts M, Verfaillie CM (2009) Adult stem and progenitor cells. Adv Biochem Eng Biotechnol 114:1–21
Ghosh Z, Huang M, Hu S, Wilson KD, Dey D, Wu JC (2011) Dissecting the oncogenic and tumorigenic potential of differentiated human induced pluripotent stem cells and human embryonic stem CellsOncogenic potential of pluripotent stem cells. Cancer Res 71(14):5030–5039
Hanahan D, Weinberg RA (2011) Hallmarks of cancer: the next generation. Cell 144(5):646–674
Heller KJ (2003) Genetically engineered food. Methods and detection, 2nd edn. Wiley-VCH Verlag GmbH & Co. KGaA, Weiheim
Hockemeyer D, Jaenisch R (2016) Induced pluripotent stem cells meet genome editing. Cell Stem Cell 18(5):573–586
Hyllner J, Mason C, Wilmut I (2015) Cells: from Robert Hooke to cell therapy—a 350 year journey, vol No. 370. The Royal Society, p 20150320
Lai RC, Yeo RWY, Tan KH, Lim SK (2013) Exosomes for drug delivery—a novel application for the mesenchymal stem cell. Biotechnol Adv 31(5):543–551
Nasr HB, Hammami ST, Zeghal K (2013) Could cancer initiate from bone marrow progenitors? Iran J Cancer Prev 6(2):108
Palma C, Tannous M, Malta T, Russo EMS, Covas DT, Picanço-Castro V (2013) Forced expression of OCT4 influences the expression of pluripotent genes in human mesenchymal stem cells and fibroblasts. Genet Mol Res 12(2):1054–1060
Park JS, Suryaprakash S, Lao Y-H, Leong KW (2015) Engineering mesenchymal stem cells for regenerative medicine and drug delivery. Methods 84:3–16
Pera MF, Reubinoff B, Trounson A (2000) Human embryonic stem cells. J Cell Sci 113(1):5–10
Rao MS (1999) Multipotent and restricted precursors in the central nervous system. Anat Rec 257(4):137–148
Rossant J, Tam PP (2022) Early human embryonic development: blastocyst formation to gastrulation. Dev Cell 57(2):152–165
Saeedi P, Halabian R, Fooladi AAI (2019) A revealing review of mesenchymal stem cells therapy, clinical perspectives and modification strategies. Stem Cell Investig 6:34
Shakhbazau A, Sevyaryn I, Goncharova N, Grinev V, Kosmacheva S, Potapnev M (2008) Viral vectors for stable transduction of human mesenchymal stem cells: systems based on adeno-associated viruses and lentiviruses. Bull Exp Biol Med 146(4):531
Shi Y, Inoue H, Wu JC, Yamanaka S (2017) Induced pluripotent stem cell technology: a decade of progress. Nat Rev Drug Discov 16(2):115–130
Slack JM (2018) The science of stem cells. John Wiley & Sons
Theunissen TW, Powell BE, Wang H, Mitalipova M, Faddah DA, Reddy J, Fan ZP, Maetzel D, Ganz K, Shi L (2014) Systematic identification of culture conditions for induction and maintenance of naive human pluripotency. Cell Stem Cell 15(4):471–487
Thomas ED, Storb R, Clift RA, Fefer A, Johnson FL, Neiman PE, Lerner KG, Glucksberg H, Buckner CD (1975a) Bone-marrow transplantation: (first of two parts). N Engl J Med 292(16): 832–843
Thomas ED, Storb R, Clift RA, Fefer A, Johnson FL, Neiman PE, Lerner KG, Glucksberg H, Buckner CD (1975b) Bone-marrow transplantation: (second of two parts). N Engl J Med 292(17):895–902

Tian XC, Kubota C, Enright B, Yang X (2003) Cloning animals by somatic cell nuclear transfer–biological factors. Reprod Biol Endocrinol 1(1):1–7

Tran C, Damaser MS (2015) Stem cells as drug delivery methods: application of stem cell secretome for regeneration. Adv Drug Deliv Rev 82:1–11

Trounson A, McDonald C (2015) Stem cell therapies in clinical trials: progress and challenges. Cell Stem Cell 17(1):11–22

Trusler O, Huang Z, Goodwin J, Laslett AL (2018) Cell surface markers for the identification and study of human naive pluripotent stem cells. Stem Cell Res 26:36–43

Ulloa-Montoya F, Verfaillie CM, Hu W-S (2005) Culture systems for pluripotent stem cells. J Biosci Bioeng 100(1):12–27

Villa-Diaz L, Ross A, Lahann J, Krebsbach P (2013) Concise review: the evolution of human pluripotent stem cell culture: from feeder cells to synthetic coatings. Stem Cells 31(1):1–7

Wang Y-C, Nakagawa M, Garitaonandia I, Slavin I, Altun G, Lacharite RM, Nazor KL, Tran HT, Lynch CL, Leonardo TR (2011) Specific lectin biomarkers for isolation of human pluripotent stem cells identified through array-based glycomic analysis. Cell Res 21(11):1551–1563

Yilmaz A, Benvenisty N (2019) Defining human pluripotency. Cell Stem Cell 25(1):9–22

Zhu H, Guo Z-K, Jiang X-X, Li H, Wang X-Y, Yao H-Y, Zhang Y, Mao N (2010) A protocol for isolation and culture of mesenchymal stem cells from mouse compact bone. Nat Protoc 5(3): 550–560

Zubkova ES, Beloglazova IB, Ratner EI, Dyikanov DT, Dergilev KV, Menshikov MY, Parfyonova YV (2021) Transduction of rat and human adipose-tissue derived mesenchymal stromal cells by adeno-associated viral vector serotype DJ. Biol Open 10(9):bio058461

Role of Stem Cells as a Protective Agent against Neurological Complications

4

Uzma Nayeem, Shaista Ahmed, Dipak Kumar, Sadaf Jahan,
Andleeb Khan, Gulzar Rather, and Arif Jamal Siddiqui

Abstract

Different degrees of paralysis and/or loss of consciousness and feeling are caused by neurodegenerative diseases that develop as neurons gradually lose their structure and function. The lack of efficient treatments for neurodegenerative diseases has a severe negative social and economic influence. Regenerative cell treatment, commonly referred to as stem cell therapy, has offered a fantastic chance to research potentially effective cutting-edge methods for treating neurodegenerative illnesses in the previous 20 years. This is because stem cells can

U. Nayeem
Department of Molecular Medicine, School of Interdisciplinary Science and Technology, Jamia Hamdard, New Delhi, India

S. Ahmed
Centre of Translational and Clinical Research, School of Chemical and Lifesciences, Jamia Hamdard, New Delhi, India

D. Kumar
Department of Zoology, KKM College, Munger University, Munger, India

S. Jahan
Department of Medical Laboratory Sciences, College of Applied Medical Sciences, Majmaah University, Al-Majmaah, Saudi Arabia

A. Khan
Department of Pharmacology and Toxicology, College of Pharmacy, Jazan University, Jazan, Saudi Arabia

G. Rather
Department of Biomedical Engineering, Sathyabama Institute of Science & Technology deemed to be University, Chennai, Tamil Nadu, India

A. J. Siddiqui (✉)
Department of Biology, College of Science, University of Hail, Hail City, Saudi Arabia

S. Jahan, A. J. Siddiqui (eds.), *Applications of Stem Cells and Derived Exosomes in Neurodegenerative Disorders*, https://doi.org/10.1007/978-981-99-3848-3_4

repair damaged neural tissue by replacing lost or damaged cells with differentiated cells, promoting a regenerative environment, or protecting already healthy neurons and glial cells from damage. To develop practical and effective treatments for neurodegenerative disease augmentation by a greater knowledge of stem cell technologies and further research in this area is inevitable. Considerably, this chapter provides an overview of the different stem cell types, stem cell-based treatments for neurodegenerative disorders and new developments in this area.

Keywords

Neurodegenerative diseases · Stem cell therapy · Amyotrophic lateral sclerosis · Parkinson's disease · Huntington's disease · Alzheimer's disease

4.1 Introduction

Neurodegenerative disorders (NDDs), such as Parkinson's disease (PD), Alzheimer's disease (AD), Huntington's disease (HD), amyotrophic lateral sclerosis (ALS) and frontotemporal dementia (FTD), are characterized by the loss of structure, form, or function of neurons in the brain or spinal cord (Enciu et al. 2011). There are many factors responsible for NDDs like age, hormonal imbalance, growth factors and chemical exposure (Heyer and Meredith 2017). It is reported that reversal of NDDs is not possible and current medical treatment is costly with side effects. Additionally, for most individuals, a primary barrier is the central nervous system's (CNS) restricted capacity for regeneration and the challenges associated with identifying and treating pervasive neuronal cell death (Heyer and Meredith 2017; Lie et al. 2004). The blood-brain barrier (BBB) makes treating these disorders even more challenging. Hence, researchers are continuously working on promising tools to cure NDDs. Although there are some traditional natural therapies available, cell therapy is also used. One of the uses of stem cells is in cellular therapy which is helpful to reverse the degeneration process. In stem cell treatment or regenerative therapy, stem cells or their derivatives are used to improve the capacity of dysfunctional and injured tissue to react to healing procedures. Stem cell treatments usually focus on either cellular substitution or environmental improvement (Kashyap et al. 2015; Sivandzade and Cucullo 2021). When used therapeutically, it has provided advantageous and alluring options for treating a range of diseases, including neurodegenerative disorders.

Specific groups of neurons lose their ability to maintain their protein balance in neurodegenerative disorders, and inclusion bodies that are composed of proteins that are not soluble or have unfavourable conformations are common (Soto 2003). The gradual loss of sensation, memory and motor neurons and paralysis are all results of this pathogenic process. Despite the fact that billions of dollars have been spent on clinical studies of these diseases, there are still no measurable biomarkers or effective treatments for their progression. Despite the fact that stem cell therapy is still in its infancy, it has already proven to be a beneficial, secure and breakthrough method to pursue in order to treat neurodegenerative diseases (Karussis et al. 2013). The specific types of neurons and neural networks that were affected by the disease are being restored through stem cell therapy for disorders that cause neurodegeneration. Enhancing the environment to support the neurons of the host by producing neurotrophic or scavenging factors that are toxic or constructing additional neural networks around the region that is impacted is another method of treating neurodegenerative diseases (Temple and Alvarez-Buylla 1999).

In recent times, several approaches have been attempted to treat neurodegenerative diseases using stem cells, and they participate in the production of glial and neuronal cells. To find the most effective and profitable strategy for utilizing stem cells to treat neurological disorders, scientists have investigated multiple stem cell sources (Ahmadian-Moghadam et al. 2020). In vivo studies are conducted to see the impact of stem cells into stem cell treatment for neurodegenerative illnesses. According to these studies, stem cells may be able to affect native cells, encourage the effective restoration of nervous tissue, develop into neurons and glial cells and lessen motor impairments (Lee et al. 2007). Several clinical studies are investigating different aspects of stem cell therapies for brain diseases. Other important stages in developing and bringing stem cell treatments from the lab to patients include selecting the right stem cell type, the supporting mechanism and the specific neural disease. For instance, when a particular sort of neuronal component has been removed in PD, replacing cellular components may be advantageous. On the other hand, ALS may profit from therapies that improve the spinal cord's surroundings to support the remaining neurons. The pathogenesis of common neurodegenerative illnesses and the process of regeneration were then covered. Therefore, in the next section, we will discuss the types of stem cells and how these stem cells are found to be promising therapeutic agents.

4.2 Classifications of Stem Cell

Based on the capacity to convert into lineages, stem cells are classified as pluripotent, multipotent and unipotent (Singh et al. 2016). Multipotent stem cells can specialize in distinct cells from specific cell lineages, but pluripotent stem cells (PSCs) have a broader variety of differentiation than these cells. One example is a stem cell that produces blood cells, these cells can develop into different types of cells. From a blood cell precursor, an oligopotent cell is created. However, because some multipotent cells have the capacity to differentiate into various cell types, the term "pluripotent cell" is more correct. Multiple cell types can be created from oligopotent stem cells. The stem cell known as a myeloid can only generate white blood cells; it cannot make red blood cells. The lowest degree of differentiation and singular specialization that defines unipotent stem cells is their capacity for division that is singular. They are a hopeful option for therapeutic application in the field of regenerative medicine because of their higher quality (Fig. 4.1).

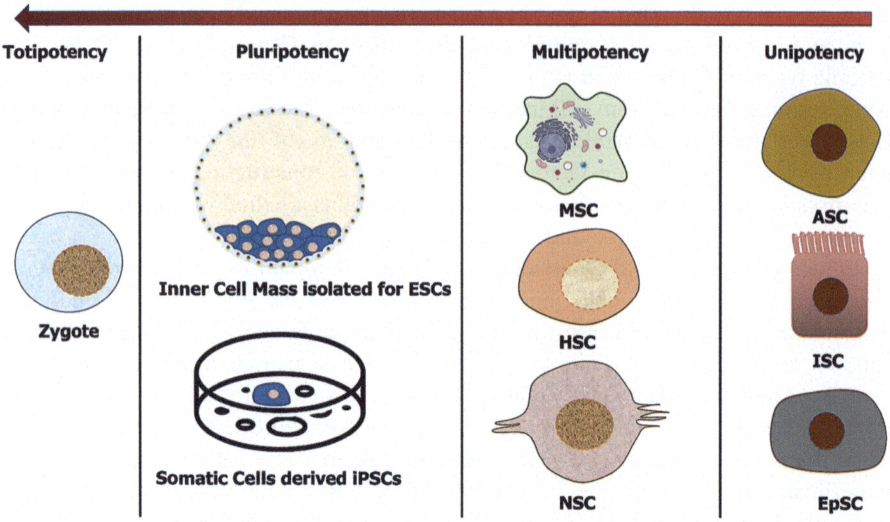

Fig. 4.1 Zygote has the highest potency which can form both embryonic and extra-embryonic tissues. As differentiation occurs, stem cells become pluripotent as embryonic stem cells (ESCs) and induced pluripotent stem cells (iPSCs), which can form embryonic tissues only. Further, neural stem cells (NSCs), haematopoietic stem cells (HSCs) and mesenchymal stem cells (MSCs) are multipotent, and adipose stem cells (ASCs), intestinal stem cells (ISCs) and epithelial stem cells are unipotent in their capabilities

4.2.1 Embryonic Stem Cells

ESCs have the potential to be intriguing research subjects because of their capacity to continuously regenerate and differentiate into all of the major cell types of the central nervous system. These cells are currently being utilized as a valuable source of human neural progenitor cells in numerous research projects regarding neurodegenerative diseases (Barker et al. 2015). Although having advantages, some disadvantages are also reported. ESCs can show that the immune rejection in the patient is in translational medicine.

4.2.2 Induced Pluripotent-Related Stem Cells (iPSCs)

iPSCs were created for the first time in Professor Shinya Yamanaka's lab in 2007 (Takahashi et al. 2007). Promoting the production of genes and transcription factors in non-pluripotent cells like fibroblasts, hepatocytes, blood T cells and keratinocytes that support the development of embryonic stem cells results in iPSCs, a type of stem cell that is similar to pluripotent stem cells (Shi et al. 2017). These altered cells now serve as a potential component of creating an endless supply of autologous neurons for transplantation in patients suffering from neurodegenerative diseases (Guan et al. 2020). Cells that can be derived from all cells in a patient's somatic tissue, such as

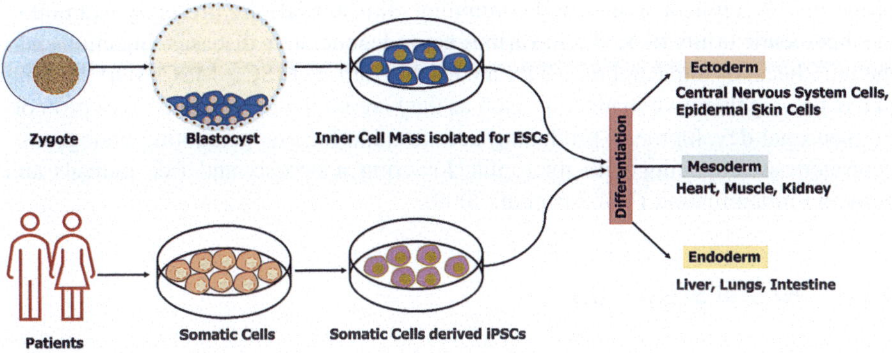

Fig. 4.2 Pluripotent stem cells as embryonic stem cells (ESCs) and induced pluripotent stem cells (iPSCs) can be derived from the blastocyst and somatic cells of patients, respectively. These stem cells have the potential to be differentiated into cells of three germ layers. Under specific conditions and according to therapeutic demands, such stem cells or thereby differentiated cells can be utilized to cure or reverse the damage caused by the neurodegenerative diseases

the cells in Fig. 4.2, are called pluripotent. The process of creating iPSCs may lead to a reliable source of cells that produce neurons affected by degenerative diseases in the brain (Sivandzade and Cucullo 2021). One of the primary benefits of using iPSCs is that they can be created without the utilization of oocytes or embryos; as a result, there is no risk of violating any religious or ethical principles. Another benefit of iPSCs is that it provides a valuable option for autologous cell transfusion without the risk of immune rejection and without the need for immunosuppressive drugs (Yasuhara et al. 2017).

The iPSCs may be advantageous for therapeutic use due to their simple harvesting procedures, reduced adverse effects and unique terminally differentiated cell phenotypes. In contrast to ESCs, the maturation process into adult neuron cells is more difficult in iPSCs.

4.2.3 Mesenchymal Stem Cells (MSCs)

MSCs are usually found in the bones, cord, fat and liver. These cells can be differentiated into a variety of cell lineages, including bone, cartilage, adipose, muscle and neurons (Sadaf Jahan et al. 2017). MSCs have substantial therapeutic promise and are considered the best source of cells for transplantation in the treatment of neurodegenerative diseases due to their extraordinary ability for self-renewal while maintaining multi-potency. Because of the simple collection techniques and fewer associated ethical, religious and immune concerns, neurons derived from MSCs have the potential to treat neurodegenerative disorders (Jahan et al. 2018). Additionally, it has been documented that unlike other primordial cells like ESCs, MSCs do not develop tumours. MSCs are beneficial cell type for investigating neurological disorders because they possess several important properties. Preclinical studies and continuing clinical trials are presently examining the therapeutic utility of MSCs in various neurodegenerative diseases. Injections into the intrathecal or intracerebral cavity are used to deliver MSCs. MSCs start working in a neuro-regenerative manner as soon as they are implanted; this involves promoting neuronal development, producing neurotrophic factors, promoting endogenous neurogenesis, engaging with microglia, lowering apoptosis and free radicals and reducing inflammation (Wilkins et al. 2009).

4.2.4 Neural Stem Cells

The brain tissues contain stem cells, which are more limited in their potential outcomes. Basically, neural stem cells develop into only a small number of various types of brain cells, such as oligodendrocytes, neurons and astrocytes, and have a limited capacity for self-renewal. NSCs can be isolated from various parts of the human prenatal and embryonic brain, as well as from brain tissues of patients who underwent surgery (Kim et al. 2013b). NSC transplantation in other brain areas is a potential medicinal approach for the management of many neurodegenerative diseases. By producing chemical substances that have a neuro-modulatory effect, such as controlling neural excitability, synaptic activity and learning, NSCs can have a gliogenic effect (Vay et al. 2016).

Unlike ESCs, NSCs are considered to have a more evolutionary stability and be less tumorigenic. The alteration of genetic material in these cells can lead to immortalized NSCs with increased proliferation potential. These cells can resolve the low capacity for self-renewal of NSCs (Martínez-Serrano and Björklund 1997).

4.3 Stem Cell-Based Therapeutic Strategies against Neurodegenerative Disorders

4.3.1 Therapy Using Stem Cells for Alzheimer's Disease (AD)

A strategy for treating terminal brain diseases like Alzheimer's has been proposed as a result of the development of stem cell technology. The primary cause of AD is neuronal degeneration throughout the brain, particularly in the basal cortex, hippocampus and prefrontal regions. Through pathological analysis, it was discovered that misfolded tau and A-(amyloid) proteins accumulated as AD progressed, and this adversely affected the neurons of the cholinergic system and their connections, ultimately leading to the degeneration of other neurons in various cerebral regions. There is currently no treatment for this illness, despite cases where acetylcholinesterase inhibitors briefly relieve some symptoms. Recent research have shown that neurons are the source of early Alzheimer's disease death that are compromised as well as dysfunctional glial cells (Zhao et al. 2021; Tajbakhsh et al. 2021). The cells that are derived from the stem cells will be used to promote natural neurogenesis and repair damaged or defective cells in the brain (Taupin 2008). All accessible stem cell sources are comprised of human ESC, MSC, iPSC and NSC produced from the umbilical cord and bone marrow, as well as directly initiated neurons (iN) derived from somatic cells (Alessandrini et al. 2019). Even though some ESC transplant experiments have shown that they can improve brain performance in rodent models of brain injury, the ability to translate these findings to humans has been limited (Titomanlio et al. 2011).

A research from 2013 showed that ESCs can grow into progenitor cells that resemble middle ganglionic eminences, a form of temporary brain cell that is located in the developing brain. Following transplantation into a rat brain injury model, these cells were able to grow into GABAergic and cholinergic neurons as well as form synapses with host networks, which caused changes in impaired learning and spatial memory (Liu et al. 2013). Despite continuing preclinical research, the development of ESC-based treatments is severely hampered by the intrinsic moral and immunogenic problems associated with using allogeneic cells as a source of cells (Tolosa et al. 2016). It has been demonstrated that the paracrine activity of NSCs has significant therapeutic promise. In a rat AD model, growth factor-secreting NSC transplantation enhanced neurogenesis and cognitive performance (Blurton-Jones et al. 2009) and aged primate brain (Kordower et al. 1994), while human NSCs overexpressing choline acetyltransferase reversed learning and memory deficits in a cholinergic neurotoxic rat model (Park et al. 2013). Recent additional studies found that NSC implantation reduced neuro-inflammation in rodent models of AD (Zhang et al. 2016), attenuated tau and AD neuropathology (Lee et al. 2015), promoted neurogenesis and synaptogenesis (Lilja et al. 2015) (Ager et al. 2015) and reversed cognitive impairments (Ager et al. 2015; Zhang et al. 2016). Although the therapeutic mechanisms underpinning these changes are unclear, paracrine neuroprotective or immunomodulatory factor release as well as direct neuronal differentiation is most likely to be involved in their mediating effects (Moghadam et al. 2009; Xuan et al. 2009).

Because of their abundance, MSCs are one of the stem cell types that are presently being researched the most. In elderly rodent models, transplanted MSCs have been shown to differentiate into various types of brain cells, which increased the amounts of the neurotransmitters acetylcholine, BDNF and NGF and improved movement and cognitive function (Park et al. 2013). With the addition of MSC-secreted compounds that promote survival, neuronal development and growth in areas that are naturally neurogenic. Paracrine activity with anti-oxidative impact of MSCs has a big role in medical field (Teixeira et al. 2015) and in cellular models of AD (Zilka et al. 2011). Similar to this, the transplantation of MSCs has been shown to prevent the death of cells associated with Aβ and tau in rodent models of AD (Lee et al. 2012); reduce Aβ deposits and plaque formation (Naaldijk et al. 2017); stimulate neurogenesis, synaptogenesis and neuronal differentiation (Oh et al. 2015; Yang et al. 2013); and rescue spatial learning and memory deficits (Lee et al. 2012). According to some studies, donated MSCs have extra anti-inflammatory and immunologically modulatory paracrine activity. This is accompanied by elevated levels of neuroprotective cytokines like IL-10 and decreased levels of inflammatory cytokines like TNF- and IL-1 (Kim et al. 2013a; Lee et al. 2012). MSCs that are intravenously administered have the capacity to travel to areas of neural damage and penetrate the blood-brain barrier without developing tumorigenicity or immunological reactions (Ra et al. 2011).

The iPSC-derived neurons have the ability to develop electro-physiologically active synaptic networks because they are physically and functionally developed (Pang et al. 2011). Additionally, it enabled the differentiation to be directed into specific neuronal groups, such as the dopamine neurons, by utilizing extra transcription factors during the induction period (X. Liu et al. 2012). Because iPSC transplantation is a novel technique, there haven't been many early animal model transplantation trials. In a rat model of an ischaemic stroke, human iPSC-derived NSCs improved diminished neurological function and lowered pro-inflammatory indicators through a neurotrophin-associated bystander impact. Another recent research discovered that human iPSC-derived cholinergic neuronal progenitors survived a transgenic AD rat model's intra-hippocampal transplant evolved into phenotypically mature cholinergic neurons and corrected spatial memory impairments (Fujiwara et al. 2013).

The iPSC technique makes it possible to produce individual pluripotent stem cells while avoiding the moral and immunological rejection issues related to sources that are not identical to the patient. Conversely, autologous iPSCs may not be very beneficial for neuro-replacement because neurons from AD patients have neuropathological features including increased Aβ levels, hyper-phosphorylated tau, shorter processes and reduced electrical excitability (Balez et al. 2016; Hossini et al. 2015). Alternatively, there are significant implications regarding the origin of AD and the detection of potential treatments when using iPSC-derived neurons to simulate the disease in vitro. As a result, as mentioned previously, they are currently the focus of extensive in vitro research (Truong et al. 2016) (Fig. 4.3).

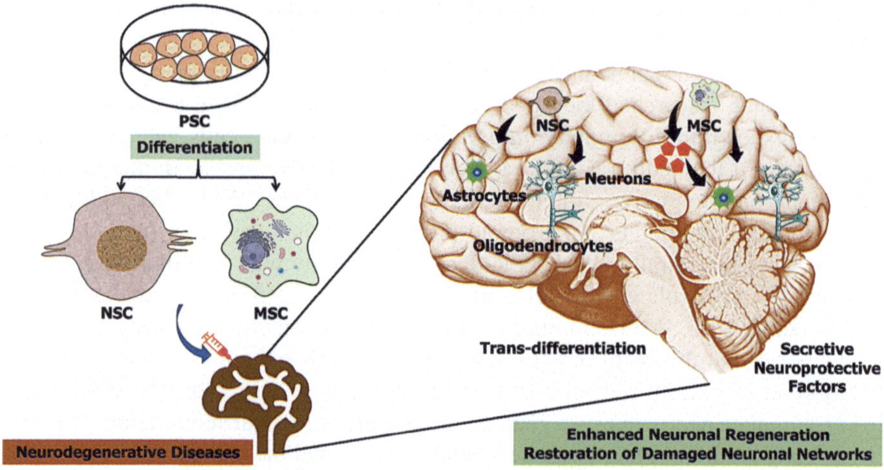

Fig. 4.3 Human pluripotent stem cells (PSCs) such as iPSCs and ESCs can be differentiated into neural stem cells (NSCs) and mesenchymal stem cells (MSCs), from which the cells of the brain like neurons, oligodendrocytes and astrocytes can be derived. MSCs have a trophic effect that includes the release of growth factors and neurotrophic chemicals which can facilitate the recovery of damaged nerve tissue by promoting angiogenesis, neurogenesis and immunomodulation

4.3.2 Therapy for Parkinson's Disease (PD) Using Stem Cells

The most prevalent neurological illness is Parkinson's disease (PD), which is second only to Alzheimer's disease (AD). Worldwide, PD affects six million people, the majority of whom are over 65. Ageing is a major risk factor for the development of the disease, and the average age at which PD shows itself is 55 (Marchetti et al. 2020). Young-onset PD, described as a diagnosis between the ages of 21 and 50, affects about 10% of PD cases (Schrag and Schott 2006), which is more likely to be inherited from family members (Poewe et al. 2017). Beyond the age of 70, PD patients are frequently identified. Clinically, PD manifests as motor dysfunction, which includes bradykinesia, muscular stiffness, resting trembling and unstable gait. The accumulation of toxic -synuclein (−syn) is what primarily drives the neurological processes connected to Parkinson's disease (PD), a presynaptic neural protein that builds up in the nervous system to create Lewy bodies (LBs) and Lewy neurites (LNs). With ageing, motor impairment becomes more prevalent in PD patients. The disease's most prevalent sign, i.e. tremor, manifests at repose but subsides with intentional movement. Additional non-motor complaints and these include cognitive loss, hyposmia, constipation, anxiety, melancholy, orthostatic hypotension, urinary disruption and disorders of rapid eye movement sleep behaviour. Only a tiny percentage of Parkinson's patients experience symptoms of mental illness like dementia and melancholy. Years after the start of motor disability, dementia-related PD is common and is likely to impact 30–80% of those who are afflicted (Maeda et al. 2017).

There is still a debate regarding the exact cause of PD. Advanced age is a major risk factor for PD. The substantia nigra (SN) can experience a series of stressors brought on by ageing that reduce neurons' receptivity to shocks in the future. It is well known that environmental and hereditary factors affect the risk and development of disease. Only 10% of instances of PD are brought on by hereditary causes (Tysnes and Storstein 2017). Vascular brain shock, repetitive head trauma, neuroleptics, pesticide exposure, nicotine, caffeine use and manganese toxicity were all linked to an increased risk of Parkinson's disease (Ayeni et al. 2022). Numerous crucial pathways have been found as potential treatment targets as a result of the increased comprehension of the origin and pathophysiology of Parkinson's disease. With the help of recent genetic research, it may be feasible to use nerve preventive treatments in people who are predisposed to MS, delaying the start and development of the illness. Traditional approaches are still used to address the signs and symptoms of PD. Researchers are looking into stem cells as potential replacements for damaged or diseased organs or synapses in addition to efforts to avoid and manage the symptoms of PD (Parmar et al. 2020).

Dopaminergic (DA) cell transfer is regarded as the most effective cell replacement therapy. A DA-depleted striatum can be transplanted with midbrain DA neurons to re-establish DA neurotransmission and aid PD patients in replacing missing neurons. Research is being done on a few stem cell methods as a potential way to restore cells through regenerative medicine. Clinical studies showed that the transfer of foetal midbrain tissue improved neurological symptoms and recovered muscle function in PD patients (Kefalopoulou et al. 2014; Olanow et al. 2003). Striatal grafts of midbrain DA neurons can be made from embryonic tissues, carotid body cells, immature retinal cells and SN neurons from pig embryos (Obeso et al. 2010). Despite the need for more research into the safety and efficacy of many cell-based therapies in humans, this is an exciting field to explore, and it is believed that they will eventually fall under the purview of PD management.

Oculomotor neurons, reticular neurons and DA neurons from the substantia nigra (SN) and ventral tegmental (VT) areas are some of the neuronal groups that make up the foetal ventral mesencephalon (VM) (Puelles 2007). Late in the 1970s and early in the 1980s, the first studies to involve patients in the open-label transfer of VM cells into the brains of PD patients were conducted (Lindvall et al. 1989, 1990). The recovery of DA neurotransmission is observed at 6 months; however, the slow recovery suggests that the cells transplanted are still developing (Piccini et al. 2000). The transferred cells of some patients who had received foetal VM transplants did, however, show signs of Lewy bodies (LB) illness upon postmortem inspection (Farag et al. 2009; Kordower et al. 1994), increasing the chance that LB pathology may be transferred from the recipient to the transplant (Brundin et al. 2008).

The special effects of embryonic VM cells for transplantation on the symptoms of PD have been documented in animals extensively. In a European, open-label, assessor-blinded trial, conducted by TRANSEURO, VM tissues were used to treat 150 individuals with younger-onset Parkinson's disease (PD) without substantial levodopa-induced dyskinesia (Z. Liu and Cheung 2020). The patients underwent regular PET and MRI scans as well as clinical monitoring for over 4 years (Bang et al. 2015). The TRANSEURO trials helped to define preclinical and clinical standards, the viability and clinical effectiveness of treating with foetal cells and the procedures for upcoming cell-based therapies using DA neurons derived from iPSCs (Fan and Ng 2020).

There are a number of technical hurdles that must be cleared before foetal VM transplantation can be used in therapeutic practice. The limited DA neuron survival and the constrained dopaminergic re-innervation in the host striatum are two important drawbacks. In order to overcome these challenges, co-transplantation with neural/para-neural supplies and neurotrophic factors such as BDNF have been employed to promote neurogenesis (Wang et al. 2020). The absence of standardization and a variety of techniques, as well as the scarcity of human foetal tissues, are two additional problems. One different cell treatment that has been studied is mesenchymal stem cells (MSCs). MSCs have been shown to be effective in the treatment of a number of illnesses, including PD (Mendes Filho et al. 2018). With low immunogenicity, MSCs are found to be appropriate for transplantation with low chances of teratoma and tumorigenicity (Jahan et al. 2017). Human umbilical cord MSCs are appropriate for use in medicinal applications due to their multi-lineage differentiation potential, autologous transplantation viability, simplicity of procurement and absence of ethical concerns (Chung et al. 2018). In contrast, the mortality rate and infection of human umbilical cord MSCs in the host are low, and cells transplanted from a vein have a chance of causing embolism in the capillaries, which would lead to uncontrolled growth. Only a handful of human umbilical cord MSCs implanted into the body have been observed in the intended location (Phinney and Prockop 2007). Nevertheless, after transplantation, it was reported that MSCs from different sources reduced PD symptoms in an animal model using the 6-OHDA drug. Some benefits were also noticed when transplanted MSCs released mitotic and angiogenic factors like fibroblast growth factor 2, endothelial growth factor and vascular endothelial growth factor, and MSC-secreted neurotrophic factors (BDNF, glial cell line-derived neurotrophic factor, nerve growth factor) can prevent DA neurons from apoptosis and promote neurogenesis (d'Angelo et al. 2020). 6-OHDA is linked with the induction of inflammation and oxidative stress, which can contribute to the denervation of dopaminergic neurons and degeneration of the nervous system (Gasparotto et al. 2017; Saeedi et al. 2019).

MSCs can be manipulated to degenerate into astrocytes and neurons in the presence of the proper conditions. Both undifferentiated MSCs and neuronal-primed MSCs had advantageous impacts following their injection into PD animal models, indicating that MSCs might not always be able to reverse PD by replacing cells directly. In neurotoxin-induced PD models, MSCs not only reduce PD symptoms but also potentially slow down the degradation of DA neurons. By preventing tau phosphorylation, MSCs preserved the assembly of microtubules in a PD model that also contained the A53T -syn mutant. As a result, the trafficking of axons was increased, which enhanced the autophagosome clearance of -syn and the neuroprotective effects on DA neurons. MSCs can come from a variety of origins and possess less racial issues rather than foetal VM tissues, but they are varied and frequently challenging to distinguish from fibroblasts. Additionally, MSCs are not expandable for numerous runs. Clinical-grade MSCs are frequently transferred prior to phase 3, yet the number of cells is usually not enough for donation at this point (dos Santos et al. 2011).

Both ESCs and iPSCs are examples of pluripotent cells (also known as PSCs), which can develop into different types of cells, including DA neurons. PSCs are capable of being renewed indefinitely, allowing an infinite supply of brain transplanting cells to be produced. Since Shinya Yamanaka's discovery of iPSC-based cell replacement therapy, it has become more widely used because iPSCs have less ethical concerns than ESCs but a comparable capacity for differentiation (Chung et al. 2018; Sonntag et al. 2018). Both 2D and 3D cultures have protocols for separating human ESCs/iPSCs to particular, localized neuronal subgroups (Golas 2018).

In PD rodent models, transplantation results in the DA neurons that were previously generated using the FP-based method were transferred, and then this technique may be successfully engrafted in vivo and functionally restore motor deficits (Liu and Cheung 2020). A behavioural investigation showed that rotational asymmetry caused by amphetamine had been restored (Kirkeby et al. 2012). Furthermore, 6-OHDA PD rats with DA progenitors implanted into their striatum displayed better forelimb use and akinesia in tests (Hargus et al. 2010; Martínez-Cerdeño et al. 2010). Midbrain DA neurons from hESCs performed similarly to human foetal DA neurons in terms of transplant survival, efficacy and ability to improve motor function in PD rats (Grealish et al. 2014). Since a single injection of transplantation is sufficient and the long-term goal is to decrease the possibility of neurosurgery, engraftment and survival of cells is paramount for cell therapy. Midbrain DA neurons derived from hESCs exhibited a lifespan of over 6 months after being transplanted into rats and mice with 6-OHDA lesioning as shown by PET and single-photon emission computed tomography (Grealish et al. 2014). The electrical and chemical activity of the transplant can now be observed and analysed in real time via the development of optogenetics. To produce DA and make effective recovery possible, the activity of engrafted cells needs to be carefully controlled. Drug-dependent activation or inhibition of human DA neuron transplants recovered motor function in PD rodents (Chen et al. 2016).

One benefit of using iPSCs for cell replacement is the absence of risk associated with immunological rejection when using native iPSC-derived cells in the donor's own body (Laperle et al. 2020). The applicability of this idea has been established in wild animals. In MPTP-lesioned cynomolgus monkey brains, human iPSC-derived DA neurons significantly improved motor function without immunosuppression (Hallett et al. 2015). Clinical studies involving human iPSC-derived DA neurons must be investigated in a primate model that involves long-term studies in nonhuman primates with post-transplantation safety and efficacy evidence. In an MPTP-induced monkey model of PD, hPSC-derived DA progenitor cells were observed to have a lifespan of at least 2 years, to grow in the absence of tumours and to function as midbrain DA neurons (Kikuchi et al. 2017). For the treatment of Parkinson's disease, clinical trials involving hESC-/iPSC-derived cell products have started in Australia (NCT02452723), China (NCT03119636) and Japan (JMA-IIA00384, UMIN000033564); however, these investigations have a higher risk than other clinical investigations that utilize MSCs (Doi et al. 2020). Other cells of the stem cell family that are also utilized in medical research include human amniotic epithelial cells (hAECs). These cells have the capacity to promote neuronal cell survival and regrowth, restore neurons that have been harmed and produce and release neurotrophic compounds and neurotransmitters (Xu et al. 2019).

4.3.3 Stem Cell Application in Huntington's Disease

Other than the cognitive and emotional declines, HD is a deadly, autosomal-dominant neurodegenerative disease marked by the loss of inhibitory, spiny neurons in the striatum of the forebrain, which affects the cortex, thalamus and hippocampus (Gan and Johnson 2014). Unintended movement, dementia and cognitive and emotional declines are all caused by the huntingtin protein's N-terminal aberrant increase of CAG repeats (Htt) (Lunn et al. 2011). With a disease path spanning 20 to 30 years, HD frequently manifests and the same occurred to her when she was in her fourth or fifth decade of existence. Effective treatments for HD are still illusive despite the illness's well-established genetic basis and disease processes.

Stem cell-based treatments have yielded encouraging results and a lot of attention for HD treatments. Repairing deceased or injured neurons while also altering faulty genes with enlarged CAG repeats is the aim of the stem cell therapy for HD. The majority of stem cells used for HD treatment, according to a recent study, are NSCs. Multiple sources, including the body and brain cells of HD patients, have been used to create and produce NSCs. Despite the fact that HD experimental and therapeutic stem cell research studies are still in the early phases, strong proof shows that stem cells or their derivatives can be infused into animal models of HD. Motor neuron integration and host circuitry growth were observed in the first stem cell treatments using ESC-derived NSCs transplanted into HD animals. The moral and theological ramifications of using foetal tissues, however, continue to be crucial problems. While unaltered neural stem cells (NSCs) had no neuroprotective advantages, GDNF-expressing neural progenitor cells (NPCs) offered nerve protection and functional recovery. Due to their capacity to minimize immune system dysfunction, augment compensatory neurogenesis, decrease apoptosis, augment mitochondrial function and enhance cell viability, MSCs are presently a hopeful supply of cells for treating HD (Connor 2018). Dey et al. demonstrated in 2010 that genetically modified MSCs that overexpressed BDNF or NGF had the opposite effect on the behaviour and death of neurons. Thus, transplanting BDNF-expressing MSCs into the striatum may create a favourable environment to arrest neurodegenerative processes (Sakthiswary and Raymond 2012). Lin et al. also showed that neural development, neurotrophic support and anti-apoptotic activities of human-derived MSCs contributed to neuroprotection and neuro-restoration. Other studies have revealed that tooth pulp stem cells may offer a therapeutic source for HD treatment that is less likely to experience post-transplant immunological resistance (Snyder et al. 2011). Injections of NPC into the striatum of HD rats produced comparable functional benefits (Lunn et al. 2014). Functional gains are brought about by NPC incorporation and relocation to secondary places related to HD sickness.

Few studies have investigated the ability of iPSC-derived NSCs to treat HD through cell replacement. In the initial investigation, when Jeon et al. attempted to transplant 72 CAG repeats from an HD patient's hiPSC-derived NSCs into an HD rodent model, the behavioural results were better, and there was no build-up of human mutant huntingtin (mHTT) in the transferred cells (Jeon et al. 2012). After that, iPSC-derived NSCs were transplanted into the brain of healthy mice, and 33 weeks later, the cells were observed to exhibit mHTT aggregation in the lateral ventricle. In a second research, An et al. produced human neural stem cells (hNSCs) after correcting an iPSC mutation obtained from HD patients, which were then implanted into HD mouse models. In addition to surviving the transfer, the cells effectively developed into motor neurons. Stem cell therapy for the management of HD is still a long way, and it will take more extensive, thorough preliminary research to show how effective it is as a treatment (Rosser and Svendsen 2014).

4.3.4 Therapy Using Amyotrophic Lateral Sclerosis Stem Cells

Motor neuron loss in the ventral horn and cortex of the spinal cord is a hallmark of Lou Gehrig's disease (ALS), a degenerative, progressive illness. As the illness worsens, symptoms like muscular weakness, twitching, rigidity and a lack of purposeful movement control become more noticeable (Gan and Johnson 2014; Sakthiswary and Raymond 2012). Typically, 20 to 48 months are needed for ALS to progress from the beginning to mortality. Investigations have uncovered additional abnormalities that may be associated with the onset of random and non-SOD1 familial ALS (Sivandzade et al. 2019). These include the deleterious and degenerative effects of non-neural cells on motor neurons, protein instability, RNA-/DNA-binding protein aggregation and disintegration and malfunction of the neuronal cytoskeleton (Lin et al. 2019).

Stem cell treatments have piqued the curiosity of many people as a potential therapeutic method. However, no definitive outcomes from preclinical or clinical research have yet been revealed. The most difficult obstacles to surmount when deciding on the most effective cell type and location for implantation have a deficiency of knowledge about the progression of the disease in the human body. Similar to other neurodegenerative illnesses, ALS can be effectively treated with stem cell treatments by replacing injured or deceased motor neurons, reducing inflammation and upregulating the expression of neurotrophic factors. An ultimate goal of the treatment is identical to the other objectives: to provide both a neural component that is integrated and the necessary contextual additions to inhibit the degeneration of existing motor neurons (Lunn et al. 2011). The purpose of the experiment which took place in 2010 at Emory University was to determine if NSCs were safe to insert into the nerves of the spinal cords of patients (Raore et al. 2011). Then, Martinez et al. looked into the safety of stem cell implantation in ALS patients' frontal motor cortex (Meamar et al. 2013).

Other than safety, the effectiveness of the treatment was necessary to be assessed. ALS patients in Spain who received MSC transplants showed an increase in motor neuron numbers and a decrease in ubiquitin deposits in their motor neurons (Blanquer et al. 2012). Because trustworthy data shows that injecting foetal NSCs into patients' spinal cords delays the onset of ALS, current clinical studies also concentrate on exogenous NSC transplantation (Mazzini et al. 2015). Research is currently being done on growth factors' critical shielding impact on surviving motor neurons. For instance, Brainstorm Cell Therapeutics is working on a method to add MSCs to the fluid surrounding ALS patients' brains and spinal cords (through intrathecal injection). The aim of the study is to investigate the beneficial impacts of neurotrophic, angiogenic, anti-inflammatory and immunomodulatory chemicals generated by MSCs (Petrou et al. 2016). Overall, despite the reality that ALS stem cell therapy is still in its early stages, experts from around the world are optimistic that therapies like this will slow the progression of the disease and augment the effectiveness of existing treatments (Lunn et al. 2014).

4.3.5 Dementia with Frontotemporal Stem Cell Therapy

The second most frequent factor contributing to progressive dementia is also the most sneaky of the diseases that impact people under the age of 65: frontotemporal dementia (FTD) (Karageorgiou and Miller 2014; Bang et al. 2015). Growing abnormalities in personality, cognition, behaviour and language skills associated with FTD are brought on by the frontotemporal cortex neurodegeneration and selective death of cerebral cortical neurons (Capozzo et al. 2017). FTD is a complex hereditary and pathological disease that can occur in families or on its own. According to statistics, up to 15–20% of FTD cases are family variations, and the MAPT gene, which produces the tau protein linked with microtubules, is mutated in up to 20% of these cases (Ferrari et al. 2019; Kim et al. 2018). FTD is a public health issue because of its rising prevalence, dearth of treatments and social burden.

Mutations in the tau (MAPT), progranulin (PGRN) and C9ORF72 genes have been found to be the most common genetic variables associated with FTD to date (Hedges et al. 2016). Recently, these genes have been the focus of research into the illness process and the creation of ground-breaking FTD pharmaceutical treatments. The mechanisms of FTD are still unknown, despite major study findings over the preceding two decades. This might be because there aren't many disease models that properly represent the various diseases associated with FTD. On the basis of clinical diagnostic factors, there are also still difficult and erroneous clinical evaluations of sporadic FTD.

It has been difficult to create suitable models for studying the molecular causes of FTD because cell lines and animal models lack the ability to replicate the intricate pattern of mutations present in the human CNS in its adult form. In addition, multiple tau overexpression experiments may result in FTD symptoms that may not accurately reflect endogenous tau expression (Lines et al. 2020). Surprisingly, the capacity to convert somatic cells into iPSCs may serve as a beneficial model for studying the cause of FTD (Karch et al. 2019). The iPSCs have proven to be a successful method for simulating the symptoms of FTD patients in order to identify the disease pathways and accelerate the creation of new treatments (Guo et al. 2017). Recent investigations have described the procedures for and characteristics of iPSCs derived from individuals with familial FTD (Freibaum et al. 2015). Individuals with FTD supplied peripheral blood mononuclear cells, according to Lee et al.

In Grn-deficient patient neurons, they successfully discovered cell-autonomous, reversible abnormalities, offering a useful model for researching and the effects of GRN-dependent processes that are pathogenic and the creation of potential treatments. Ehrlich et al. also generated iPSCs from individuals with mutations in the MAPT gene in order to develop an in vitro model for the detection of particular degenerative changes in frontotemporal dementia with parkinsonism. These genes are associated with FTD and then differentiated the neurons into an adult phenotype (FTDP-17) (Ehrlich et al. 2015). In order to develop patient-specific iPSC lines and assess the expression of disease indicators like FTD-tau, TDP-43, active caspase-3 and fused in sarcoma, all of which have been associated with FUS, were all identified in PBMCs from two different patients with FTD (Kim et al. 2020). They discovered that the active caspase-3 expression was considerably greater than in the controls in their immunocytochemical and immunoblot analyses. This neurodegenerative component of FTD has the potential to be utilized as a means of identifying pathways associated with the disease and screening for therapeutics. To construct a genuine FTD model that selectively degenerates the frontal and temporal regions, Raitano et al. also investigated the specific neurons and neurons of the cortex that were derived from the iPSCs of patients with FTD (Raitano et al. 2015). Despite the fact that the use of stem cell technology for FTD modelling is still in its infancy, researchers anticipate using iPSCs as a helpful instrument to support the investigation of disease causes and the creation of treatment approaches.

4.4 Summary and Conclusion

Because of lack of proper treatment, neurodegenerative disorders previously were considered complicated to cure. Nowadays, there are many available natural therapies including cell-based therapy. Studies have shown that stem cell therapies are successful in treating neurological illnesses. Alzheimer's disease, amyotrophic lateral sclerosis, amyotrophic lateral degeneration, Huntington's disease, spinal muscular atrophy and other diseases are being treated with stem cells. Early investigations of stem cell therapy for ALS in humans have made significant progress towards ensuring safety, despite still being in its infancy. Determining treatment efficacy is now critically important, while in simultaneously recognising the most effective cell dose, stem cell source, delivery location and method. Additionally, if treatment is to be most effective early in the course of the illness, the therapeutic window of opportunity must be precisely defined, and this field of study may need to collaborate with early ALS diagnosis techniques.

Stem cell therapy for ALS offers tremendous confidence and is one of the disease's early warning signs. While it is being developed from the lab to the bedside, it will be important to protect its validity and the safety of the patients, promoting sound clinical study design to accurately gauge therapy effectiveness. Another study on Huntington's disease was reported for the experimental drug Cellavita HD's dose-response analysis following intravenous administration in patients with Huntington's disease. Huntington's disease stem cell treatment Cellavita HD is now available. Therefore, we can state that stem cell therapy currently offers powerful, promising treatments for many diseases. There is still much to be done, even though some of these drugs are presently undertaking clinical studies and preclinical research is becoming more and more important. By replacing injured neurons and offering neuroprotection and neuro-restoration, stem cell therapy is anticipated to play a major role in the development of future therapies for neurodegenerative disorders. The transport of medicines and regenerative treatments has also been improved by recent developments in hydrogels and nanoparticles with integrated stem cells. As a consequence, it is anticipated that replacement and regenerative treatments for the brain will be successfully integrated into therapeutic practice in the near future.

References

Ager RR, Davis JL, Agazaryan A, Benavente F, Poon WW, LaFerla FM, Blurton-Jones M (2015) Human neural stem cells improve cognition and promote synaptic growth in two complementary transgenic models of Alzheimer's disease and neuronal loss. Hippocampus 25(7):813–826

Ahmadian-Moghadam H, Sadat-Shirazi M-S, Zarrindast M-R (2020) Therapeutic potential of stem cells for treatment of neurodegenerative diseases. Biotechnol Lett 42:1073–1101

Alessandrini M, Preynat-Seauve O, De Bruin K, Pepper MS (2019) Stem cell therapy for neurological disorders. South African Medical Journal *109*(8 Supplement 1):S71–S78

Ayeni EA, Aldossary AM, Ayejoto DA, Gbadegesin LA, Alshehri AA, Alfassam HA, Tawfik EA et al (2022) Neurodegenerative diseases: implications of environmental and climatic influences on neurotransmitters and neuronal hormones activities. Int J Environ Res Public Health 19(19): 12495

Balez R, Steiner N, Engel M, Muñoz SS, Lum JS, Wu Y, O'Connor M et al (2016) Neuroprotective effects of apigenin against inflammation, neuronal excitability and apoptosis in an induced pluripotent stem cell model of Alzheimer's disease. Sci Rep 6(1):1–16

Bang J, Spina S, Miller BL (2015) Frontotemporal dementia. Lancet 386(10004):1672–1682

Barker RA, Drouin-Ouellet J, Parmar M (2015) Cell-based therapies for Parkinson disease—past insights and future potential. Nat Rev Neurol 11(9):492–503

Blanquer M, Moraleda JM, Iniesta F, Gómez-Espuch J, Meca-Lallana J, Villaverde R, Bleda P et al (2012) Neurotrophic bone marrow cellular nests prevent spinal motoneuron degeneration in amyotrophic lateral sclerosis patients: a pilot safety study. Stem Cells 30(6):1277–1285

Blurton-Jones M, Kitazawa M, Martinez-Coria H, Castello NA, Müller F-J, Loring JF, LaFerla FM et al (2009) Neural stem cells improve cognition via BDNF in a transgenic model of Alzheimer disease. Proc Natl Acad Sci 106(32):13594–13599

Brundin P, Li J-Y, Holton JL, Lindvall O, Revesz T (2008) Research in motion: the enigma of Parkinson's disease pathology spread. Nat Rev Neurosci 9(10):741–745

Capozzo R, Sassi C, Hammer MB, Arcuti S, Zecca C, Barulli MR, Seripa D et al (2017) Clinical and genetic analyses of familial and sporadic frontotemporal dementia patients in southern Italy. Alzheimers Dement 13(8):858–869

Chen Y, Xiong M, Dong Y, Haberman A, Cao J, Liu H, Zhang S-C et al (2016) Chemical control of grafted human PSC-derived neurons in a mouse model of Parkinson's disease. Cell Stem Cell 18(6):817–826

Chung T-H, Hsu S-C, Wu S-H, Hsiao J-K, Lin C-P, Yao M, Huang D-M (2018) Dextran-coated iron oxide nanoparticle-improved therapeutic effects of human mesenchymal stem cells in a mouse model of Parkinson's disease. Nanoscale 10(6):2998–3007

Connor B (2018) Concise review: the use of stem cells for understanding and treating Huntington's disease. Stem Cells 36(2):146–160

d'Angelo M, Cimini A, Castelli V (2020) Insights into the effects of mesenchymal stem cell-derived secretome in Parkinson's disease. Int J Mol Sci 21(15):5241

Doi D, Magotani H, Kikuchi T, Ikeda M, Hiramatsu S, Yoshida K, Morizane A et al (2020) Pre-clinical study of induced pluripotent stem cell-derived dopaminergic progenitor cells for Parkinson's disease. Nat Commun 11(1):3369

Ehrlich M, Hallmann A-L, Reinhardt P, Araúzo-Bravo MJ, Korr S, Röpke A, Oblak AL et al (2015) Distinct neurodegenerative changes in an induced pluripotent stem cell model of frontotemporal dementia linked to mutant TAU protein. Stem Cell Rep 5(1):83–96

Enciu AM, Nicolescu MI, Manole CG, Mureşanu DF, Popescu LM, Popescu BO (2011) Neuroregeneration in neurodegenerative disorders. BMC Neurol 11:1–7

Fan Y, Ng S-Y (2020) Replacing what's lost: a new era of stem cell therapy for Parkinson's disease. Transl Neurodegener 9(1):1–10

Farag ES, Vinters HV, Bronstein J (2009) Pathologic findings in retinal pigment epithelial cell implantation for Parkinson disease. Neurology 73(14):1095–1102

Ferrari R, Manzoni C, Hardy J (2019) Genetics and molecular mechanisms of frontotemporal lobar degeneration: an update and future avenues. Neurobiol Aging 78:98–110

Freibaum BD, Lu Y, Lopez-Gonzalez R, Kim NC, Almeida S, Lee K-H, Wong PC et al (2015) GGGGCC repeat expansion in C9orf72 compromises nucleocytoplasmic transport. Nature 525(7567):129–133

Fujiwara N, Shimizu J, Takai K, Arimitsu N, Saito A, Kono T, Suzuki T et al (2013) Restoration of spatial memory dysfunction of human APP transgenic mice by transplantation of neuronal precursors derived from human iPS cells. Neurosci Lett 557:129–134

Gan L, Johnson JA (2014) Oxidative damage and the Nrf2-ARE pathway in neurodegenerative diseases. Biochim Biophys Acta 1842(8):1208–1218

Gasparotto J, Ribeiro CT, Bortolin RC, Somensi N, Rabelo TK, Kunzler A, Gelain DP et al (2017) Targeted inhibition of RAGE in substantia nigra of rats blocks 6-OHDA–induced dopaminergic denervation. Sci Rep 7(1):8795

Golas MM (2018) Human cellular models of medium spiny neuron development and Huntington disease. Life Sci 209:179–196

Grealish S, Diguet E, Kirkeby A, Mattsson B, Heuer A, Bramoulle Y, Björklund A et al (2014) Human ESC-derived dopamine neurons show similar preclinical efficacy and potency to fetal neurons when grafted in a rat model of Parkinson's disease. Cell Stem Cell 15(5):653–665

Guan X, Xu W, Zhang H, Wang Q, Yu J, Zhang R, Wang D et al (2020) Transplantation of human induced pluripotent stem cell-derived cardiomyocytes improves myocardial function and reverses ventricular remodeling in infarcted rat hearts. Stem Cell Res Ther 11(1):1–11

Guo W, Fumagalli L, Prior R, Van Den Bosch L (2017) Current advances and limitations in modeling ALS/FTD in a dish using induced pluripotent stem cells. Front Neurosci 11:671

Hallett PJ, Deleidi M, Astradsson A, Smith GA, Cooper O, Osborn TM, Brownell A-L et al (2015) Successful function of autologous iPSC-derived dopamine neurons following transplantation in a non-human primate model of Parkinson's disease. Cell Stem Cell 16(3):269–274

Hargus G, Cooper O, Deleidi M, Levy A, Lee K, Marlow E, Hallett PJ et al (2010) Differentiated Parkinson patient-derived induced pluripotent stem cells grow in the adult rodent brain and reduce motor asymmetry in parkinsonian rats. Proc Natl Acad Sci 107(36):15921–15926

Hedges EC, Mehler VJ, Nishimura AL (2016) The use of stem cells to model amyotrophic lateral sclerosis and frontotemporal dementia: from basic research to regenerative medicine. Stem Cells Int 2016:9279516

Heyer DB, Meredith RM (2017) Environmental toxicology: sensitive periods of development and neurodevelopmental disorders. Neurotoxicology 58:23–41

Hossini AM, Megges M, Prigione A, Lichtner B, Toliat MR, Wruck W, Makrantonaki E et al (2015) Induced pluripotent stem cell-derived neuronal cells from a sporadic Alzheimer's disease donor as a model for investigating AD-associated gene regulatory networks. BMC Genomics 16:1–22

Jahan S, Singh S, Srivastava A, Kumar V, Kumar D, Pandey A, Pant AB et al (2018) PKA-GSK3β and β-catenin signaling play a critical role in trans-resveratrol mediated neuronal differentiation in human cord blood stem cells. Mol Neurobiol 55:2828–2839

Jahan S, Kumar D, Kumar A, Rajpurohit CS, Singh S, Srivastava A, Pant AB et al (2017) Neurotrophic factor mediated neuronal differentiation of human cord blood mesenchymal stem cells and their applicability to assess the developmental neurotoxicity. Biochem Biophys Res Commun 482(4):961–967

Jeon I, Lee N, Li J-Y, Park I-H, Park KS, Moon J, Kwon J et al (2012) Neuronal properties, in vivo effects, and pathology of a Huntington's disease patient-derived induced pluripotent stem cells. Stem Cells 30(9):2054–2062

Karageorgiou E, Miller BL (2014) Frontotemporal lobar degeneration: a clinical approach. Paper presented at the Seminars in neurology. Semin Neurol 34:189

Karch CM, Kao AW, Karydas A, Onanuga K, Martinez R, Argouarch A, Bowles KR et al (2019) A comprehensive resource for induced pluripotent stem cells from patients with primary tauopathies. Stem cell Rep 13(5):939–955

Karussis D, Petrou P, Kassis I (2013) Clinical experience with stem cells and other cell therapies in neurological diseases. J Neurol Sci 324(1–2):1–9

Kashyap MP, Kumar V, Singh AK, Tripathi VK, Jahan S, Pandey A, Pant AB et al (2015) Differentiating neurons derived from human umbilical cord blood stem cells work as a test system for developmental neurotoxicity. Mol Neurobiol 51:791–807

Kefalopoulou Z, Politis M, Piccini P, Mencacci N, Bhatia K, Jahanshahi M, Björklund A et al (2014) Long-term clinical outcome of fetal cell transplantation for Parkinson disease: two case reports. JAMA Neurol 71(1):83–87

Kikuchi T, Morizane A, Doi D, Magotani H, Onoe H, Hayashi T, Inoue H et al (2017) Human iPS cell-derived dopaminergic neurons function in a primate Parkinson's disease model. Nature 548(7669):592–596

Kim E-J, Kim Y-E, Jang J-H, Cho E-H, Na DL, Seo SW, Park KH (2018) Analysis of frontotemporal dementia, amyotrophic lateral sclerosis, and other dementia-related genes in 107 Korean patients with frontotemporal dementia. Neurobiol Aging 72:186. e181–186. e187

Kim K-S, Kim HS, Park J-M, Kim HW, Park M-k, Lee H-S, Moon J et al (2013a) Long-term immunomodulatory effect of amniotic stem cells in an Alzheimer's disease model. Neurobiol Aging 34(10):2408–2420

Kim M, Kim HJ, Koh W, Li L, Heo H, Cho H, Nakanishi M et al (2020) Modeling of frontotemporal dementia using iPSC technology. Int J Mol Sci 21(15):5319

Kim SU, Lee HJ, Kim YB (2013b) Neural stem cell-based treatment for neurodegenerative diseases. Neuropathology 33(5):491–504

Kirkeby A, Grealish S, Wolf DA, Nelander J, Wood J, Lundblad M, Parmar M et al (2012) Generation of regionally specified neural progenitors and functional neurons from human embryonic stem cells under defined conditions. Cell Rep 1(6):703–714

Kordower JH, Winn SR, Liu Y-T, Mufson EJ, Sladek Jr John R, Hammang JP, Emerich DF et al (1994) The aged monkey basal forebrain: rescue and sprouting of axotomized basal forebrain neurons after grafts of encapsulated cells secreting human nerve growth factor. Proc Natl Acad Sci 91(23):10898–10902

Laperle AH, Sances S, Yucer N, Dardov VJ, Garcia VJ, Ho R, Avalos P et al (2020) iPSC modeling of young-onset Parkinson's disease reveals a molecular signature of disease and novel therapeutic candidates. Nat Med 26(2):289–299

Lee HJ, Lee JK, Lee H, Carter JE, Chang JW, Oh W, Jin HK et al (2012) Human umbilical cord blood-derived mesenchymal stem cells improve neuropathology and cognitive impairment in an Alzheimer's disease mouse model through modulation of neuroinflammation. Neurobiol Aging 33(3):588–602

Lee I-S, Jung K, Kim I-S, Lee H, Kim M, Yun S, In Park K et al (2015) Human neural stem cells alleviate Alzheimer-like pathology in a mouse model. Mol Neurodegener 10:1–16

Lee J-P, Jeyakumar M, Gonzalez R, Takahashi H, Lee P-J, Baek RC, Clarke J et al (2007) Stem cells act through multiple mechanisms to benefit mice with neurodegenerative metabolic disease. Nat Med 13(4):439–447

Lie DC, Song H, Colamarino SA, Ming G-l, Gage FH (2004) Neurogenesis in the adult brain: new strategies for central nervous system diseases. Annu Rev Pharmacol Toxicol 44:399–421

Lilja AM, Malmsten L, Röjdner J, Voytenko L, Verkhratsky A, Ögren SO, Marutle A et al (2015) Neural stem cell transplant-induced effect on neurogenesis and cognition in Alzheimer Tg2576 mice is inhibited by concomitant treatment with amyloid-lowering or cholinergic 7 nicotinic receptor drugs. Neural Plast 2015:370432

Lin CY, Wu CL, Lee KZ, Chen YJ, Zhang PH, Chang CY, Tsai HJ et al (2019) Extracellular Pgk1 enhances neurite outgrowth of motoneurons through Nogo66/NgR-independent targeting of NogoA. elife 8:e49175

Lindvall O, Brundin P, Widner H, Rehncrona S, Gustavii B, Frackowiak R, Marsden CD et al (1990) Grafts of fetal dopamine neurons survive and improve motor function in Parkinson's disease. Science 247(4942):574–577

Lindvall O, Rehncrona S, Brundin P, Gustavii B, Åstedt B, Widner H, Rothwell JC et al (1989) Human fetal dopamine neurons grafted into the striatum in two patients with severe Parkinson's disease: a detailed account of methodology and a 6-month follow-up. Arch Neurol 46(6): 615–631

Lines G, Casey JM, Preza E, Wray S (2020) Modelling frontotemporal dementia using patient-derived induced pluripotent stem cells. Mol Cell Neurosci 109:103553

Liu X, Li F, Stubblefield EA, Blanchard B, Richards TL, Larson GA, Zhang D et al (2012) Direct reprogramming of human fibroblasts into dopaminergic neuron-like cells. Cell Res 22(2): 321–332

Liu Y, Weick JP, Liu H, Krencik R, Zhang X, Ma L, Zhang S-C et al (2013) Medial ganglionic eminence–like cells derived from human embryonic stem cells correct learning and memory deficits. Nat Biotechnol 31(5):440–447

Liu Z, Cheung H-H (2020) Stem cell-based therapies for Parkinson disease. Int J Mol Sci 21(21): 8060

Lunn JS, Sakowski SA, Feldman EL (2014) Concise review: stem cell therapies for amyotrophic lateral sclerosis: recent advances and prospects for the future. Stem Cells 32(5):1099–1109

Lunn JS, Sakowski SA, Hur J, Feldman EL (2011) Stem cell technology for neurodegenerative diseases. Ann Neurol 70(3):353–361

Maeda T, Shimo Y, Chiu S-W, Yamaguchi T, Kashihara K, Tsuboi Y, Saiki H et al (2017) Clinical manifestations of nonmotor symptoms in 1021 Japanese Parkinson's disease patients from 35 medical centers. Parkinsonism Relat Disord 38:54–60

Marchetti B, Tirolo C, L'Episcopo F, Caniglia S, Testa N, Smith JA, Serapide MF et al (2020) Parkinson's disease, aging and adult neurogenesis: Wnt/β-catenin signalling as the key to unlock the mystery of endogenous brain repair. Aging Cell 19(3):e13101

Martínez-Cerdeño V, Noctor SC, Espinosa A, Ariza J, Parker P, Orasji S, Kriegstein AR et al (2010) Embryonic MGE precursor cells grafted into adult rat striatum integrate and ameliorate motor symptoms in 6-OHDA-lesioned rats. Cell Stem Cell 6(3):238–250

Martínez-Serrano A, Björklund A (1997) Immortalized neural progenitor cells for CNS gene transfer and repair. Trends Neurosci 20(11):530–538

Mazzini L, Gelati M, Profico DC, Sgaravizzi G, Massimo PP, Muzi G, Giorgi C et al (2015) Human neural stem cell transplantation in ALS: initial results from a phase I trial. J Transl Med 13:1–16

Meamar R, Nasr-Esfahani MH, Mousavi SA, Basiri K (2013) Stem cell therapy in amyotrophic lateral sclerosis. J Clin Neurosci 20(12):1659–1663

Mendes Filho D, de Ribeiro P, Oliveira LF, de Paula DRM, Capuano V, de Assunção TSF, da Silva VJD (2018) Therapy with mesenchymal stem cells in Parkinson disease: history and perspectives. Neurologist 23(4):141–147

Moghadam FH, Alaie H, Karbalaie K, Tanhaei S, Esfahani MHN, Baharvand H (2009) Transplantation of primed or unprimed mouse embryonic stem cell-derived neural precursor cells improves cognitive function in Alzheimerian rats. Differentiation 78(2–3):59–68

Naaldijk Y, Jaeger C, Fabian C, Leovsky C, Blüher A, Rudolph L, Stolzing A et al (2017) Effect of systemic transplantation of bone marrow-derived mesenchymal stem cells on neuropathology markers in APP/PS 1 Alzheimer mice. Neuropathol Appl Neurobiol 43(4):299–314

Obeso JA, Rodriguez-Oroz MC, Goetz CG, Marin C, Kordower JH, Rodriguez M, Halliday G et al (2010) Missing pieces in the Parkinson's disease puzzle. Nat Med 16(6):653–661

Oh SH, Kim HN, Park H-J, Shin JY, Lee PH (2015) Mesenchymal stem cells increase hippocampal neurogenesis and neuronal differentiation by enhancing the Wnt signaling pathway in an Alzheimer's disease model. Cell Transplant 24(6):1097–1109

Olanow CW, Goetz CG, Kordower JH, Stoessl AJ, Sossi V, Brin MF, Godbold J et al (2003) A double-blind controlled trial of bilateral fetal nigral transplantation in Parkinson's disease. Ann Neurol 54(3):403–414

Pang ZP, Yang N, Vierbuchen T, Ostermeier A, Fuentes DR, Yang TQ, Südhof TC et al (2011) Induction of human neuronal cells by defined transcription factors. Nature 476(7359):220–223

Park D, Yang Y-H, Bae DK, Lee SH, Yang G, Kyung J, Kim GH et al (2013) Improvement of cognitive function and physical activity of aging mice by human neural stem cells over-expressing choline acetyltransferase. Neurobiol Aging 34(11):2639–2646

Parmar M, Grealish S, Henchcliffe C (2020) The future of stem cell therapies for Parkinson disease. Nat Rev Neurosci 21(2):103–115

Petrou P, Gothelf Y, Argov Z, Gotkine M, Levy YS, Kassis I, Abramsky O et al (2016) Safety and clinical effects of mesenchymal stem cells secreting neurotrophic factor transplantation in patients with amyotrophic lateral sclerosis: results of phase 1/2 and 2a clinical trials. JAMA Neurol 73(3):337–344

Phinney DG, Prockop DJ (2007) Concise review: mesenchymal stem/multipotent stromal cells: the state of transdifferentiation and modes of tissue repair—current views. Stem Cells 25(11): 2896–2902

Piccini P, Lindvall O, Björklund A, Brundin P, Hagell P, Ceravolo R, Rehncrona S et al (2000) Delayed recovery of movement-related cortical function in Parkinson's disease after striatal dopaminergic grafts. Ann Neurol 48(5):689–695

Poewe W, Seppi K, Tanner CM, Halliday GM, Brundin P, Volkmann J et al (2017) Parkinson disease. Nat Rev Dis Primers 3:17013

Puelles E (2007) Genetic control of basal midbrain development. J Neurosci Res 85(16):3530–3534

Ra JC, Shin IS, Kim SH, Kang SK, Kang BC, Lee HY, Choi HJ et al (2011) Safety of intravenous infusion of human adipose tissue-derived mesenchymal stem cells in animals and humans. Stem Cells Dev 20(8):1297–1308

Raitano S, Ordovàs L, De Muynck L, Guo W, Espuny-Camacho I, Geraerts M, Voets T et al (2015) Restoration of progranulin expression rescues cortical neuron generation in an induced pluripotent stem cell model of frontotemporal dementia. Stem Cell Rep 4(1):16–24

Raore B, Federici T, Taub J, Wu MC, Riley J, Franz CK, Johe K et al (2011) Cervical multilevel intraspinal stem cell therapy: assessment of surgical risks in Gottingen minipigs. Spine 36(3): E164–E171

Rosser A, Svendsen CN (2014) Stem cells for cell replacement therapy: a therapeutic strategy for HD? Mov Disord 29(11):1446–1454

Saeedi P, Halabian R, Fooladi AAI (2019) A revealing review of mesenchymal stem cells therapy, clinical perspectives and modification strategies. Stem Cell Investig 6:34

Sakthiswary R, Raymond AA (2012) Stem cell therapy in neurodegenerative diseases: from principles to practice. Neural Regen Res 7(23):1822

dos Santos F, Andrade PZ, Abecasis MM, Gimble JM, Chase LG, Campbell AM, Cabral JMS et al (2011) Toward a clinical-grade expansion of mesenchymal stem cells from human sources: a microcarrier-based culture system under xeno-free conditions. Tissue Eng Part C Methods 17(12):1201–1210

Schrag A, Schott JM (2006) Epidemiological, clinical, and genetic characteristics of early-onset parkinsonism. Lancet Neurol 5(4):355–363

Shi Y, Inoue H, Wu JC, Yamanaka S (2017) Induced pluripotent stem cell technology: a decade of progress. Nat Rev Drug Discov 16(2):115–130

Singh VK, Saini A, Kalsan M, Kumar N, Chandra R (2016) Describing the stem cell potency: the various methods of functional assessment and in silico diagnostics. Front Cell Develop Biol 4: 134

Sivandzade F, Cucullo L (2021) Regenerative stem cell therapy for neurodegenerative diseases: an overview. Int J Mol Sci 22(4):2153

Sivandzade F, Prasad S, Bhalerao A, Cucullo L (2019) NRF2 and NF-κB interplay in cerebrovascular and neurodegenerative disorders: molecular mechanisms and possible therapeutic approaches. Redox Biol 21:101059

Snyder BR, Cheng P-H, Yang J, Yang S-H, Huang AHC, Chan AWS (2011) Characterization of dental pulp stem/stromal cells of Huntington monkey tooth germs. BMC Cell Biol 12(1):1–8

Sonntag K-C, Song B, Lee N, Jung JH, Cha Y, Leblanc P, Schweitzer J et al (2018) Pluripotent stem cell-based therapy for Parkinson's disease: current status and future prospects. Prog Neurobiol 168:1–20

Soto C (2003) Unfolding the role of protein misfolding in neurodegenerative diseases. Nat Rev Neurosci 4(1):49–60

Tajbakhsh A, Read M, Barreto GE, Ávila-Rodriguez M, Gheibi-Hayat SM, Sahebkar A (2021) Apoptotic neurons and amyloid-beta clearance by phagocytosis in Alzheimer's disease: pathological mechanisms and therapeutic outlooks. Eur J Pharmacol 895:173873

Takahashi K, Tanabe K, Ohnuki M, Narita M, Ichisaka T, Tomoda K, Yamanaka S (2007) Induction of pluripotent stem cells from adult human fibroblasts by defined factors. Cell *131*(5):861–872

Taupin P (2008) Adult neurogenesis, neuroinflammation and therapeutic potential of adult neural stem cells. Int J Med Sci 5(3):127

Teixeira FG, Carvalho MM, Neves-Carvalho A, Panchalingam KM, Behie LA, Pinto L, Salgado AJ et al (2015) Secretome of mesenchymal progenitors from the umbilical cord acts as modulator of neural/glial proliferation and differentiation. Stem Cell Rev Rep 11:288–297

Temple S, Alvarez-Buylla A (1999) Stem cells in the adult mammalian central nervous system. Curr Opin Neurobiol 9(1):135–141

Titomanlio L, Kavelaars A, Dalous J, Mani S, El Ghouzzi V, Heijnen C, Gressens P et al (2011) Stem cell therapy for neonatal brain injury: perspectives and challenges. Ann Neurol 70(5): 698–712

Tolosa L, Pareja E, Gómez-Lechón MJ (2016) Clinical application of pluripotent stem cells: an alternative cell-based therapy for treating liver diseases? Transplantation 100(12):2548–2557

Truong A, Si E, Duncan T, Valenzuela M (2016) Modeling neurodegenerative disorders in adult somatic cells: a critical review. Front Biol 11:232–245

Tysnes O-B, Storstein A (2017) Epidemiology of Parkinson's disease. J Neural Transm 124:901–905

Vay SU, Blaschke S, Klein R, Fink GR, Schroeter M, Rueger MA (2016) Minocycline mitigates the gliogenic effects of proinflammatory cytokines on neural stem cells. J Neurosci Res 94(2): 149–160

Wang J, Hu W-W, Jiang Z, Feng M-J (2020) Advances in treatment of neurodegenerative diseases: perspectives for combination of stem cells with neurotrophic factors. World J Stem Cells 12(5): 323

Wilkins A, Kemp K, Ginty M, Hares K, Mallam E, Scolding N (2009) Human bone marrow-derived mesenchymal stem cells secrete brain-derived neurotrophic factor which promotes neuronal survival in vitro. Stem Cell Res 3(1):63–70

Xu H, Zhang J, Tsang KS, Yang H, Gao W-Q (2019) Therapeutic potential of human amniotic epithelial cells on injuries and disorders in the central nervous system. Stem Cells Int *2019*:5432301

Xuan AG, Luo M, Ji WD, Long DH (2009) Effects of engrafted neural stem cells in Alzheimer's disease rats. Neurosci Lett 450(2):167–171

Yang H, Xie ZH, Wei LF, Yang HN, Yang SN, Zhu ZY, Bi JZ et al (2013) Human umbilical cord mesenchymal stem cell-derived neuron-like cells rescue memory deficits and reduce amyloid-beta deposition in an AβPP/PS1 transgenic mouse model. Stem Cell Res Ther 4:1–14

Yasuhara T, Kameda M, Sasaki T, Tajiri N, Date I (2017) Cell therapy for Parkinson's disease. Cell Transplant 26(9):1551–1559

Zhang Q, Wu H-h, Wang Y, Gu G-j, Zhang W, Xia R (2016) Neural stem cell transplantation decreases neuroinflammation in a transgenic mouse model of Alzheimer's disease. J Neurochem 136(4):815–825

Zhao M, Jiang X-F, Zhang H-Q, Sun J-H, Pei H, Ma L-N, Li H et al (2021) Interactions between glial cells and the blood-brain barrier and their role in Alzheimer's disease. Ageing Res Rev 72: 101483

Zilka N, Zilkova M, Kazmerova Z, Sarissky M, Cigankova V, Novak M (2011) Mesenchymal stem cells rescue the Alzheimer's disease cell model from cell death induced by misfolded truncated tau. Neuroscience 193:330–337

5 Mesenchymal Stem Cells and their Applications against Neurodegenerative Disorders

Insaf Bahrini

Abstract

Intense research has tremendously increased our knowledge about molecular and cellular mechanisms of neuronal pathogenesis and has yielded new insights into potential therapeutic targets that may be applicable in neurodegenerative diseases. For instance, stem cell-based therapy has over the last two decades emerged as a promising tool for neurodegenerative disorder's therapy. One approach is using mesenchymal stem cells (MSCs) which had intrigued a great interest, thanks to their availability, high differentiation potency, and lack of ethical concern. MSCs are able to be differentiated into nerve and glia cells. This property makes this type of cells a perfect candidate to replace cells that were damaged or lost during neurodegenerative disorders. In this chapter, we have outlined potential therapeutic mechanisms of MSCs with respect to neurodegenerative diseases, in particular amyotrophic lateral sclerosis, multiple sclerosis, and Parkinson's, Huntington's, and Alzheimer's diseases. Then, we reviewed on recent advances relevant to pre-clinical and clinical trials using this technique and lastly, we discussed the limitations that hamper the application of this approach for neurodegenerative diseases.

Keywords

Mesenchymal stem cells · Mechanisms · Neurodegenerative diseases · Treatment · Clinical trials

I. Bahrini (✉)
Department of Biology, College of Science, University of Ha'il, Ha'il, Saudi Arabia

S. Jahan, A. J. Siddiqui (eds.), *Applications of Stem Cells and Derived Exosomes in Neurodegenerative Disorders*, https://doi.org/10.1007/978-981-99-3848-3_5

Abbreviations

AD	Alzheimer's disease
Aβ	Amyloid-β
ALS	Amyotrophic lateral sclerosis
ASCs	Mesenchymal stem cells isolated from adipose tissue
BDNF	Brain-derived neurotrophic factor
BM-MSCs	Mesenchymal stem cells isolated from bone marrow
BNDF	Brain-derived neurotrophic factor
ChAT	Choline acetyltransferase
CNS	Central nervous system
DA	Dopamine
DPSCs	Dental pulp stem cells
EAE	Encephalomyelitis
EVs	Extracellular vesicles
GDNF	Glial cell-derived neurotrophic factor
HD	Huntington's disease
HTT	Huntingtin gene or protein
IL-10	Interleukin-10
IL-6	Interleukin-6
MCP-1	Monocyte chemoattractant protein-1
MS	Multiple sclerosis
MSA	Multiple system atrophy
MSCs	Mesenchymal stem cells
NDs	Neurodegenerative diseases
NTFs	Neurotrophic factors
PD	Parkinson's disease
SN	Substantia nigra
SOD1	Superoxide dismutatse-1
TGF-β	Transforming growth factor-β
TH	Tyrosine hydroxylase
TNF-α	Tumor necrosis factor-α
UC-MSCs	Mesenchymal stem cells isolated from the umbilical cord
WJ-MSCs	Mesenchymal stem cells isolated from Wharton's jelly
α-Syn	Alpha-synuclein

5.1 Introduction

Neurodegenerative diseases (NDs) include a wide range of nervous system disorders mainly described by progressive neuronal degeneration, decrease in neuronal activity, cerebral shrinkage, and pathological accumulation of protein aggregates. The most common NDs are multiple system atrophy (MSA), Huntington's disease (HD), amyotrophic lateral sclerosis (ALS), Parkinson's disease (PD), Alzheimer's disease (AD), spinal muscular atrophy, prion disease, multiple sclerosis (MS), and spinocerebellar ataxia. Given that the brain has low capacity to self-repairing and neurogenesis, conventional treatments for NDs have failed to achieve satisfactory results. Hence, available treatments can only help control the disease's symptoms in order to ameliorate patients' well-being. Over the last two decades, cell-based approaches have risen up as a potential efficient strategy in ND therapy. Particularly, marrow stromal cells or mesenchymal stem cells (MSCs) are considered as a highly efficient candidate for cell-based therapy, thanks to their abundance, accessibility in the body, regeneration capacity, ability to regulate the immune system, associated low risk to initiate tumorigenesis, and limited related ethical concerns (Aleynik et al. 2014). MSCs were initially characterized by Friedenstein et al. (1976) as "colony forming units-fibroblastic" because of their capacity to form colonies from a singular cell. Later on, in the 2000s, they were named as "multipotent mesenchymal stromal cells" (Krampera et al. 2013).

MSCs are multipotent cell type that can self-renew and differentiate into cells of the mesodermal lineage, including adipocytes, osteocytes, tenocytes, chondrocytes, skeletal myocytes, and cardiomyocytes (Pittenger et al. 1999). Furthermore, studies have indicated that cells from ectodermal and endodermal origin, including functional liver cells, pancreatic islet β-like cells, renal epithelial lineage, and photoreceptor-like cells, can be differentiated from MSCs when specific culture conditions and stimuli are applied (Morigi et al. 2004). MSCs can also give rise to neuronal cells with neuron-like functions (Hermann et al. 2004). However, it is still uncertain whether MSCs that have undergone differentiation or undifferentiated MSCs can integrate the host neuronal circuitry and establish new synaptic connections with host cells (Zeng et al. 2015). In the following, we overviewed the therapeutic mechanisms of MSCs and explored their potential use as novel targets for ND treatment. Then, we reviewed recent progress in using MSCs in preclinical and clinical applications in ND therapy. Finally, we discussed the underlying limitations of the use of stem-based approach for neurodegenerative disorders.

5.2 Therapeutic Mechanisms of MCSs

MSCs have attracted considerable interest for many years, thanks to their high self-renewal capacity, differentiation potency, and low immunogenicity. The therapeutic potential of MSCs implies different mechanisms including immunomodulation, paracrine secretion, and migration toward damaged areas in response to cytokines, chemokines, and other signaling molecules secreted from lesions (Fig. 5.1). Being initially used in tissue repair and regenerative medicine because of their multipotent differentiation capacity, MSCs are currently investigated for the treatment of many diseases including neurodegenerative disorders, thanks to their distinctive migratory property and trophic and immunosuppressive effects.

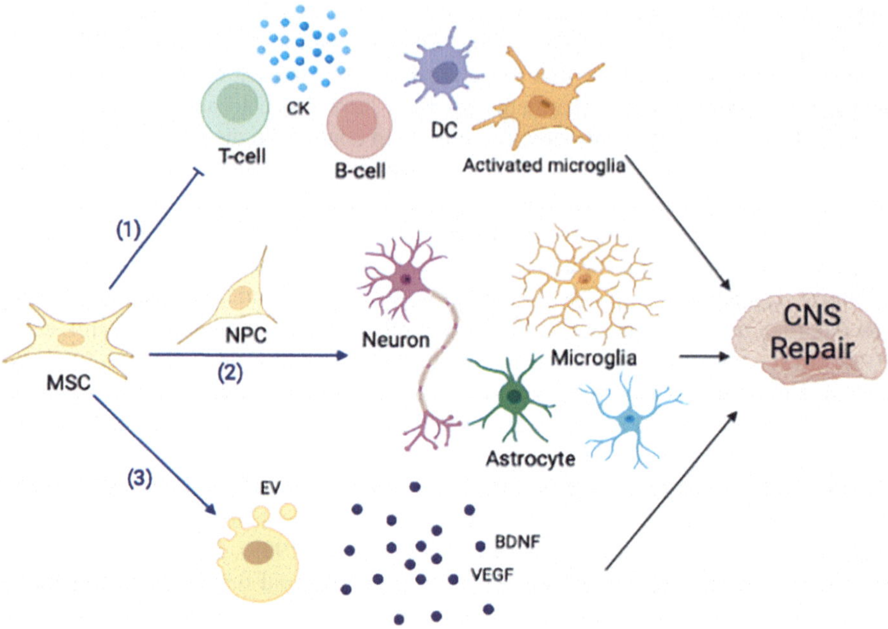

Fig. 5.1 Potential therapeutic mechanisms of MSCs during neurodegenerative disorders. (1) Immune regulation, (2) migration and differentiation, (3) paracrine secretion. NPC, neuronal progenitor cell; DC, dendritic cell; VEGF, vascular endothelial growth factor

5.2.1 Migration into Damaged Tissue

MSCs can spontaneously migrate to the damaged areas after transplantation, a process known as homing (Rombouts and Ploemacher 2003). Previous histopathological studies have indicated an efficient migration of MSCs to the damaged tissue of the central nervous system (CNS) and their differentiation into neuronal and glial cells, resulting in persisting enhancement of motor function after acute spinal cord injury (Chopp et al. 2000). Significant homing capacity over long distance in the brain was also demonstrated after transplantation of MSCs previously labeled with iron nanoparticles, which migrated from the site of injection in the subventricular zone toward damaged areas in the olfactory bulb (Delcroix et al. 2009). In addition, when transfused intravenously, MSCs were reported to engraft into the bone marrow compartment of allogeneic recipients without significant toxicity (Devine et al. 2001). However, effectiveness of MSC homing varies greatly depending on many factors including age and number of cells, protocol of cell isolation and culture, pretreatment of cells, route of administration, and time of injection (Rombouts and Ploemacher 2003). Moreover, studies indicated that MSC migration, proliferation, and differentiation markedly decreased in age-related disorders including ALS (Violatto et al. 2015).

5.2.2 Paracrine Secretion

Latest findings have indicated that therapeutic effect of engrafted MSCs is mainly due to their secretory function (Merimi et al. 2021). Stem cells are able to secrete a large number of molecules, including cytokines, chemokines, antioxidants, and nucleic acids packaged into extracellular vesicles (EVs), that can modulate cell migration, apoptosis, angiogenesis, differentiation, proliferation, and inflammation in damaged tissue (Boomsma and Geenen 2012). The paracrine effect of MSCs and its potential applications in treating human diseases have been reported in several studies (Osugi et al. 2012). Furthermore, evidence has indicated that MSCs could also act through a paracrine signaling mechanism without direct cell interaction. For example, it was reported that the conditioned medium obtained by culturing MSCs had a protective effect on neurons by reducing inflammation in the absence of engraftment (Sun et al. 2013). In addition, EVs derived from MSCs were shown to exert similar beneficial effects as MSCs (Nawaz et al. 2016). Their role in the treatment of NDs has been well documented in many studies (Gratpain et al. 2021), as they can easily cross the blood-brain barrier.

5.2.3 Immunomodulatory and Anti-Inflammatory Effects

Given their immunoregulatory and anti-inflammatory effects that are primarily mediated by intercellular interactions and secretion of a large number of molecules, MSCs may play a significant role in several diseases (Aggarwal and Pittenger 2005). Previous research findings have indicated that MSCs could prevent cytokine secretion, proliferation, and activation of cells of the immune system, particularly B and T lymphocytes and natural killer cells, and inhibit the maturation and activation of dendritic cells (Aggarwal and Pittenger 2005; Spaggiari et al. 2008). In the CNS, immune response deficiency is a key process in nervous system disorders. Neuroinflammation in the CNS results from the activation of glial cells and/or infiltration of lymphocytes which contribute to neurodegenerative process. Several studies have indicated that MSCs may play important immunomodulatory roles in NDs, mostly by modulating microglia-mediated neuroinflammation (Liu et al. 2019). Importantly, it was shown that MSCs' immunoregulatory actions may vary depending on the disease condition and the microenvironment (Gao et al. 2016).

5.2.4 MCSs Application in Neurodegenerative Disorders

Given their plasticity and high multipotency to differentiate into various neuronal cells, MSCs represent a promising approach for treatment of NDs. Researches have shown that MSCs can regulate neuroinflammation (Lee et al. 2012), promote axonal growth (Sasaki et al. 2009), stimulate endogenous neurogenesis (Kan et al. 2011), activate astroglia (Vigo et al. 2021), induce functional modifications of microglia (Giunti et al. 2012), enhance synaptic connection and interhemispheric cortical communication, and decrease apoptosis and oxidative stress ((Hirota et al. 2022, Angeloni et al. 2020).

5.2.4.1 Parkinson's Disease

Parkinson's disease is the second most frequent neurodegenerative disorder among older adults worldwide. PD pathology has been mainly associated with a dopamine (DA) depletion in the striatum resulting from the loss of DA-secreting neuronal cells in the substantia nigra (SN) (Buddhala et al. 2015). Neuronal degeneration is usually associated with the accumulation of intracellular inclusions, named Lewy bodies, which are composed mainly of alpha-synuclein (α-Syn) (Gómez-Benito et al. 2020). Common symptoms include bradykinesia, resting tremor, muscular rigidity, sleep disorder, and olfactory and cognitive impairment. Existing treatments for PD are only used to relieve the symptoms and are unable to slow or stop the disease progression. To date, several studies are investigating the therapeutic actions of MSCs for PD treatment. Particularly, MSCs can be differentiated into dopaminergic neurons and can produce a large number of trophic factors that stimulate tissue regeneration (Shetty et al. 2013). It was reported that administration of MSCs into the SN has been found to reconstitute damaged dopaminergic neuronal circuitry and decrease DA deficiency in PD animal models (Danielyan et al. 2011). Moreover, research data indicated a rescue of dopaminergic neurons resulting in an improvement in motor functions and reduction in behavioral deterioration following MSC treatment in PD animal models (Mostafavi et al. 2019). More research findings demonstrated that MSCs exert neuroprotective actions by modulating α-syn transmission in Parkinsonian models (Shin and Lee 2020). Furthermore, data suggested that MSCs protect dopaminergic neurons by decreasing pro-inflammatory cytokines and modulating activation of microglia (Kim et al. 2009a). MSCs have also been used to mediate gene transfer in experimental PD models. For example, transplantation of engineered MSCs overexpressing the gene coding for a marker for DA, tyrosine hydroxylase (TH), induced DA level in damaged striatal region of the rat's brain (Lu et al. 2005). In addition, administration of glial cell-derived neurotrophic factors (GDNF)-overexpressing MSCs significantly rescued the dopaminergic neurons from neurotoxicity in Parkinsonian rat models (Hoban et al. 2015) and MPTP-treated nonhuman primates (Ren et al. 2013).

5.2.4.2 Multiple Sclerosis

Multiple sclerosis is an inflammatory and chronic neurodegenerative disorder characterized by a progressive degeneration of axons and loss of myelin sheath. It is also known as the most frequent disabling neurological disorder of younger adults (Dimitrov and Turner 2014). In addition, many MS patients eventually develop secondary-progressive MS, resulting in a steady progression of symptoms (Inojosa et al. 2021). Available therapeutic strategies are predicated on immunomodulation, which can merely delay the disease progression and relieve symptoms. During the past decade, evidence suggested a high advantage of using MSCs for MS treatment using animal models of MS, including experimental autoimmune encephalomyelitis (EAE), the common animal model for inflammatory demyelinating disease (Martin

1997; Karussis et al. 2010). Studies have indicated that MSCs injected intravenously could migrate toward the CNS and lymphatic system and thereafter decreased neuroinflammation, demyelination, and disease severity in EAE mice model (Zappia et al. 2005). Moreover, as reported by Mohajeri et al. (2011), engrafted MSCs were able to induce immune tolerance by increasing T regulatory cell frequency in EAE animal model. Nevertheless, the use of MSCs to treat MS is not restricted to its immunomodulatory activity. Their mechanism of action in MS-based therapy was also related to their ability to differentiate and their paracrine function. Interestingly, human BM-MSCs showed efficient differentiation into myelinogenic mature oligodendrocytes in vivo (Lam et al. 2019). Also, remyelination of mouse brain by MSCs significantly extended animal lifespan and improved motor function (Lam et al. 2019). Moreover, MSC secretome was found to promote axonal myelination and stem cell oligodendrogenesis and integration into the neuronal circuitry (Jadasz et al. 2018). Further studies by Mohammadi (2020) showed that systemic injection of exosomes secreted by MSCs resulted in sustained recovery and improved motor function of EAE mice. This recovery was associated with reduction in neuroinflammation (Mohammadi 2020).

5.2.4.3 Alzheimer's Disease

Alzheimer's disease is a progressive neurologic disorder in older adults and the most prevalent form of dementia, a syndrome characterized by a loss or decline in memory and other cognitive and behavioral abilities such as conversation, attention, memory, judgment, and social skills. The disease was initially recorded in 1907 by Dr. Alois Alzheimer. The aggregation and accumulation of Tau proteins and amyloid-β (Aβ) in the brain are key players of AD pathophysiology. AD is also characterized by a progressive neuronal loss and synapse damage throughout the brain (DeTure and Dickson 2019). Available pharmacological treatments do not cure AD, but are used to temporarily manage symptoms of dementia and mental disorders such as depression. MSCs derived from the bone marrow (BM-MSCs) were broadly investigated in animal models with AD, including DAL mice, APP mice, and scopolamine-induced rats (Qin et al. 2020). Evidence have indicated that MSC administration could attenuate neuropathology, cognitive deficit, and memory loss (Qin et al. 2020; Lee et al. 2012). Moreover, MSCs manifested remarkable ability to reduce Aβ plaques and Tau tangle accumulation, enhance synaptic transmission, and reduce neuronal apoptosis (Bae et al. 2013). MSCs were reported to have neuroprotective functions due to their paracrine effects. For example, BM-MSC-derived EVs were found to enhance neurogenesis in the subventricular region and in the hippocampus, relieve cognitive deficits (Reza-Zaldivar et al. 2019), and improve neuronal survival by reducing synapse damage and oxidative stress (Bodart-Santos et al. 2019). Another study by Nakano et al. (2020) proposed that exosomal transfer by MSC led to restoration of astrocytic function which enhanced synaptogenesis and correction of cognitive decline in mice model of AD.

5.2.4.4 Amyotrophic Lateral Sclerosis

Also called Charcot or Lou Gehrig's syndrome, amyotrophic lateral sclerosis is a late-onset fatal disease that results from a progressive motor nerve cell degeneration in the CNS. The accumulation of ubiquitinated inclusions into motor neurons and surrounding glial cells is usually observed in ALS (Blokhuis et al. 2013). Muscle weakness, crumps, loss of coordination, slurred speech or dysarthria, twitching, and muscle wasting (myopenia), which can eventually lead to fatal paralysis and even to death, are hallmark signs of ALS (Nowicka et al. 2019). Despite the significant research efforts in clinical trials for ALS therapy, no effective treatment is available to alleviate progression of the disease. Instead, patients are treated to partially alleviate the symptoms (Nowicka et al. 2019). It has been reported that MSCs isolated from multiple tissues could differentiate into motor neurons expressing specific markers such as Pax-6, choline acetyltransferase (ChAT), and nestin (Abdullah et al. 2016). In this regard, many studies were conducted to assess the advantage of using MSCs in ALS animal models, particularly superoxide dismutatse-1 (SOD1) transgenic models (Philips and Rothstein 2015). Research data indicated significant delay of disease onset and disease progression, protection of motor neurons, and enhanced motor function recovery, leading to increased viability in ALS rodent model (Kim et al. 2010). Authors indicated a dose-dependent effect of MSCs (Kim et al. 2010). In addition, it was demonstrated that MSC transplantation can slow down disease progression, improve motor performance, prevent neuroinflammation, reduce microglial activation and astrogliosis, and increase motor neuronal survival in pre-symptomatic SOD1 mice (Sun et al. 2014) and after the occurrence of clinical symptoms (Sun et al. 2014; Sironi et al. 2017). Furthermore, evidence has shown that repeated MSC injections were more advantageous in controlling ALS pathology than single injection (Sironi et al. 2017; Sun et al. 2014), during early stage of the disease and after symptom onset.

5.2.4.5 Huntington's Disease

HD is an inherited rare disease that results from a selective degeneration of striatal neurons, leading to a more widespread brain atrophy. The disease occurs after inheritance of an autosomal dominant mutation in Huntingtin (*HTT*) gene. The movement disorder is distinctive and is the hallmark of HD, but symptoms may get progressively worse, leading to cognitive decline such as dementia, dystonia, memory lapses, and psychiatric disorders. Available treatments for HD can only help reduce some of the problems it causes but are ineffective to slow down or reverse the disease progression. Therefore, many preclinical trials were carried out in order to investigate potential benefit of using MSCs in HD therapy. Studies showed that transplantation of MSCs isolated from human adipose tissues (ASCs) in the ipsilateral striatal border rescued HD mice models from striatal degeneration and behavioral deterioration and increased their survival rate (Lee et al. 2009). A possible role of secreted factors by transplanted MSCs was proposed (Lee et al. 2009). Similarly, other studies showed that intraperitoneal injection of MSC extract increased motor impairment recovery in HD rodent models (Im et al. 2013). The treatment also decreased the level of *HTT* aggregates and associated atrophy in the striatal region

(Im et al. 2013). Another study showed that human BM-MSC intrastriatal transplantation attenuated motor impairment and prolonged survival of mouse model with HD, as administrated MSCs appeared to activate proliferation and differentiation of neurons and migration of neuroblasts and microglia into lesioned areas (Lin et al. 2011). Authors suggested angiogenic and anti-apoptotic effects of MSC transplants (Lin et al. 2011). Migratory property of MSCs over a long distance in the CNS was also demonstrated in quinolinic acid model with HD by magnetic resonance imaging and histological study. Engrafted MSCs that are induced to secrete neurotrophic factors (NTFs) could regenerate damaged dopaminergic neurons in the striatum (Sadan et al. 2009). Moreover, several studies have demonstrated that MSCs may have an immunomodulatory effects through secretion of cytokines and NTFs including brain-derived neurotrophic factor (BDNF). It was reported that intranasally administrated BM-MSCs increased TH expression in the striatum, microglia shift into alternative M2 phenotype, and suppressed monocyte chemoattractant protein-1 (MCP-1), interleukin-6 (IL-6), and tumor necrosis factor-α (TNF-α) which are usually upregulated in the striatum during HD (Yu-Taeger et al. 2019). In addition, MSCs used as vectors for gene transfer to target tissue were investigated in HD experimental models. For example, administration of MSCs overexpressing BDNF, known to promote neuronal survival and function, enhanced motor recovery in HD transgenic mice (Dey et al. 2010).

5.3 Clinical Trials Using MSC Therapies for Neurodegenerative Diseases

Preclinical success using cell-based therapies has supported their approval for human trials. MSCs are the most frequently used stem cells for human medical applications, given their unique therapeutic potentials. Since their first clinical use in 1995 (Lazarus et al. 1995), several studies using MSCs were conducted to treat many diseases including cardiovascular diseases, bone and cartilage diseases, autoimmune diseases, diabetes, liver diseases, cancer, and neurodegenerative disorders in clinical setting (Petrou et al. 2016; Chulpanova et al. 2018) for decades. Over 900 clinical studies, either completed or in the process, that report using MSCs are listed at ClinicalTrials.gov, 218 studies of which are related to the nervous system. Registered clinical trials for AD, PD, ALS, MS, and HD are overviewed in Table 5.1.

5.3.1 ALS

Several delivery methods including direct transplantation into the spinal cord and intravenous, intramuscular, and intrathecal injection have been reported to test the tolerability of MSC injection in ALS patients (Petrou et al. 2016; Mazzini et al. 2012). Outcomes of human pilot clinical trials using MSC approach suggested some

Table 5.1 Clinical research studies registered in ClinicalTrials.gov for neurodegenerative disorders (as of July 2022)

Disease	Clinical phase	Source of MSCs	Site of injection	Location
ALS	I/II	WJ-MSCs	Intrathecal	Poland
	I	UC-MSCs	Intrathecal	Antigua and Barbuda
	II	ASCs	Intrathecal	USA
	I	BM-MSCs	Intravenous	Iran
	I	BM-MSCs	Intrathecal	Iran
	II	UC-MSCs	Intrathecal	China
	I/II	BM-MSCs	Intrathecal	Brazil
	III	BM-MSCs	Intrathecal	USA
	I/II	ASCs	Intravenous	Spain
	II	ASCs	Intrathecal	USA
	II	BM-MSCs	Intramuscular	Israel
	I/II	BM-MSCs	Intrathecal and intramuscular	Israel
PD	I	UC-MSCs	Intravenous	China
	I/II	BM-MSCs	Intravenous	China
	I	UC-MSCs	Intravenous	Antigua and Barbuda
	I	BM-MSCs	Intravenous	USA
	II	BM-MSCs	Intravenous	USA
	I/II	UC-MSCs	Intravenous	Jordan
	II/III	BM-MSCs	Intravenous	Belarus
MS	I/II	BM-MSCs	Intravenous	Iran
	I/II	BM-MSCs	Intravenous	Italy
	I/II	BM-MSCs	Intravenous	Sweden
	I	BM-MSCs	Intravenous	Sweden
	I/II	BM-MSCs	Intrathecal	Jordan
	I	ASCs	Intravenous	Cayman Islands
	I/II	BM-MSCs	Intravenous	France
	I/II	UC-MSCs	Intravenous	USA
	II	BM-MSCs	Intravenous	Spain
	I/II	BM-MSCs	Intravenous	UK
	I/II	UC-MSCs	Intravenous	China
	I/II	BM-MSCs	Intravenous	Spain
	I/II	BM-MSCs	Intrathecal	Norway
	I/II	BM-MSCs	Intravenous	UK
	I/II	BM-MSCs	Intravenous	Spain
	I/II	UC-MSCs	Intrathecal	Jordan
	I	BM-MSCs	Intrathecal	USA
	I	UC-MSCs	Intravenous	Antigua and Barbuda
	II	BM-MSCs	Intravenous	Canada
	II	BM-MSCs	Intrathecal	USA
	I	BM-MSCs	Intravenous	USA

(continued)

Table 5.1 (continued)

Disease	Clinical phase	Source of MSCs	Site of injection	Location
	I/II	UC-MSCs	Intravenous	Trinidad and Tobago
	I/II	UC-MSCs	Intravenous	Spain
	II	UC-MSCs	Intravenous	Israel
	I/II	BM-MSCs	Intravenous	Spain
AD	II	BM-MSCs	Intravenous	USA
	I/II	UC-MSCs	Intraventricular	Korea
	I	BM-MSCs	Intravenous	USA
	I	UC-MSCs	Intraventricular	Korea
	I/II	UC-MSCs	Intravenous	China
	I/II	UC-MSCs	Intraventricular	Korea
HD	I	DPSCs	Intravenous	Brazil
	II	DPSCs	Intravenous	Brazil
	II/III	DPSCs	Intravenous	Brazil

stabilization in ALS pathology (Kim et al. 2009b, 2014; Mazzini et al. 2012). The healing effect of MSC transplantation was associated with its immunomodulatory effect (Kwon et al. 2014) and paracrine secretion of MSCs (Kim et al. 2014). A phase I research study showed that after MSC injection, secretion level of IL-10, IL-6, and TGF-β cytokines was increased, while production of MCP-1, which has a synergistic effect on ALS pathogenesis, was reduced (Oh et al. 2015). Authors have indicated that multiple MSC injections increased their beneficial effects. Moreover, the advantageous effect of MSCs that are induced to secrete NTFs was also demonstrated (Gothelf et al. 2014; Petrou et al. 2016). Additionally, MSCs' mode of administration appeared to be important for the observation of therapeutic effects in patients with ALS. Intrathecal method was less invasive and showed good migration of MSCs to the CNS, leading to more efficacy to reduce ALS progression (Petrou et al. 2016; Karussis et al. 2010). In contrast, intramuscular injection resulted in only little local impact on ALS disease (Petrou et al. 2016).

5.3.2 PD

Human clinical trials have demonstrated good safety and some level of efficacy of using MSCs in controlling PD pathogenesis. Hence, a pilot study by Venkataramana et al. (2010) using MSC transplantation was shown effective in improving functional recovery including dyskinesia and had no serious clinical side effects in patients with PD. More recently, single intravenous infusion of MSCs was reported to be tolerated with no immunogenic response in PD patients with mild to moderate symptoms (Schiess et al. 2021). This study indicated a downregulation of inflammatory cytokines along with an increased level of BDNF, when MSCs were administrated at highest doses. All patients sustained motor improvement except for one patient who suffered from dyskinesias within 3 weeks following MSC infusion (Schiess et al. 2021).

5.3.3 MS

Phase I and phase II clinical studies were conducted to evaluate whether MSC administration is safe and effective in slowing down MS progression in human (Karussis et al. 2010). Trial studies have indicated that MSCs intravenously or intrathecally injected over several days in subjects with MS were tolerated with no serious complication or evidence of disease activation (Bonab et al. 2012; Riordan et al. 2018). Many benefits from using this method including a decrease in magnetic resonance imaging lesions, walking and upper extremity motor recovery, control of bladder and bowel dysfunction, fatigue and sexual impotence, and enhanced life quality were observed. Improvement in symptoms was recorded 1 month (for some cases, 1 year) after MSC transplantation (Riordan et al. 2018). Moreover, a recent study by Lu and colleagues suggested that combining intrathecal and intravenous transplantation of human MSCs derived from the umbilical cord (UC-MSCs) at low dosage was safe and feasible (Lu et al. 2020).

5.3.4 AD

Though preclinical trials in AD animal models demonstrated neuroprotective effects of MSC transplantation, they did not translate clinically (Drummond and Wisniewski 2017; Duncan and Valenzuela 2017). Several studies were carried out in AD patients in order to assess the effects of MSCs isolated from different tissues, but majority of trials have not been completed and have failed to publish results (Kim et al. 2015a; Oliva Jr et al. 2020). Recently, it was showed that skin amyloid deposition disappeared after intravenous administration of ASCs in AD patients (Shigematsu et al. 2021). Authors suggested a possible secretion of neprilysin, an Aβ peptide-degrading enzyme, by MSCs. In addition, it was shown that single stereotactic injection of human UC-MSCs into the brain was generally safe and well tolerated and did not report any significant side effects between patients with mild to moderate dementia recruited in a phase I clinical trial. However, this technique did not reduce or slow the progression of AD symptoms over a 2-year follow-up (Kim et al. 2015a). Although EVs isolated from MSCs have proven their efficacy in AD experimental models, only one clinical study received approval to evaluate their tolerability and efficacy in subjects having mild to moderate dementia so far (Wang et al. 2020).

5.3.5 HD

MSCs have been largely investigated in animal models with HD and have shown success in alleviating the disease pathology; nonetheless, not much clinical trials has been undertaken so far, due to the complexity of HD syndrome in human. Initial clinical trials in HD patients aimed to investigate the tolerability as well as effectiveness of fetal striatal MSC administration (Bachoud-Lévi et al. 2006; Rosser et al. 2002). Findings reported improved motor and cognitive performance maintained for several years post-transplantation. Nevertheless, benefits of MSC administration were not consistent in all studies, and some safety concerns were addressed, notably in advanced stage of HD (Barker et al. 2013). More recently, Deng and colleagues have engineered BDNF-overexpressing MSCs to be potentially tested in a phase I clinical trial for safety and efficacy in patients with early stage of HD (Deng et al. 2016). At present, three clinical studies using MSCs are ongoing in order to assess their efficacy and safety to treat patients with HD (NCT02728115, NCT03252535, and NCT04219241). Another study was designed to characterize small molecule markers of HD in cerebrospinal fluid and plasma, including BDNF (NCT01937923).

5.4 Limitations and Future Directions in MSC Application for ND Treatment

Although the preclinical and early clinical findings were promising, several limitations need to be overcome before MSC-based treatments for neurodegenerative disorders become widely available. For example, many studies reported that transplanted MSCs have short lifespan and limited migration capacity (Preda et al. 2021; Eggenhofer et al. 2012). Moreover, outcomes of preclinical studies varied largely with the cell donor, age of cells, number of MSC injection, cell dosage, administration method, and time of intervention (Lu et al. 2020; Schiess et al. 2021; Sun et al. 2014; Kim et al. 2015b). Some strategies have been proposed so far to improve cell survival or potency, including preconditioning of MSCs or priming (Sart et al. 2014) and genetic manipulation (Nakano et al. 2020). However, despite the stable and intensive potency of engineered MSCs in the treatment of neurodegenerative disorders, critical concerns about their safety might hamper their utilization in clinical setting. Furthermore, preclinical studies in ND outcomes were frequently model-dependent and poorly reflected in clinical research studies performed on patients. Notably, transgenic animal models for NDs are mainly based on gene mutations associated with familial forms of the diseases, while the large majority of human neurodegenerative disorders occurs sporadically (Kalinderi et al. 2016), thereby limiting the effectiveness of translational research. Furthermore, key questions including mechanisms of action of MSCs, preferred source, suitable dose, best route of administration, time and frequency of intervention, and long-term safety are critical to exploit full potency of MSCs in ND therapy.

References

Abdullah RH, Yaseen NY, Salih SM, Al-Juboory AA, Hassan A, Al-Shammari AM (2016) Induction of mice adult bone marrow mesenchymal stem cells into functional motor neuron-like cells. J Chem Neuroanat 77:129–142

Aggarwal S, Pittenger MF (2005) Human mesenchymal stem cells modulate allogeneic immune cell responses. Blood 105(4):1815–1822

Aleynik A, Gernavage KM, Mourad YS, Sherman LS, Liu K, Gubenko YA, Rameshwar P (2014) Stem cell delivery of therapies for brain disorders. Clin Transl Med 3(1):1–10

Angeloni C, Gatti M, Prata C, Hrelia S, Maraldi T (2020) Role of mesenchymal stem cells in counteracting oxidative stress—related neurodegeneration. Int J Mol Sci 21(9):3299

Bachoud-Lévi AC, Gaura V, Brugières P, Lefaucheur JP, Boissé MF, Maison P, Baudic S, Ribeiro MJ, Bourdet C, Remy P, Cesaro P (2006) Effect of fetal neural transplants in patients with Huntington's disease 6 years after surgery: a long-term follow-up study. Lancet Neurol 5(4):303–309

Bae JS, Jin HK, Lee JK, Richardson JC, Carter JE (2013) Bone marrow-derived mesenchymal stem cells contribute to the reduction of amyloid-β deposits and the improvement of synaptic transmission in a mouse model of pre-dementia Alzheimer's disease. Curr Alzheimer Res 10(5):524–531

Barker RA, Mason SL, Harrower TP, Swain RA, Ho AK, Sahakian BJ, Mathur R, Elneil S, Thornton S, Hurrelbrink C, Armstrong RJ (2013) The long-term safety and efficacy of bilateral transplantation of human fetal striatal tissue in patients with mild to moderate Huntington's disease. J Neurol Neurosurg Psychiatry 84(6):657–665

Blokhuis AM, Groen EJ, Koppers M, van den Berg LH, Pasterkamp RJ (2013) Protein aggregation in amyotrophic lateral sclerosis. Acta Neuropathol 125(6):777–794

Bodart-Santos V, de Carvalho LR, de Godoy MA, Batista AF, Saraiva LM, Lima LG, Abreu CA, De Felice FG, Galina A, Mendez-Otero R, Ferreira ST (2019) Extracellular vesicles derived from human Wharton's jelly mesenchymal stem cells protect hippocampal neurons from oxidative stress and synapse damage induced by amyloid-β oligomers. Stem Cell Res Ther 10(1):1–13

Bonab MM, Sahraian MA, Aghsaie A, Karvigh SA, Hosseinian SM, Nikbin B, Lotfi J, Khorramnia S, Motamed MR, Togha M, Harirchian MH, Moghadam NB, Alikhani K, Yadegari S, Jafarian S, Gheini MR (2012) Autologous mesenchymal stem cell therapy in progressive multiple sclerosis: an open label study. Curr Stem Cell Res Ther 7(6):407–414

Boomsma RA, Geenen DL (2012) Mesenchymal stem cells secrete multiple cytokines that promote angiogenesis and have contrasting effects on chemotaxis and apoptosis. PLoS One 7(4):e35685

Buddhala C, Loftin SK, Kuley BM, Cairns NJ, Campbell MC, Perlmutter JS, Kotzbauer PT (2015) Dopaminergic, serotonergic, and noradrenergic deficits in Parkinson disease. Ann Clin Transl Neurol 2(10):949–959

Chopp M, Zhang XH, Li Y, Wang L, Chen J, Lu D, Lu M, Rosenblum M (2000) Spinal cord injury in rat: treatment with bone marrow stromal cell transplantation. Neuroreport 11(13):3001–3005

Chulpanova DS, Kitaeva KV, Tazetdinova LG, James V, Rizvanov AA, Solovyeva VV (2018) Application of mesenchymal stem cells for therapeutic agent delivery in anti-tumor treatment. Front Pharmacol 9:259

Danielyan L, Schäfer R, von Ameln-Mayerhofer A, Bernhard F, Verleysdonk S, Buadze M, Lourhmati A, Klopfer T, Schaumann F, Schmid B, Koehle C (2011) Therapeutic efficacy of intranasally delivered mesenchymal stem cells in a rat model of Parkinson disease. Rejuvenation Res 14(1):3–16

Delcroix GJR, Jacquart M, Lemaire L, Sindji L, Franconi F, Le Jeune JJ, Montero-Menei CN (2009) Mesenchymal and neural stem cells labeled with HEDP-coated SPIO nanoparticles: in vitro characterization and migration potential in rat brain. Brain Res 1255:18–31

Deng P, Torrest A, Pollock K, Dahlenburg H, Annett G, Nolta JA, Fink KD (2016) Clinical trial perspective for adult and juvenile Huntington's disease using genetically-engineered mesenchymal stem cells. Neural Regen Res 11(5):702–705

DeTure MA, Dickson DW (2019) The neuropathological diagnosis of Alzheimer's disease. Mol Neurodegener 14(1):1–18

Devine SM, Bartholomew AM, Mahmud N, Nelson M, Patil S, Hardy W, Sturgeon C, Hewett T, Chung T, Stock W, Sher D (2001) Mesenchymal stem cells are capable of homing to the bone marrow of non-human primates following systemic infusion. Exp Hematol 29(2):244–255

Dey ND, Bombard MC, Roland BP, Davidson S, Lu M, Rossignol J, Sandstrom MI, Skeel RL, Lescaudron L, Dunbar GL (2010) Genetically engineered mesenchymal stem cells reduce behavioral deficits in the YAC 128 mouse model of Huntington's disease. Behav Brain Res 214(2):193–200

Dimitrov LG, Turner B (2014) What's new in multiple sclerosis? Br J Gen Pract 64(629):612–613

Drummond E, Wisniewski T (2017) Alzheimer's disease: experimental models and reality. Acta Neuropathol 133(2):155–175

Duncan T, Valenzuela M (2017) Alzheimer's disease, dementia, and stem cell therapy. Stem Cell Res Ther 8(1):1–9

Eggenhofer E, Benseler V, Kroemer A, Popp FC, Geissler EK, Schlitt HJ, Baan CC, Dahlke MH, Hoogduijn M (2012) Mesenchymal stem cells are short-lived and do not migrate beyond the lungs after intravenous infusion. Front Immunol 3:297

Friedenstein AJ, Gorskaja JF, Kulagina N (1976) Fibroblast precursors in normal and irradiated mouse hematopoietic organs. Exp Hematol 4(5):267–274

Gao F, Chiu SM, Motan DAL, Zhang Z, Chen L, Ji HL, Tse HF, Fu QL, Lian Q (2016) Mesenchymal stem cells and immunomodulation: current status and future prospects. Cell Death Dis 7(1):e2062–e2062

Giunti D, Parodi B, Usai C, Vergani L, Casazza S, Bruzzone S, Mancardi G, Uccelli A (2012) Mesenchymal stem cells shape microglia effector functions through the release of CX3CL1. Stem Cells 30(9):2044–2053

Gothelf Y, Abramov N, Harel A, Offen D (2014) Safety of repeated transplantations of neurotrophic factors-secreting human mesenchymal stromal stem cells. Clin Transl Med 3(1): 1–11

Gómez-Benito M, Granado N, García-Sanz P, Michel A, Dumoulin M, Moratalla R (2020) Modeling Parkinson's disease with the alpha-synuclein protein. Front Pharmacol 11:356

Gratpain V, Mwema A, Labrak Y, Muccioli GG, Van Pesch V, des Rieux A (2021) Extracellular vesicles for the treatment of central nervous system diseases. Adv Drug Deliv Rev 174:535–552

Hermann A, Gastl R, Liebau S, Popa MO, Fiedler J, Boehm BO, Maisel M, Lerche H, Schwarz J, Brenner R, Storch A (2004) Efficient generation of neural stem cell-like cells from adult human bone marrow stromal cells. J Cell Sci 117(19):4411–4422

Hirota R, Sasaki M, Kataoka-Sasaki Y, Oshigiri T, Kurihara K, Fukushi R, Oka S, Ukai R, Yoshimoto M, Kocsis JD, Yamashita T (2022) Enhanced network in corticospinal tracts after infused mesenchymal stem cells in spinal cord injury. J Neurotrauma, (ja) 39:1665

Hoban DB, Howard L, Dowd E (2015) GDNF-secreting mesenchymal stem cells provide localized neuroprotection in an inflammation-driven rat model of Parkinson's disease. Neuroscience 303: 402–411

Im W, Ban J, Lim J, Lee M, Lee ST, Chu K, Kim M (2013) Extracts of adipose derived stem cells slows progression in the R6/2 model of Huntington's disease. PLoS One 8(4):e59438

Inojosa H, Proschmann U, Akgün K, Ziemssen T (2021) A focus on secondary progressive multiple sclerosis (SPMS): challenges in diagnosis and definition. J Neurol 268(4):1210–1221

Jadasz JJ, Tepe L, Beyer F, Samper Agrelo I, Akkermann R, Spitzhorn LS, Silva ME, Oreffo RO, Hartung HP, Prigione A, Rivera FJ (2018) Human mesenchymal factors induce rat hippocampal-and human neural stem cell dependent oligodendrogenesis. Glia 66(1):145–160

Kalinderi K, Bostantjopoulou S, Fidani L (2016) The genetic background of Parkinson's disease: current progress and future prospects. Acta Neurol Scand 134(5):314–326

Kan I, Barhum Y, Melamed E, Offen D (2011) Mesenchymal stem cells stimulate endogenous neurogenesis in the subventricular zone of adult mice. Stem Cell Rev Rep 7(2):404–412

Karussis D, Karageorgiou C, Vaknin-Dembinsky A, Gowda-Kurkalli B, Gomori JM, Kassis I, Bulte JW, Petrou P, Ben-Hur T, Abramsky O, Slavin S (2010) Safety and immunological effects of mesenchymal stem cell transplantation in patients with multiple sclerosis and amyotrophic lateral sclerosis. Arch Neurol 67(10):1187–1194

Kim HJ, Seo SW, Chang JW, Lee JI, Kim CH, Chin J, Choi SJ, Kwon H, Yun HJ, Lee JM, Kim ST (2015a) Stereotactic brain injection of human umbilical cord blood mesenchymal stem cells in patients with Alzheimer's disease dementia: a phase 1 clinical trial. Alzheimers Dement (N Y) 1(2):95–102

Kim YS, Kwon OR, Choi YJ, Suh DS, Heo DB, Koh YG (2015b) Comparative matched-pair analysis of the injection versus implantation of mesenchymal stem cells for knee osteoarthritis. Am J Sports Med 43(11):2738–2746

Kim KS, Lee HJ, An J, Kim YB, Ra JC, Lim I, Kim SU (2014) Transplantation of human adipose tissue-derived stem cells delays clinical onset and prolongs life span in ALS mouse model. Cell Transplant 23(12):1585–1597

Kim H, Kim HY, Choi MR, Hwang S, Nam KH, Kim HC, Han JS, Kim KS, Yoon HS, Kim SH (2010) Dose-dependent efficacy of ALS-human mesenchymal stem cells transplantation into cisterna magna in SOD1-G93A ALS mice. Neurosci Lett 468(3):190–194

Kim YJ, Park HJ, Lee G, Bang OY, Ahn YH, Joe E, Kim HO, Lee PH (2009a) Neuroprotective effects of human mesenchymal stem cells on dopaminergic neurons through anti-inflammatory action. Glia 57(1):13–23

Kim HY, Paek JY, Park HK, Choi MR, Yoon HS, Kim KS, Kim SH (2009b) Efficacy and safety of autologous bone marrow-derived mesenchymal stem cell treatment in patients with amyotrophic lateral sclerosis. J Korean Neurol Assoc:163–169

Krampera M, Galipeau J, Shi Y, Tarte K, Sensebe L (2013) Immunological characterization of multipotent mesenchymal stromal cells—the International Society for Cellular Therapy (ISCT) working proposal. Cytotherapy 15(9):1054–1061

Kwon MS, Noh MY, Oh KW, Cho KA, Kang BY, Kim KS, Kim YS, Kim SH (2014) The immunomodulatory effects of human mesenchymal stem cells on peripheral blood mononuclear cells in ALS patients. J Neurochem 131(2):206–218

Lam G, Shea GKH, Wu LK, Sui MLT, Hei KC, Kuen ZCC, Yin YWC, Tsui AYP, Shum DKY, Chan YS (2019) Derivation of oligodendrocyte precursor cells from human bone marrow stromal cells and use for re-myelination in the congenitally dysmyelinated brain. Cells 10(8): 2166

Lazarus H, Haynesworth SE, Gerson SL, Rosenthal NS, Caplan AI (1995) Ex vivo expansion and subsequent infusion of human bone marrow-derived stromal progenitor cells (mesenchymal progenitor cells): implications for therapeutic use. Bone Marrow Transplant 16(4):557–564

Lee HJ, Lee JK, Lee H, Carter JE, Chang JW, Oh W, Yang YS, Suh JG, Lee BH, Jin HK, Bae JS (2012) Human umbilical cord blood-derived mesenchymal stem cells improve neuropathology and cognitive impairment in an Alzheimer's disease mouse model through modulation of neuroinflammation. Neurobiol Aging 33(3):588–602

Lee ST, Chu K, Jung KH, Im WS, Park JE, Lim HC, Won CH, Shin SH, Lee SK, Kim M, Roh JK (2009) Slowed progression in models of Huntington disease by adipose stem cell transplantation. Ann Neurol 66(5):671–681

Lin YT, Chern Y, Shen CKJ, Wen HL, Chang YC, Li H, Cheng TH, Hsieh-Li HM (2011) Human mesenchymal stem cells prolong survival and ameliorate motor deficit through trophic support in Huntington's disease mouse models. PLoS One 6(8):e22924

Liu Y, Zeng R, Wang Y, Huang W, Hu B, Zhu G, Zhang R, Li F, Han J, Li Y (2019) Mesenchymal stem cells enhance microglia M2 polarization and attenuate neuroinflammation through TSG-6. Brain Res 1724:146422

Lu Z, Zhu L, Liu Z, Wu J, Xu Y, Zhang CJ (2020) IV/IT hUC-MSCs infusion in RRMS and NMO: a 10-year follow-up study. Front Neurol 11:967

Lu L, Zhao C, Liu Y, Sun X, Duan C, Ji M, Zhao H, Xu Q, Yang H (2005) Therapeutic benefit of TH-engineered mesenchymal stem cells for Parkinson's disease. Brain Res Protocol 15(1): 46–51

Martin R (1997) Immunological aspects of experimental allergic encephalomyelitis and multiple sclerosis and their application for new therapeutic strategies. J Neural Transm Suppl 49:53–67

Mazzini L, Mareschi K, Ferrero I, Miglioretti M, Stecco A, Servo S, Carriero A, Monaco F, Fagioli F (2012) Mesenchymal stromal cell transplantation in amyotrophic lateral sclerosis: a long-term safety study. Cytotherapy 14(1):56–60

Merimi M, El-Majzoub R, Lagneaux L, Moussa Agha D, Bouhtit F, Meuleman N, Fahmi H, Lewalle P, Fayyad-Kazan M, Najar M (2021) The therapeutic potential of mesenchymal stromal cells for regenerative medicine: current knowledge and future understandings. Front Cell Dev Biol 9:1625

Mohajeri M, Farazmand A, Bonab MM, Nikbin B, Minagar A (2011) FOXP3 gene expression in multiple sclerosis patients pre-and post mesenchymal stem cell therapy. Iran J Allergy Asthma Immunol 10:155–161

Mohammadi R (2020) Exosomes secreted by mesenchymal stem cell to treat autoimmune disorders. University of California, Irvine

Morigi M, Imberti B, Zoja C, Corna D, Tomasoni S, Abbate M, Rottoli D, Angioletti S, Benigni A, Perico N, Alison M (2004) Mesenchymal stem cells are renotropic, helping to repair the kidney and improve function in acute renal failure. J Am Soc Nephrol 15(7):1794–1804

Mostafavi H, Ghassemifard L, Rostami A, Alipour M, Nadri S (2019) Trabecular meshwork mesenchymal stem cell transplantation improve motor symptoms of parkinsonian rat model. Biologicals 61:61–67

Nakano M, Kubota K, Kobayashi E, Chikenji TS, Saito Y, Konari N, Fujimiya M (2020) Bone marrow-derived mesenchymal stem cells improve cognitive impairment in an Alzheimer's disease model by increasing the expression of microRNA-146a in hippocampus. Sci Rep 10(1):1–15

Nawaz M, Fatima F, Vallabhaneni KC, Penfornis P, Valadi H, Ekström K, Kholia S, Whitt JD, Fernandes JD, Pochampally R, Squire JA (2016) Extracellular vesicles: evolving factors in stem cell biology. Stem Cells Int 2016

Nowicka N, Juranek J, Juranek JK, Wojtkiewicz J (2019) Risk factors and emerging therapies in amyotrophic lateral sclerosis. Int J Mol Sci 20(11):2616

Oh KW, Moon C, Kim HY, Oh SI, Park J, Lee JH, Chang IY, Kim KS, Kim SH (2015) Phase I trial of repeated intrathecal autologous bone marrow-derived mesenchymal stromal cells in amyotrophic lateral sclerosis. Stem Cells Transl Med 4(6):590–597

Oliva AA Jr, Baumel B, Brody M, Agronin M, Herskowitz B, Bookheimer SY, McClain-Moss L, Hitchinson B, Ramdas K, Pena A, Diaz LA (2020) Clinical evaluation of allogeneic mesenchymal stem cells for Alzheimer's disease: human/human trials: inflammation. Alzheimers Dement 16:e046634

Osugi M, Katagiri W, Yoshimi R, Inukai T, Hibi H, Ueda M (2012) Conditioned media from mesenchymal stem cells enhanced bone regeneration in rat calvarial bone defects. Tissue Eng A 18(13–14):1479–1489

Petrou P, Gothelf Y, Argov Z, Gotkine M, Levy YS, Kassis I, Vaknin-Dembinsky A, Ben-Hur T, Offen D, Abramsky O, Melamed E (2016) Safety and clinical effects of mesenchymal stem cells secreting neurotrophic factor transplantation in patients with amyotrophic lateral sclerosis: results of phase 1/2 and 2a clinical trials. JAMA Neurol 73(3):337–344

Philips T, Rothstein JD (2015) Rodent models of amyotrophic lateral sclerosis. Curr Protocol Pharmacol 69(1):5–67

Pittenger MF, Mackay AM, Beck SC, Jaiswal RK, Douglas R, Mosca JD, Moorman MA, Simonetti DW, Craig S, Marshak DR (1999) Multilineage potential of adult human mesenchymal stem cells. Science 284(5411):143–147

Preda MB, Neculachi CA, Fenyo IM, Vacaru AM, Simionescu M, Burlacu A (2021) Short lifespan of syngeneic transplanted MSC is a consequence of in vivo apoptosis and immune cell recruitment in mice. Cell Death Dis 12(6):1–12

Qin C, Lu Y, Wang K, Bai L, Shi G, Huang Y, Li Y (2020) Transplantation of bone marrow mesenchymal stem cells improves cognitive deficits and alleviates neuropathology in animal models of Alzheimer's disease: a meta-analytic review on potential mechanisms. Transl Neurodegener 9(1):1–20

Ren Z, Wang J, Wang S, Zou C, Li X, Guan Y, Chen Z, Zhang YA (2013) Autologous transplantation of GDNF-expressing mesenchymal stem cells protects against MPTP-induced damage in cynomolgus monkeys. Sci Rep 3(1):1–11

Reza-Zaldivar EE, Hernández-Sapiéns MA, Gutiérrez-Mercado YK, Sandoval-Ávila S, Gomez-Pinedo U, Márquez-Aguirre AL, Vázquez-Méndez E, Padilla-Camberos E, Canales-Aguirre AA (2019) Mesenchymal stem cell-derived exosomes promote neurogenesis and cognitive function recovery in a mouse model of Alzheimer's disease. Neural Regen Res 14(9):1626

Riordan NH, Morales I, Fernández G, Allen N, Fearnot NE, Leckrone ME, Markovich DJ, Mansfield D, Avila D, Patel AN, Kesari S (2018) Clinical feasibility of umbilical cord tissue-derived mesenchymal stem cells in the treatment of multiple sclerosis. J Transl Med 16(1):1–12

Rombouts WJC, Ploemacher RE (2003) Primary murine MSC show highly efficient homing to the bone marrow but lose homing ability following culture. Leukemia 17(1):160–170

Rosser AE, Barker RA, Harrower T, Watts C, Farrington M, Ho AK, Burnstein RM, Menon DK, Gillard JH, Pickard J, Dunnett SB (2002) Unilateral transplantation of human primary fetal tissue in four patients with Huntington's disease: NEST-UK safety report ISRCTN no 36485475. J Neurol Neurosurg Psychiatry 73(6):678–685

Sadan O, Shemesh N, Cohen Y, Melamed E, Offen D (2009) Adult neurotrophic factor-secreting stem cells: a potential novel therapy for neurodegenerative diseases. Isr Med Assoc J 11(4): 201–204

Sart S, Ma T, Li Y (2014) Preconditioning stem cells for in vivo delivery. BioRes Open Access 3(4):137–149

Sasaki M, Radtke C, Tan AM, Zhao P, Hamada H, Houkin K, Honmou O, Kocsis JD (2009) BDNF-hypersecreting human mesenchymal stem cells promote functional recovery, axonal sprouting, and protection of corticospinal neurons after spinal cord injury. J Neurosci 29(47): 14932–14941

Schiess M, Suescun J, Doursout MF, Adams C, Green C, Saltarrelli JG, Savitz S, Ellmore TM (2021) Allogeneic Bone Marrow–Derived Mesenchymal Stem Cell Safety in Idiopathic Parkinson's Disease. Mov Disord 36(8):1825–1834

Shetty P, Thakur AM, Viswanathan C (2013) Dopaminergic cells, derived from a high efficiency differentiation protocol from umbilical cord derived mesenchymal stem cells, alleviate symptoms in a Parkinson's disease rodent model. Cell Biol Int 37(2):167–180

Shigematsu K, Takeda T, Komori N, Tahara K, Yamagishi H (2021) Hypothesis: intravenous administration of mesenchymal stem cells is effective in the treatment of Alzheimer's disease. Med Hypotheses 150:110572

Shin JY, Lee PH (2020) Mesenchymal stem cells modulate misfolded α-synuclein in parkinsonian disorders: a multitarget disease-modifying strategy. Stem Cell Res 47:101908

Sironi F, Vallarola A, Violatto MB, Talamini L, Freschi M, De Gioia R, Capelli C, Agostini A, Moscatelli D, Tortarolo M, Bigini P (2017) Multiple intracerebroventricular injections of human umbilical cord mesenchymal stem cells delay motor neurons loss but not disease progression of SOD1G93A mice. Stem Cell Res 25:166–178

Spaggiari GM, Capobianco A, Abdelrazik H, Becchetti F, Mingari MC, Moretta L (2008) Mesenchymal stem cells inhibit natural killer–cell proliferation, cytotoxicity, and cytokine production: role of indoleamine 2, 3-dioxygenase and prostaglandin E2. Blood 111(3):1327–1333

Sun H, Hou Z, Yang H, Meng M, Li P, Zou Q, Yang L, Chen Y, Chai H, Zhong H, Yang ZZ (2014) Multiple systemic transplantations of human amniotic mesenchymal stem cells exert therapeutic effects in an ALS mouse model. Cell Tissue Res 357(3):571–582

Sun H, Bénardais K, Stanslowsky N, Thau-Habermann N, Hensel N, Huang D, Claus P, Dengler R, Stangel M, Petri S (2013) Therapeutic potential of mesenchymal stromal cells and MSC conditioned medium in amyotrophic lateral sclerosis (ALS)-in vitro evidence from primary motor neuron cultures, NSC-34 cells, astrocytes and microglia. PLoS One 8(9):e72926

Venkataramana NK, Kumar SK, Balaraju S, Radhakrishnan RC, Bansal A, Dixit A, Rao DK, Das M, Jan M, Gupta PK, Totey SM (2010) Open-labeled study of unilateral autologous bone-marrow-derived mesenchymal stem cell transplantation in Parkinson's disease. Transl Res 155(2):62–70

Vigo T, Voulgari-Kokota A, Errede M, Girolamo F, Ortolan J, Mariani MC, Ferrara G, Virgintino D, Buffo A, Kerlero de Rosbo N, Uccelli A (2021) Mesenchymal stem cells instruct a beneficial phenotype in reactive astrocytes. Glia 69(5):1204–1215

Violatto MB, Santangelo C, Capelli C, Frapolli R, Ferrari R, Sitia L, Tortarolo M, Talamini L, Previdi S, Moscatelli D, Salmona M (2015) Longitudinal tracking of triple labeled umbilical cord derived mesenchymal stromal cells in a mouse model of amyotrophic lateral sclerosis. Stem Cell Res 15(1):243–253

Wang G, Gao X, Hospital R (2020) The safety and the efficacy evaluation of allogenic adipose MSC-exos in patients with Alzheimer's disease. Identifier NCT04388982

Yu-Taeger L, Stricker-Shaver J, Arnold K, Bambynek-Dziuk P, Novati A, Singer E, Lourhmati A, Fabian C, Magg J, Riess O, Schwab M (2019) Intranasal administration of mesenchymal stem cells ameliorates the abnormal dopamine transmission system and inflammatory reaction in the R6/2 mouse model of Huntington disease. Cell 8(6):595

Zappia E, Casazza S, Pedemonte E, Benvenuto F, Bonanni I, Gerdoni E, Giunti D, Ceravolo A, Cazzanti F, Frassoni F, Mancardi G (2005) Mesenchymal stem cells ameliorate experimental autoimmune encephalomyelitis inducing T-cell anergy. Blood 106(5):1755–1761

Zeng X, Qiu XC, Ma YH, Duan JJ, Chen YF, Gu HY, Wang JM, Ling EA, Wu JL, Wu W, Zeng YS (2015) Integration of donor mesenchymal stem cell-derived neuron-like cells into host neural network after rat spinal cord transection. Biomaterials 53:184–201

iPSCs and their Role in Amelioration of Neurodegenerative Disorders

6

Dipak Kumar

Abstract

Since the birth of human-induced pluripotent stem cells (iPSCs) in 2007, its applications have been widened to solve the challenges of metabolic and neurodegenerative disorders. The developments of pioneering techniques and novel treatment methods for the neurodegenerative disorders like amyotrophic lateral sclerosis (ALS) and Alzheimer's and Parkinson's diseases using innovative systems such as 3D in vitro cell culture and clustered regularly interspaced short palindromic repeat-associated protein 9 (CRISPR-Cas9) have opened a novel vista of hope in the field of medical sciences. As iPSC technique produces the pluripotent stem cells from one's own somatic cells, its application in the field of stem cell therapy has been immensely potentiated. There are various iPSC-based disease models for drug development and therapeutic interventions having vast implications in the field of modern neurobiology. In this chapter, we will elaborate how iPSCs can be employed to ameliorate the neurodegenerative disorders.

Keywords

Human-induced pluripotent stem cells (iPSCs) · Neurodegenerative disorders · CRISPR-Cas9 · Drug development · 3D in vitro cell culture

D. Kumar (✉)
Department of Zoology, KKM College, Munger University, Munger, India

S. Jahan, A. J. Siddiqui (eds.), *Applications of Stem Cells and Derived Exosomes in Neurodegenerative Disorders*, https://doi.org/10.1007/978-981-99-3848-3_6

6.1 Introduction

The human brain is a composite and prime organ. It is the center of control and coordination of the nervous system. Besides the brain, the spinal cord contributes to the functioning of the central nervous system (CNS). The peripheral nervous system (PNS) further extends into the extremities and organs of the individual and establishes a network of communication. Lately, discovery of neurogenesis and regenerative capacity of the brain has given a new tool to the neurobiologist to explore the cellular and molecular mechanisms governing brain structures and functions (Lepousez et al. 2015). A scalable amount of stem cells and defined in vitro conditions to differentiate them into desired lineages including brain cells, viz., neuron, glia and oligodendrocyte, have further accelerated the voyage to explore this master organ (Conley et al. 2004).

6.1.1 Stem Cells

Naïve cells, which can self-renew and have differentiation potential forming cells of various lineages, are termed as stem cells. Basically, they can proliferate and differentiate as per their niche and meet the requirements of time and space of a multicellular organism's life (Moore and Lemischka 2006). All multicellular organisms start their development from a single cell, zygote, which forms the three germ layers and extraembryonic membranes. Subsequently, the capacity of proliferation and lineage specification become limited as individuals mature and cells get terminally differentiated (Zakrzewski et al. 2019).

6.1.2 Stem Cell Types and Potency

Stem cells have been categorised and named as per their differentiation potential and origin. The potency of stem cells is their ability to differentiate in various lineages. Totipotent stem cells can form the cells of both embryonic and extraembryonic tissues, whereas unipotent stem cells have potential only to differentiate into a particular cell type. The zygote is a totipotent stem cell as it can form the foetus as well as the placenta which connects the foetus to its mother. At the blastula stage of embryonic development, inner cell mass (ICM) has the capability to form all the germ layers; these cells are denoted as pluripotent stem cells (PSCs). Moreover, they are also named as embryonic stem cells (ESCs), which means being derived from the embryo. In 2006, Takahashi and Yamanaka characterised a group of genes whose expression can induce pluripotency in somatic cells. They termed these cells as induced pluripotent stem cells (iPSCs) (Yu et al. 2007). Multipotent stem cells and oligopotent stem cells have potency that ranges between pluripotency and unipotency. For instance, haematopoietic stem cells (HSCs) can differentiate into various blood cells and mesenchymal stem cells (MSCs) have been differentiated into several somatic cells (Bacigalupo 2004). These multipotent cells have been named as per their origins. Similarly, neural stem cells (NSCs) (Gage 2000), adipocyte stem cells (ASCs) (Bunnell et al. 2008), endothelial stem cells (Chambers et al. 2021), mammary stem cells (Tharmapalan et al. 2019), intestinal stem cells (Santos et al. 2018), etc. have their respective origin and sometimes referred to as somatic or adult stem cells.

6.1.3 Sources of Stem Cells

The discovery and characterization of stem cells have been made mainly from four sources: embryonic tissues, foetal tissues, adult organ tissues and differentiated somatic cells. The pluripotent stem cells have been isolated from embryonic tissues, whereas the multipotent and unipotent stem cells have been purified from foetal tissues and fully developed organ tissues, respectively. iPSCs have been developed from various terminally differentiated somatic cells. ICM isolated from preimplantation blastocyst can be cultured in in vitro conditions. These cells can be retained in undifferentiated and pluripotent forms in the presence of leukaemia inhibitory factor (LIF) or embryonic fibroblast for virtually indefinite time (Mossahebi-Mohammadi et al. 2020). Further, in specific conditions (using growth factors, cytokines, etc.), the cells can be differentiated as per need.

Adult stem cells have varied sources. Almost every tissue has been described to contain the resident stem cells residing in specific niches. The umbilical cord (UC) has been extensively utilized as a mesenchymal stem (stromal) cell (MSC) source. Distinct anatomical regions of human UC like Wharton's jelly, umbilical arteries, vein, cord lining and the blood from it have been described in literature for isolating MSCs (Iwatani et al. 2019). These cells have been differentiated into adipocytes, hepatocytes, osteocytes, chondrocytes and cardiac and neural cells, suggesting their multipotency (Mebarki et al. 2021).

6.1.4 Safety and Ethics

Although stem cells have tremendous potential in the areas of regenerative medicine and drug development, their clinical applications must be thoroughly assessed (Doss and Sachinidis 2019). Safety of donors from any infection must be assured after the somatic cell isolation for the iPSC preparation or taking tissue for the isolation of adult stem cells (Garreta et al. 2018). Various procedures of clinical applications require compliance with ethical guidelines (Volarevic et al. 2018). Terms and conditions of consent related to a particular procedure of isolation and therapy must be understood and duly signed by the concerned subject (Barker et al. 2018).

6.2 Induced Pluripotent Stem Cells

Since the 1950s, researchers have been able to demonstrate that differentiated cells have the same genetic information as the cells of developing embryo, thanks to development like the somatic cell nuclear transfer technique (SCNT). Subsequent isolation, culture and characterization of PSCs and the demonstrations that transcription factors govern the cell fate whose manipulation can transfer one mature cell into another provided the platform for the creation of iPSCs. Moreover, the development of iPSCs is legally and ethically more feasible and technically more achievable than the SCNT. So, in order to utilize the iPSC potential, suitable technique of factor delivery, identification of appropriate reprogrammed cell and establishment of clinical procedures were required (Rowe and Daley 2019).

6.2.1 Origin of iPSC

The induced pluripotency was first achieved in 2006 by Kazutoshi Takahashi and Shinya Yamanaka. Their laboratory devised an elegant method for screening pluripotency-associated genes (Yamanaka 2012). A pool of 24 genes could activate ESC-specific allele present on Fbxo15 locus which otherwise was dormant and drug resistant. The drug-resistant mouse fibroblast colony had characteristic ESC morphology. The laboratory eliminated these genes in their subsequent experiments and reached on conclusion of minimally required essential set of four genes: Klf4, Sox2, c-Myc and Oct4 (Takahashi et al. 2007). Later, these sets of genes were known as Yamanaka factors. iPSCs are then created by reprogramming mouse or human somatic cells with Yamanaka factors and other combinations of factors. Such pluripotent cells expressed the markers like SSEA-1 and Nanog and engendered subcutaneous teratomas when inoculated in immunocompromised mice (Ohnuki et al. 2009). Although iPSCs fulfilled the various criteria of being pluripotent, they were not the same as ESCs. The expression level of key genes was found to be lower when compared with the pluripotency genes of ESCs. Specifically, iPSCs demonstrated incomplete promoter demethylation (Liang and Zhang 2013) for genes like Oct4 and could neither generate postnatal chimera nor form germline (Puri and Nagy 2012). Inferring that these first generations of iPSCs are partially reprogrammed, several laboratories improved upon these outcomes. For instance, some essential pluripotency genes such as Nanog and Oct4 were selected for reactivation instead of Fbxo15. The case of X chromosome inactivation and reactivation of human female iPSCs through the reprogramming process was also taken into consideration to remove any abnormality that can hamper the authenticity of pluripotency and subsequent clinical applications.

Pluripotency has been induced in the somatic cells of various species including human (Ohnuki et al. 2009), rat (Liao et al. 2009) and rhesus monkey (Fang et al. 2014) by the expression of Yamanaka factors. Hence, it can be concluded that the transcriptional machinery controlling pluripotency has evolved while retaining its essential features. Similarly, there have been various cell populations such as neural cells, keratinocytes, melanocytes, the liver and stomach cells, as well as genetically marked β cells from the pancreas and terminally differentiated lymphocytes being utilized for the reprogramming into iPSCs, proving the ubiquity of induced pluripotency (Rowe and Daley 2019). Now, iPSCs from any source is identical to the ESCs, and in vitro they have capability to differentiate into every cells of all germ layers and essentially can originate from any cell of an adult organism (Wu and Hochedlinger 2011). iPSCs have the capability to aggregate and incorporate the ICM and can develop into an embryo when injected into host blastocyst followed by transfer into a foster pregnant female. Moreover, iPSCs contributing to the germline formation (Zhao et al. 2021) have potential to produce viable and fertile organisms (Yamashiro et al. 2018).

6.2.2 Genetics

As evident, some genes play indispensable function in the preservation of stemness in the cells. Ectopic expression of these genes opened direct route to pluripotency. Pioneer studies on iPSCs constitutively utilized retroviral vectors to stably integrate genes like c-Myc, Klf4, Oct4 and Sox2 into the host cell genome (Malik and Rao 2013). Generally, the retroviral transgenes are silenced at the end of reprogramming, but the process is often incomplete due to activation of DNA and histone methyltransferases (Haridhasapavalan et al. 2020). This results into incompletely reprogramed cell lines that need exogenous factors corresponding to endogenous genes for pluripotency characterization.

Often, the viral transgenes have residual activities or get reactivated in iPSC-originated somatic cells that can hamper their developmental capabilities or can subsequently form tumours in vivo. The issue gets escalated when iPSCs are produced using constitutive lentiviral vectors because lentiviral vectors are even less proficiently silenced than that of retroviral vectors (Chan et al. 2009). This leads to differentiation block. Using the inducible lentiviral vectors provides an alternative as their expression can be controlled by the inert drug doxycycline. This permits selection of fully reprogrammed iPSCs, as the cells requiring exogenous factors stop proliferation in the absence of doxycycline (Sim et al. 2014). Moreover, including all four essential reprogramming factors in lentiviral vectors, the efficiency of reprogramming increases by transferring cells from various sources to the lentiviral vectors (Bar-Nur et al. 2014).

Conventional reprogramming methods are based on integrating vectors having issues of reactivation, cell death, immunogenicity, residual expression, insertional mutagenesis and uninhibited silencing of transgenes (Krause et al. 2016). These issues have been addressed by numerous alternative methods. These methods are generally divided into two categories: transgene and chemical reprogramming. Transgene reprogramming can be further classified as protein transduction and RNA- and DNA-based reprogramming. Protein transduction involves direct delivery of gene product, while RNA programming is achieved by transduction or transfection of either synthetic mRNAs, miRNAs, RNA viruses or synthetic RNA replicons.

DNA-based methods are the most common, and they mostly take three forms: transposons, plasmids and viral particles. These viruses may be retro- or DNA viruses (Malik and Rao 2013) as summarized in Table 6.1.

6.2.3 Physiology

Morphologically, the reprogrammed cells form compact colonies, have distinct margins with well-defined boundaries and made up of scanty cytoplasm, large nucleus and large nucleoli. Although many such colonies appear morphologically like ESCs, only a few have analogous physiology in terms of their molecular and functional characteristics. The bona fide iPSCs can be distinguished from that of partially reprogrammed cells only by using molecular hallmark, i.e. pluripotency genes such as OCT4, SOX2 and NANOG in the level comparable to that of ESCs. At the molecular level, it can be noticed that telomerase gene expression is reactivated, THY1 is downregulated and SSEA1 is upregulated (Spyrou et al. 2019). Earlier, cells staining for alkaline phosphatase were considered as pluripotent, but as intermediately reprogrammed cells also stained positive, this method for testing pluripotency was discouraged. In virus-mediated reprogrammed cells, the pro-viral genes get silenced when endogenous genes for pluripotency are activated, and this event can be paired with the expression of embryonic antigens such as DNA methyltransferase 3β (DNMT3β), TRA-1-60, TRA-1-81, REX1 and SSEA3 (Loh et al. 2009).

Moreover, the epigenetic reprogramming throughout the genome is crucial for confirming cells as fully reprogrammed, and notably, the extent of success is assessed by the evaluation of methylation of the promoter region of the genes accountable for maintenance of pluripotency as well as at the genes essential for controlling differentiation (Stadtfeld et al. 2008). The reactivation of X chromosome is an equally important epigenetic phenomenon which occurs later during the reprogramming and represents assurance of basal state pluripotency (Payer et al. 2011). If reprogrammed iPSCs show all these characteristic features at the physiological level, they can be said to be like ESCs. However, disparities in epigenetic reprogramming in terms of methylation extent, persistence integrated provirus expression and other factors can modify iPSC's differentiation potentials (Maherali et al. 2007).

6.2.4 In Vitro Culture and Consideration

Somatic cell reprogramming for producing iPSCs is extremely inefficient. Only 0.01% to 0.1% of cell population gets converted into fully pluripotent cells because the process takes a long time to complete, i.e. around 2 weeks (Robinton and Daley 2012). Even if the somatic cells express the factors homogenously, the fibroblast reprogramming efficiency mostly remains between 1% and 5% (Table 6.1). This observation has been explained by two opposing and mutually nonexclusive models.

Table 6.1 Reprogramming methods for somatic cells into iPSCs (Robinton and Daley 2012)

Vector type		Cell types	Advantages	Disadvantages	Efficiency (%)
Integrating	**Retroviral**	Fibroblasts, neural stem cells, keratinocytes, stomach cells, amniotic cells, liver cells, blood cells and adipose cells	Judiciously effective	Genomic integration, incomplete pro-viral silencing and slow kinetics	~0.001–1
	Lentiviral	Fibroblasts and keratinocytes	Transduces both proliferating and non-proliferating cells with reasonable efficiency	Genomic integration and incomplete pro-viral silencing	~0.1–1.1
	Inducible lentiviral	Fibroblasts, β cells, keratinocytes, blood cells and melanocytes	Factors can be controlled and are reasonably efficient	Genomic integration and requirement for trans-activator expression	~0.1–2
Excisable	**Transposon**	Fibroblasts	Judiciously effective and no genomic integration	Labor-intensive screening of excised lines	~0.1
	loxP-flanked lentiviral	Fibroblasts	Judiciously effective and no genomic integration	Labor-intensive screening of excised lines and loxP sites retained in the genome	~0.1–1
Non-integrating	**Adenoviral**	Fibroblast and liver cells	No genomic integration	Low efficacy	~0.001
	Plasmid	Fibroblast	Only occasional genomic integration	Low efficacy and random vector genomic integration	~0.001
DNA free	**Sendai virus**	Fibroblast	No genomic integration	Sequence-sensitive RNA replicase and difficulty in eliminating the cells of replicating virus	~1
	Protein	Fibroblast	The genome is not integrated, transcription factors are directly delivered and no complications relating to DNA are involved	Low efficacy, short half-life and large amount of pure proteins and multiple applications of protein needed	~0.001

(continued)

Table 6.1 (continued)

Vector type		Cell types	Advantages	Disadvantages	Efficiency (%)
	Modified mRNA	Fibroblast	Reprogramming kinetics are faster, controllable and highly efficient, and there is no genomic integration	Multiple rounds of transfection needed	~1–4.4
	MicroRNA	Adipose stromal cells and dermal fibroblasts	In contrast to lentiviral vectors and retroviral vectors, this vector is more efficient and faster and does not require exogenous transcription factors and no integration	Lower efficacy than other commonly used methods	~0.1

The first one is the "elite" or "deterministic" model which explains that only rare somatic cells in the culture are receptive to reprogramming, which is why the reprogramming efficiency is low. The second one, in contrast to the first one, states that all somatic cells in the culture are equally responsive to the factor-driven reprogramming, but they must pass across a sequence of epigenetic incidents for acquiring pluripotency. Here, only a limited number of somatic cells successfully go through the roadblocks and hence the efficiency is low (Omole and Fakoya 2018).

6.2.4.1 Co-culture

Studying neurodegenerative disorders (NDDs) in vitro has always been challenging due to lack of proper models that could mimic the human brain. Development of iPSCs and other techniques has enabled successful differentiation of iPSCs into numerous cell types including the cells of the nervous system such as dopaminergic neurons and astrocytes. One of the advantages of using iPSCs is the fact that iPSC-differentiated cells have the same genetic mark as the donor which allows researchers to investigate the cause of NDDs. There are various protocols to differentiate iPSCs into neurons, astrocytes and midbrain-like neural progenitor cells (NPCs) (Denham and Dottori 2011). Astrocytes carry its core properties along with their specialized functions influenced by their local identity which may have a role to play during the development of NDDs. Therefore, it becomes imperative to produce neurons with astrocytes which have originated from the same midbrain-like NPC population to study NDDs such as Parkinson's disease (Laperle et al. 2020). Here, interaction and communication at both physical and chemical levels between astrocytes and neurons become crucial for the maintenance and disruption of neuronal health. Notably, dysfunctional astrocytes have been observed causing NDDs (Li et al. 2019). That's why several culture systems have been developed for studying communication between astrocyte and neuron such as co-culture of dopaminergic neuron and astrocyte, NPC differentiating into astrocyte and dopaminergic neuron and proper dopaminergic neuron and astrocyte (Kuijlaars et al. 2016).

6.2.4.2 3D Culture

Culturing cells on flat surface is always different from the cells in a tissue. A cell in natural milieu can grow, divide and communicate in all directions. In a dynamic environment, extracellular matrix (ECM) plays its role by letting the cell to migrate, home around and form gradients of different biochemical factors (Ahmed 2016). Hence, there are number of physiological functions such as proliferation, differentiation, survival, apoptosis, growth and neurotrophic and other factors that can be influenced by the 3D micro-environmental conditions. Keeping the differences between flat surfaces (2D) and 3D biomimetic conditions in consideration, researchers have been trying to better understand the behaviour of cells in 3D conditions. To establish the 3D organotypic culture system, compatibility of cells with biomaterials has to be checked first. So, different combinations of cells with biomaterials have been investigated to meet the demand of optimum proliferation rate and desired differentiation lineage (Schaap-Oziemlak et al. 2014). Among all other cells, pluripotent stem cells have advantages in terms of both differentiation

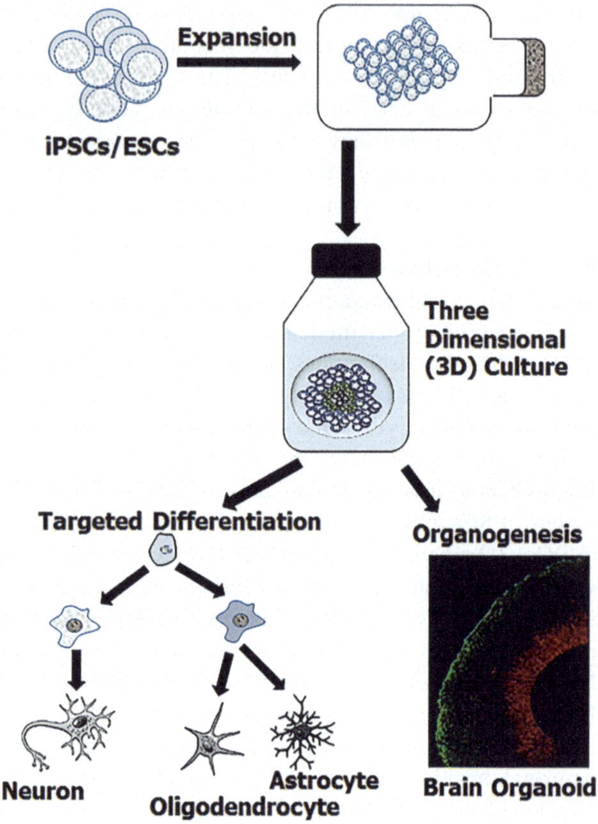

Fig. 6.1 Pluripotent stem cells (PSCs) can be expanded on conventional flat surface culture and characterized for the purity of pluripotency. Under suitable conditions of growth factors and biomaterial, the stem cells can further be differentiated into neuronal lineage cells such as neurons, oligodendrocytes or astrocytes. Further, to recapitulate the in vivo three-dimensional milieu, stem cells may be utilized to form brain organoids

and proliferation capabilities (Fig. 6.1). In this connection, bulk proliferation and differentiation of human pluripotent stem cells (HPSCs) have been done. Others have indicated towards specific needs of cells when proliferating and differentiating them (Dawson et al. 2008). Alternatively, it has been advocated that different ECMs should be used for these two purposes.

iPSCs have edge over other pluripotent stem cells in relation to ethical clearance and wide applicability. iPSCs from patients under in vitro conditions can directly be observed and manipulated as and when required. The deficiencies can be corrected and differentiated into required cells or organs or organoids in 3D culture conditions. Subsequently, they can be grafted in the body of the patient without immunological consequences. HPSCs, both iPSCs and ESCs, have been well characterized for their proliferation and differentiation in 3D suspension as well as matrix (hydrogel)-assisted 3D cultures to form human brain organoid. The study on gene-environment interactions has addressed the use of iPSCs in time-dependent interaction of chemicals to interfere with epigenetic mechanisms. Additionally, poly (3-hydroxybutyrateco-3-hydroxyvalerate) (PHB-HV) was found to be an attractive option for spinal cord repair. Co-culture of tissues can also guide organogenesis by manipulating epithelial-mesenchymal interactions (Sasai 2013).

HPSCs have been either differentiated into the specific lineage cells such as dopamine neuron, striatal neuron and glial cells or directed into 3D neuronal tissue formation, for instance, cerebral organoid (Lancaster et al. 2013), inner ear sensory epithelia (Koehler et al. 2013), optic cup (Eiraku et al. 2011) and adenohypophysis (Suga et al. 2011). In vitro maintenance and specific differentiation of the stem cells have resulted in different neuronal forms. One of the examples is the use of Rho-associated protein kinase (ROCK) inhibitor in serum-free chemically defined medium which greatly improved the isolated survival of HPSCs. The efficiency and speed of neuronal induction have been enhanced in manifold ways by the use of inhibitors of transforming growth factor (TGF) and signalling governed by bone morphogenetic protein (BMP) (Lei and Schaffer 2013). Using precise patterning strategies combined with the above in vitro-modified tools, one can develop CNS and PNS tissues in 3D environment. Having achieved the above feats, the in vivo process of brain development which starts from neurogenesis and is followed by migration, dendrite and axon elongation, myelination, synaptogenesis and apoptosis can be accomplished in 3D in vitro systems.

6.2.4.3 Brain Organoids

Fine-tuned in vitro system and procedures required for in vitro organogenesis forming a functional 3D organ-like tissue have been established. Apart from other human organ development, the development of brain tissues has been successfully achieved by different researchers. No report yet claims the development of whole functional human brain but parts of it have been successfully formed in 3D in vitro culture conditions (refer to Table 6.2). As a general rule, when one starts from the pluripotent stem cells, the lineage specification is required to form a particular organ or tissue. For instance, the neural differentiation takes place in ESCs when these cells are grown in the absence of mesoderm or endoderm inductive signals. Following this, neural progenitors are specified along the dorso-ventral and anterior-posterior axes by RA, Shh and Wnt patterns. It is believed that there are different "organizer" regions which impart signals along the wide areas of the embryonic tissues for pattern formation, creating morphogenetic gradients. The gradient forms the committed regions of cells, giving rise to specific organs during organogenesis.

Table 6.2 Neurodevelopmental/neurodegenerative 3D organoid models

Year	3D organoid models	References
2017	Inner ear organoid	(Koehler et al. 2017)
2016	Alzheimer's disease model	(Choi et al. 2016)
2013	Inner ear sensory epithelium	(Koehler et al. 2013)
2013	Cerebral organoid and microcephaly	(Lancaster et al. 2013)
2011	Self-organised optic cup	(Eiraku et al. 2011)
2011	Functional adenohypophysis	(Suga et al. 2011)
2008	Self-organized polarised cortical tissue	(Eiraku et al. 2008)

The expansion of neuro-epithelium generating radial glial (RG) stem cells marks the start of mammalian brain organogenesis. The RGs proliferate at the apical surface of the ventricular zone to generate intermediate progenitors and neurons. Subsequently, intermediate progenitors populate forming adjacent subventricular zone (SVZ) along with the neuronal migration, making other layers and zones of cortical plate. Corticogenesis, forming elaborate structure of the brain cortex, makes humans different from other mammals (Rakic 1995). In humans, SVZ is split by an inner fibre layer (IFL) into an inner and an outer SVZ. The outer SVZ is markedly formed by intermediate progenitors and a unique set of stem cells called outer radial glia. The IFL and outer SVZ are not present in mice, a difference that emphasizes the development of functional in vitro human brain model (Grinand and Matsuzaki 2012).

For understanding the step-by-step developmental procedure of the human brain and its interaction with the surrounding environment, cerebral organoid model developed by Lancaster et al. (2013) stands a milestone. The cerebral organoid model displays distinct regions of the brain with interdependency among them. Several regions in the developed organoid can be identified by histochemical studies, including the forebrain (FOXG1 and SIX3), hindbrain (KROX20 and ISL1), dorsal cortex (EMX1), hippocampus (NRP2, FZD9 and PROX1), prefrontal cortex (AUTS2), ventral forebrain (NKX2–1) and choroid plexus (TTR). Moreover, the model recapitulates the fundamental organizational events and functional interconnections of in vivo human brain development. Particularly, the cortical organization and functions have been well studied in the said model. Earlier in 2011, Suga's laboratory taking mouse ESCs developed a protocol for Rathke's pouch formation. The 3D aggregate culture system supported the efficient self-formation of 3D adenohypophysis tissue via the formation of pituitary primordium (Rathke's pouch). The developed pituitary gland contained active endocrine cells which secreted adrenocorticotrophic hormone (ACTH) and responded for the positive and negative regulators as in in vivo endocrine homoeostasis. In addition, the researchers could also rescue the hypo-pituitary mouse after transplanting it under the kidney capsule of the mouse. This puts the 3D culture system firm at its applicability for generation of in vitro fully functional organ (Suga et al. 2011).

Studies of neurodegenerative diseases are among the principal forces behind the advancement of in vitro models of the human brain. A successful attempt has been made by Choi and his group to formulate 3D culture system depicting familial Alzheimer's disease (FAD). The neural stem cell-derived in vitro model allowed the expression of FAD mutations in beta-APP and presenilin 1, forming the extracellular depositions of amyloid beta as well as amyloid beta plaques. Notably, the model also exhibited neuronal cells expressing high levels of silver-positive and detergent-resistant aggregates of phosphorylated tau along with filamentous tau in the cell body as well as neurites. In this way, the model successfully recapitulated amyloid beta and tauopathy in a 3D neural cell culture system (Raja et al. 2016).

Generating sensory organs is secondary to the generation of the brain; however, it further proves the worth of 3D culture system for studying neuron and its activity. Various results have been coming forth in relation to the generation of inner ear sensory epithelium, self-organizing optic cup and skin. Eiraku et al. (2011) reported the morphogenesis of self-organizing optic cup in 3D culture. The model not only formed the optic cup but also the stratified neural retina tissue resembling the natural eye. A similar level of success was achieved by Koehler et al. (2013) when they tried to form inner ear sensory epithelium taking pluripotent stem cell. After deciphering the non-neural ectoderm origin of the sensory epithelium of the ear, they recapitulated the step-by-step induction of pluripotent stem cells into the formation of neuro-ectoderm. The ear organoid was found to be mimicking the structural and biochemical aspects of natural vestibular organ.

6.3 Neurodegenerative Disorders

Studying neurodegenerative disorders in vitro has been difficult due to the absence of accurate model which can recapitulate the human brain. The roadblock has been gradually removed after the successful differentiation of iPSCs into brain cells such as astrocytes and dopaminergic neurons. The analysis of NDDs got further elaborated after the development of 3D cell culture system, gene-modifying technologies such as CRISPR-Cas9 and other tools. This section of the chapter deals with the application of iPSCs for ameliorating NDDs.

6.3.1 Origin and Definition

In the second decade of the nineteenth century, Parkinson's disease (PD) was described as a neurological syndrome, while Alzheimer's disease (AD) was reported in the first decade of the twentieth century. Subsequently, such diseases were called neurodegenerative diseases and defined as age-related conditions where brain functions started to decline due to loss of neurons in contrast to static loss of neuron because of toxic or metabolic shock. Various symptoms pertaining to these incurable and devastating diseases were then outlined such as decline in cognitive abilities and/or loss of locomotor functions. Worldwide, the number of NDD patients is increasing due to ageing population posing a challenge before the healthcare system while patients' quality of life is compromised (Jennekens 2014).

Various characteristic conditions are associated with different NDDs, including cognitive deficits in Alzheimer's disease, frontotemporal dementia, vascular dementia, mixed dementia, dementia with Lewy bodies, impaired locomotor activity in amyotrophic lateral sclerosis (ALS), Parkinson's disease (PD), multiple sclerosis (MS) and spinocerebellar ataxia (SA). NDDs are classified based on either primary clinical features or anatomical distribution of neuron loss or principally involved molecular abnormalities. Based on clinical features, it may be identified either as dementia, motor neuron disease or parkinsonism; on anatomical distribution of neurodegeneration basis, either as frontotemporal degeneration, extrapyramidal disorder or spinocerebellar degeneration; and on the involvement of principal molecular abnormalities, it may either be amyloidosis, taupathies, α-synucleinopathies or TDP-43 proteinopathies (Dugger and Dickson 2017).

6.3.2 Blood-Brain Barrier

All the mammals including humans have well-developed blood-brain barrier (BBB) which is created by the endothelial cells from the wall of capillaries, epithelial cells of the choroid plexus and the avascular arachnoid epithelium (Daneman and Prat 2015). Capillaries forming BBB provide huge area for blood-brain exchange which may range from 12 to 18 m^2 for an average adult. Epithelial cells of the choroid plexus forms the second interface facing cerebrospinal fluid (CSF), while the brain extracellular fluid, interstitial fluid (ISF), is secreted across the first interface, i.e. the capillary endothelium of the BBB. The CSF and ISF secretions are controlled by the osmotic and ionic gradient formed by Na^+, K^+-ATPase pump present on the abluminal membrane of the endothelium and the apical membrane of the choroid plexus of the BBB. The third interface, the avascular arachnoid epithelium, underlying the dura, completely encloses the CNS and acts as a seal between extracellular fluid of the CNS and the rest of the body. All three interfaces of the BBB function as barrier systems through a combination of physical barriers in the form of tight junctions, transport barriers in the form of specificity and metabolic barriers in which specific enzymes stop molecules from travelling between the two. The barrier function is dynamic in nature which can be regulated and modulated as per physiological and pathological demands (Wu et al. 2021).

6.3.3 Cells of CNS

The central nervous system comprises three types of cells: neurons, glial cells and vascular cells with pericytes. While neurons sense the environment and communicate with other neurons, glial cells support them to perform their functions and remove metabolic wastes. Vascular cells along with pericytes regulate the interactions between blood vessels and brain parenchyma. Subtypes of glial cells are astrocytes, microglia, oligodendrocytes and ependymal cells.

6.3.3.1 Neural Progenitor Cells and Neuron

Progenitor cells are present in the developing embryos as well as neonatal and adult brain which differentiate into brain cells. Neural progenitor cells (NPCs) give rise to many, if not all, of the glial and neuronal cell types. These cells are resident adult stem cells which can replace the brain cells as and when required. Out of many brain cells, neurons are considered as functional units as they communicate among themselves as well as with other cells such as muscle and gland cells. Neurons are specialized in their structure and physiology to perform a specific task. Structurally, neurons have cell body or soma with cytoplasm containing Nissl substance, many dendrites and a long axon. Nissl granules contain rough endoplasmic reticulum and free ribosomes where protein synthesis occurs. Schwann cells in the peripheral nervous system (PNS) and oligodendrocytes in CNS, present along the length of axon with the myelin sheath, provide insulation which facilitates the rapid transmission of nerve impulse. The axon ends are called axon terminals which form synapses with other cells and are site of interneuron signal transmission.

6.3.3.2 Astrocytes

Star-shaped glial cells located in the brain and spinal cord are referred to as astrocytes. They have a wide range of functions including metabolic support to neurons, extracellular ionic environment regulation, removal of neurotransmitter from the extracellular space, modulation of synaptic transmission and assistance to oligodendrocytes in myelination process. In the CNS, their number ranges from 20% to 40% of the total glial cells (Li et al. 2019).

6.3.3.3 Microglia

Microglial cells have originated from the mesoderm unlike other CNS cells which have originated from the ectoderm of embryonic germ layers. They make 10%–15% of the total glial cells and form resident immune system of the brain. Their phagocytic activity removes foreign material when tissue damage is detected by them. These cells can also act as antigen presenting cell (Hickman et al. 2018).

6.3.3.4 Oligodendrocytes

The myelin sheath produced by the oligodendrocytes enwraps the axons of neurons. There may be myelinated and unmyelinated axons, and the way of signal transmission is different in both types of axons. In myelinated axons, gaps are present between the sheaths, and these are known as nodes of Ranvier. Chemically, the myelin is a lipid-rich substance, and it works as an insulator. The oligodendrocytes in the CNS are the representatives of Schwann cells in the PNS. A single oligodendrocyte can myelinate up to 50 axonal segments. The process of myelination starts in utero—early in the third trimester—increases rapidly during infancy in line with the development of cognitive and motor skills and continues throughout the adolescence and early adulthood (Kuhn et al. 2019).

6.3.3.5 Pericytes and Vascular Cells

Pericytes are vascular lining cells implanted in the basement membrane of fine blood vessels present throughout the body including the brain. Their processes are extended around the capillaries including pre-capillary arterioles and post-capillary venules. They are strategically placed between astrocytes, endothelial cells and neurons in the neurovascular unit of the CNS. Pericytes perform various vital functions by integrating, coordinating and processing the signals received from the neighbouring cells. They regulate BBB permeability, angiogenesis, capillary haemodynamic (blood flow) responses, removal of toxic metabolites, stem cell activities and neuroinflammation. The structure, function and expression profile of pericytes are location specific in the brain along the vascular bed. Dysfunctional pericytes have been associated with the development of vascular diseases like stroke and neurodegenerative diseases like AD. Recently, pericytes have been extensively studied for their association with various neural disorders and as a prospective therapeutic target (Brown et al. 2019).

6.3.4 AD

Alzheimer's disease is a severe and progressive neurodegenerative disease elucidated by ample loss of cognitive and behavioural abilities, leading to the decline in patient's quality of life. Its pathogenesis is not yet completely understood. However, AD's two main forms have been described: early-onset familial AD (EOFAD) and late-onset AD (LOAD). The first one is less common, accounting for the 5% total cases, and is generally diagnosed before 65 years of age. Its inheritance shows autosomal dominant pattern with 1 of 200 mutations till date reported in the three key genes, viz. presenilin 1 (PSEN1), presenilin 2 (PSEN2) and amyloid precursor protein (APP). It occurs after the age of 65 without any history of dementia in a family but after 35 with a family history of AD. Sporadic AD, which is also known as late-onset AD, is the most common form of dementia. After 1992, AD's pathogenesis and aetiology have been described by the amyloid cascade hypothesis. According to the hypothesis, APP-derived pathogenic amyloid β (Aβ) gets accumulated in the neurons inducing a vicious cycle that further triggers build-up of neurofibrillary tangles (NFTs), leading to the death of neurons and eventually dementia (Weller and Budson 2018).

6.3.5 PD

Parkinson's disease is basically a movement disorder having a variable aetiology. In PD, neurons of the substantia nigra have been observed getting destroyed, causing loss of deep grey matter, leading to deficiency of dopamine in the basal ganglia. PD is the second most common neurodegenerative disease, which is age related after AD, and affects 1–2% of persons aged above 65 years. This disease is also progressive in terms of loss of nigrostriatal dopaminergic neurons, which in normal circumstances present unmyelinated axons forming several synapses. This is indicated by disrupted motor functions such as bradykinesias, resting tremors, rigidity and postural instability in the affected individuals. Some non-motor symptoms like sleep disturbances, olfactory deficit, constipation, cognitive decline and dementia have also been noticed in the PD patients which reduce the quality of life of patients severely. Neuropathologically, protein inclusions are deposited in the neuronal soma and neuronal processes which are known as Lewy bodies and Lewy neurites, respectively. The Lewy bodies are mostly composed of insoluble and misfolded aggregates of α-synuclein protein at the presynaptic neurons. Genetic form of the disease has provided the information on neuropathological mechanisms, although it accounts for only 5–10% of the total PD cases (Vázquez-Vélez and Zoghbi 2021).

6.3.6 HD

Huntington's disease is the most common inherited neurodegenerative disorder, which is genotypically autosomal dominant and phenotypically identified by a triad of cognitive, motor and psychiatric descriptions. HD's onset typically takes place in the mid-life of the individual which progresses irreversibly over the next 10–15 years. All reported cases of HD have been found to be caused by the abnormal expansion of CAG repeat near the N terminal of the huntingtin gene (HTT) which when translated produces mutant huntingtin protein (mHTT). Since 1993, when the mutation was first reported, the exhaustive research has described various neuropathological mechanisms, leading to the development of the disease. Almost all these mechanisms are driven by mHTT protein's ubiquitous presence where predominantly toxic gain of function occurs. The abnormally long polyglutamine (polyQ) expansion in mHTT protein confers more taxic functions, leading to neurodegeneration (Barnat et al. 2020).

6.3.7 ALS

A motor neurodegenerative disease called amyotrophic lateral sclerosis results in a gradual loss of both upper and lower motor neurons in the brain and spinal cord, which leads to feebleness of the muscles, fasciculation, muscle atrophy, spasticity and eventually paralysis. ALS has been subtyped as "progressive muscle atrophy" which involves primarily lower motor neurons and "primary lateral sclerosis" where relatively upper motor neurons are involved. Its onset is generally focal but subsequently it can spread to various body regions; specifically the failure of respiratory muscles limits the survival of the individual to 2–5 years after the onset (Masrori and Van Damme 2020).

In about 50% of the reported cases, extra-motor indications like change in behaviour, language difficulty and executive dysfunction have been observed. In other 10% of the patients, these issues may become severe enough to be clinically considered as frontotemporal dementia (FTD). 10% of ALS patients have familial history suggesting pattern of autosomal dominant inheritance, while the other 90% of the ALS cases have no such history and are categorized as sporadic ALS. The causes of ALS are only partially described as they are heterogenous. More than 20 genes have been linked with the onset of the disease. Out of these, C9orf72 gene's hexanucleotide repeat expansion has been responsible for 30%–50% of familial and 7% of sporadic ALS. Initially in 1869, the disease was originally defined by Jean-Martin Charcot as a pure motor disease but subsequently as a multi-system neurodegenerative disorder having genetic, neuropathological and clinical heterogeneity. Age, site and rate of progression of ALS are highly variable (Brotman et al. 2021).

6.3.8 MS

Multiple sclerosis is a neurodegenerative, demyelinating and chronic inflammatory disease which affects young adults between the ages of 20 and 40 years. Women are more likely to develop multiple sclerosis than men. Neuropathology of MS includes proliferating astrocytes, activated microglia, infiltrating macrophages and lymphocytes and appearance of focal plaques which contain demyelinated axons. Reduced population of oligodendrocytes located in the surrounding area of post-capillary venules labelled by the BBB breakdown has also been noticed. Dysfunctional motor abilities, disturbed speech and vision, loss of balance or coordination, acute paralysis, numbness, fatigue, nystagmus (uncontrolled eye movement), tremor and diminishing cognitive abilities are the major clinical symptoms of MS. The disease generally begins with a relapsing-remitting phase (RRMS) that is intermittent, which progresses into a progressive phase at about 15 years after the initial onset. Out of cases yet reported, about 15% are categorized as primary progressive phase (PPMS) where the disease progresses relentlessly right from the onset (Dobson and Giovannoni 2019).

MS is a multifactorial, heterogenous and immune-mediated disease which is influenced by both environmental and genetic factors. The pathogenesis has been described by two main hypotheses: "outside-in" and "inside-out". The first one is mainly based on the findings of experiments done on animal models of MS which postulates that the activated auto-reactive T cells invade the CNS, thereby inciting neuroinflammation, tissue damage and BBB leakage. The second one hypothesizes that MS is primarily a degenerative disease whose severity increases on amplification of immune responses. In this case, the trigger is inside the CNS in the form of defective oligodendrocytes resulting from mutation that leads to their death and consequently activates microglial cells. Later, demyelination and damaged axons contribute to neurodegenerative and neuropathological symptoms (McGinley et al. 2021).

6.4 Application of iPSC Technique

It is evident that neurodegenerative diseases are progressive in nature and culminate into the loss of neurons. Most of them are poorly understood, incurable and devastatingly affecting the lives of millions of individuals globally. Therefore, they pose major health challenge with stern propositions on the wellbeing of individual as well as society. Various in vivo and in vitro disease models have been developed to study the pathophysiological mechanisms involved in the progression of these diseases. A gap of genetic and/or physiological constituents between human and the disease models have resulted into the failure of therapeutic interventions. To bridge this gap, 3D cell culture systems have been developed where iPSCs are being utilized to form biomimetic organoids. This advanced model system replicates aspects of brain development and brain physiology, mimics neuronal and glial cell interactions and incorporates the effects of blood flow in the brain enabling more accurate findings. Moreover, the iPSCs are easily accessible, originated from human and are expandable, capable of differentiating into any cells type, free from ethical constraints and useful in making personalized medicines utilizing patient-specific cells. Gene editing techniques, including CRISPR-Cas9, have made it possible for genetically defined human iPSCs to be used as disease models.

6.4.1 The iPSC Technique

Integrating viral vectors like lentiviral or retroviral vectors were initially used for cell reprogramming to generate iPSCs. Then, there was a concern about the iPSC's clinical applications owing to the potential of integration of transgenes into the host cell genome, leading to insertional mutagenesis. To get rid of this mutagenesis and these genetic modifications correlated with the lentiviral and retroviral transduction-mediated insertion of reprogramming factors, various non-integrating methods have been developed. Synthetic mRNA, episomal DNA and Sendai virus are among the various other approaches being applied to develop integration-free iPSCs which are simple and have high efficiency. The progress in the iPSC technology, along with improvement of gene editing tool such as CRISPR-Cas9, has resulted in standard clinical applications such as regenerative medicine and drug development (Yamanaka 2020).

6.4.2 Stem Cell Therapy

In 1950s, bone marrow transplantation was performed for the first time which demonstrated that cell therapy can be useful in replenishing the damaged tissue or cell. Then, one of the identical twins was suffering from leukaemia whose blood-forming cells were replaced with the healthy twin. In this case, genetic matching was not an issue, but that could only be realized after almost 10 years when bone marrow transplants were carried out between non-twin-related and unrelated donors. Today, the bone marrow transplants are commonly performed for the patients whose blood-forming stem cells have been damaged due to either some disease or high radiation or anti-cancer drug treatment. However, an obstacle of host-versus-graft disease stands before the clinicians to tackle with. This challenge can be conquered with the help of iPSC technology and latest gene editing procedures (Sharma et al. 2019).

iPSC technology has been flourishing rapidly since its establishment and has ushered in a promising new era in the fields of stem cell biology, regenerative medicine, disease modelling and drug discovery. The possibility of replacing damaged tissue/cell by iPSC-derived differentiated cells/tissues/organs has been exhibited in several trials. Some of them have been successfully advanced towards clinical trials. The first trial was aimed at treatment of macular degeneration utilizing human iPSC-derived retinal pigment epithelium (RPE) (Li et al. 2016).

6.4.3 CRISPR-Cas9

Clustered regularly interspaced short palindromic repeats (CRISPR)-CRISPR-associated (Cas) system contains single-guide RNA (sgRNA) and RNA-guided endonuclease, Cas9. Target-specific sgRNA, composed of a CRISPR RNA (crRNA) and a transactivating CRISPR RNA (tracrRNA), guides the Cas9 protein for double-stranded break (DSB) at a target-specific site. Many bacteria employ this mechanism as a defence system against phages and conjugative plasmids by exhibiting site-specific DSB in the target DNA. To recognize the target site, there must be short protospacer adjacent motif (PAM) flanking the target site and formation of R-loop accompanied by the strand scission which is driven by the complementary base pairing between the sgRNA and target DNA resulting in Cas9-DNA interactions and related conformational changes. The sgRNA is the synthetic analogue of natural tracrRNA-crRNA structure found in the bacteria. Synthetic sgRNA mimics the function of natural tracrRNA-crRNA and simplifies the gene editing procedure in vitro, ex vivo and in vivo delivery systems (Yip 2020). Other gene editing methods have also been developed which are summarized in Table 6.3.

6.4.4 Application in Neurodegenerative Disorders

Human iPSCs are specifically utilized in developing disease models whose genetic causes are well defined. Somatic cells containing disease-causing mutation are acquired to derive iPSCs and then differentiated into disease-relevant cells. Various studies relating to pathophysiological mechanisms, therapeutic strategies and drug development for the selected diseases are then carried out. The studies can be performed on either conventional two-dimensional (2D) culture system or 3D platform like organoids.

There are several reports of development of disease models using iPSCs (Fig. 6.2). Prominently, AD and PD models have been devised using neurons derived from iPSCs. The pathology of ALS was modelled utilizing a co-culture of neurons with astrocyte derived from the iPSCs. NPCs have the capability to differentiate into neural subtypes like neurons and astrocytes. These cells are patterned towards defined regional identity producing neurons of specific brain region. iPSC differentiation into NPCs is initiated by plating them on a monolayer followed by neuralization via dual SMAD inhibition. The culture is maintained for 11 days and then passaged and re-cultured for 7 days for acquisition of neuronal characters. Further, these NPCs can de differentiated into desired subtypes such as astrocytes and neurons. Similar protocols are available to model different neurodevelopmental diseases summarized in Table 6.4.

Table 6.3 Other gene editing technologies

System	Enzyme	Mode of action	Reference
ZFN	Zinc-finger nucleases	DSBs are induced by zinc-finger proteins fused to FokI; DNA repair occurs by non-homologous end joining (NHEJ) to construct small indels, or homology directed repair (HDR) to introduce precise nucleotide modifications	(Paschon et al. 2019)
TALEN	Transcription activator-like effector nucleases	DNA-binding transcription activator-like effector (TALE) protein modules fused to FokI induce site-specific DSBs; DNA repair proceeds through NHEJ or HDR to introduce indels or specific mutations	(Nakano et al. 2019)
CRISPR-Cas9	Wild-type Cas9, Cas9 nickase	The NHEJ or HDR process is triggered by RNA-guided site-specific DNA cleavage that creates indels or introduces precise modifications to DNA	(Leal and Alméciga-Díaz 2022)
	Cas9 nickase	It generates targeted DSBs by combining a Cas9 nickase with paired sgRNA. The paired nicking significantly reduces off-target activity (by about 50-fold)	(Li and Margolis 2021)
	eSpCas9	A structure-guided protein engineering approach produced SpCas9 variants with reduced off-target effects, but robust activity on target	(Wang et al. 2021)
	Cas9-VRER variant	This platform allows DNA modification to be introduced precisely in monoallelic or biallelic forms, called "consecutive re-guide or re-Cas steps to erase CRISPR-Cas9-blocked targets" (CORRECT)	(Cai et al. 2022)
CRISPR-Cas9-cytidine deaminase	Fusions of CRISPR-Cas9 and a cytidine deaminase	DSBs or donor DNA templates are not required with this base editing approach	(Huang et al. 2019)

6.5 Prospects and Challenges

Drug candidates seemingly successful in pre-clinical findings have failed in phase II and III clinical trials owing to our partial knowledge of pathophysiology, aetiology and genetic and epigenetic factors involved in the onset of NDDs. Appropriate investigational tools such as 3D in vitro culture models are therefore required for selection of promising candidate compounds to get rid of shortcomings at elementary stages. This 3D model system must recapitulate the in vivo milieu to eliminate

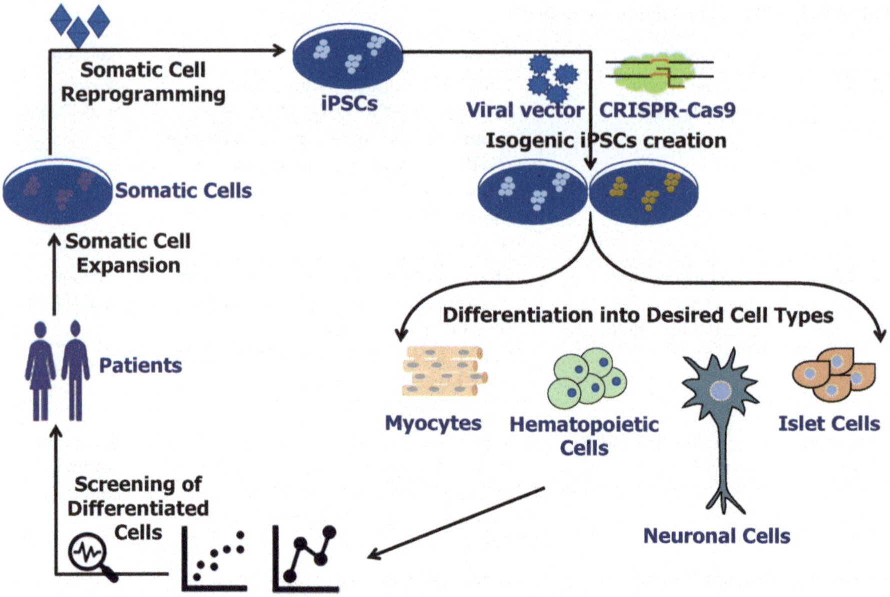

Fig. 6.2 Somatic cells isolated from patients are cultured in appropriate conditions. These cells can be reprogrammed using Yamanaka factors (Klf4, Sox2, c-Myc and Oct4) into induced pluripotent stem cells (iPSCs). With the help of viral vectors or gene editing methods such as CRISPR-Cas9, disease-specific mutation in patient-derived iPSCs can be corrected or therapeutic genes can be inserted. The modified iPSCs then can be differentiated into either neuronal lineage cells for therapeutic use in neurodegenerative diseases or myocytes or haematopoietic cells or islet cells. Before administering into the subject, the differentiated cells are screened for its purity, presence of any residual factors, expression level of therapeutic gene, etc

Table 6.4 Neurodevelopmental disease modelling using iPSCs

Disease	Molecular defect of donor cell	Cell type differentiated from iPSCs	Drug or functional tests	Reference
ALS	Heterozygous Leu144Phe mutation in SOD1	Motor neurons and glial cells	No	(Myszczynska and Ferraiuolo 2016)
Spinal muscular atrophy (SMA)	Mutations in SMN1	Neurons and astrocytes and mature motor neurons	Yes	(Fuller et al. 2016)
PD	Multifactorial, mutations in LRRK2 and/or SNCA	Dopaminergic neurons	Yes	(Beevers et al. 2013)
HD	72 CAG repeats in the huntingtin gene	None	No	(Jeon et al. 2014)
AD	A246E mutation in PSEN1	Neurons	No	(Hernández-Sapiéns et al. 2020)

any chances of deviation from in vivo drug interactions. iPSCs utilized in the 3D culture system have shown the pluripotency but they are not the same as ESCs. The expression levels of key genes have been found to be lower when compared with the pluripotency genes of ESCs. Initially, iPSCs demonstrated incomplete promoter demethylation for genes like Oct4 and could neither generate postnatal chimera nor form germline. Such observations must be noted for rectification of issues that may arise. Similarly, the case of X chromosome inactivation and reactivation of human female iPSCs during the reprogramming process must be considered to remove any abnormality that may lower the authenticity of pluripotency and subsequent clinical success.

Clinical experiences have underlined several issues in the field of regenerative medicine. Improvement in the survival and viability of transplanted cells is immediately required. We also need to find ways for facilitating cell navigation to the target location, i.e. homing, engraftment and retention of stem cells at the desired site to fully realize the potential of stem cells. Pluripotent stem cells including iPSCs have unlimited proliferation potential and innate capability of forming tumours, and this poses a risk of malignancy. Several stem cell therapies involve the usage of differentiated cells derived from iPSCs; here the differentiated population of transplanted cells must be free from any undifferentiated iPSCs. Additionally, there are few reports of stem cell line obtaining genetic alterations associated with the human cancer. In the provided scenario, an exhaustive guideline is required to monitor and guide the clinical applications of stem cells (Fig. 6.3).

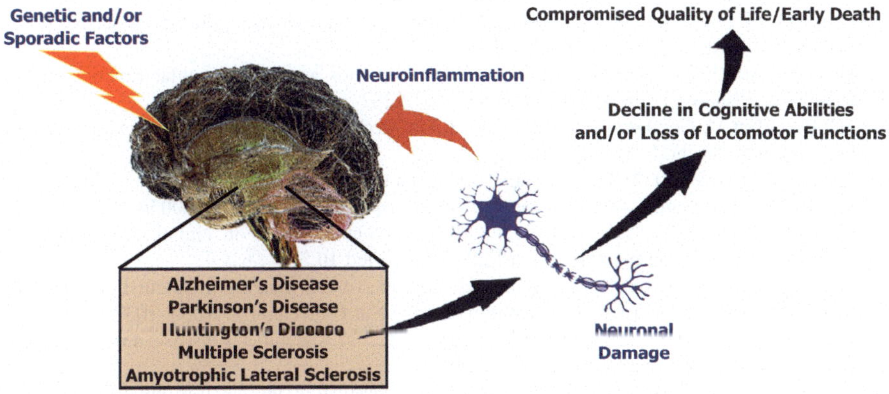

Fig. 6.3 All neurodegenerative diseases whether caused by genetic factor or occur sporadically lead to neuronal damage that can trigger neuroinflammatory reaction in the central nervous system. The damaged neurons are unable to maintain the structural and functional integrity of the brain causing decline in cognitive abilities and/or loss of locomotor functions that ultimately compromise the patient's life quality and may sometimes lead to early death

6.6 Summary

Underlined features of iPSCs such as unlimited proliferation capability, desired differentiation potential, genetic similarity with the subject and ethical clearance powered by the tools like 3D culture and CRISPR-Cas9 put us in a position where we can expect to treat diseases like neurodegenerative diseases. The powerful combination of iPSCs, 3D cultures and CRISPR-Cas9 has enabled several clinical studies and trials being carried at various phases. These trials are running at various stages. Clinical experiences from these trials have pointed out various limitations of regenerative medicine.

References

Ahmed M (2016) Extracellular matrix regulation of stem cell behavior. Curr Stem Cell Rep 2:197–206

Bacigalupo A (2004) Mesenchymal stem cells and haematopoietic stem cell transplantation. Best Pract Res Clin Haematol 17:387–399

Barker RA, Carpenter MK, Forbes S et al (2018) The challenges of first-in-human stem cell clinical trials: what does this mean for ethics and institutional review boards? Stem Cell Rep 10:1429–1431

Barnat M, Capizzi M, Aparicio E et al (2020) Huntington's disease alters human neurodevelopment. Science 369:787–793

Bar-Nur O, Brumbaugh J, Verheul C et al (2014) Small molecules facilitate rapid and synchronous iPSC generation. Nat Methods 11:1170–1176

Beevers JE, Caffrey TM, Wade-Martins R (2013) Induced pluripotent stem cell (iPSC)-derived dopaminergic models of Parkinson's disease. Biochem Soc Trans 41:1503–1508

Brotman RG, Moreno-Escobar MC, Joseph J, Pawar G (2021) Amyotrophic lateral sclerosis. In: StatPearls [Internet]. StatPearls Publishing

Brown LS, Foster CG, Courtney J-M et al (2019) Pericytes and neurovascular function in the healthy and diseased brain. Front Cell Neurosci 13:282

Bunnell BA, Flaat M, Gagliardi C et al (2008) Adipose-derived stem cells: isolation, expansion and differentiation. Methods 45:115–120

Cai G, Lin Z, Shi S (2022) Development and expansion of the CRISPR/Cas9 toolboxes for powerful genome engineering in yeast. Enzym Microb Technol 110056:110056

Chambers SE, Pathak V, Pedrini E et al (2021) Current concepts on endothelial stem cells definition, location, and markers. Stem Cells Transl Med 10:S54–S61

Chan EM, Ratanasirintrawoot S, Park I-H et al (2009) Live cell imaging distinguishes bona fide human iPS cells from partially reprogrammed cells. Nat Biotechnol 27:1033–1037

Choi SH, Kim YH, Quinti L et al (2016) 3D culture models of Alzheimer's disease: a road map to a "cure-in-a-dish.". Mol Neurodegener 11:1–11

Conley BJ, Young JC, Trounson AO, Mollard R (2004) Derivation, propagation and differentiation of human embryonic stem cells. Int J Biochem Cell Biol 36:555–567

Daneman R, Prat A (2015) The blood–brain barrier. Cold Spring Harb Perspect Biol 7:a020412

Dawson E, Mapili G, Erickson K et al (2008) Biomaterials for stem cell differentiation. Adv Drug Deliv Rev 60:215–228

Denham M, Dottori M (2011) Neural differentiation of induced pluripotent stem cells. In: Neurodegeneration. Springer, pp 99–110

Dobson R, Giovannoni G (2019) Multiple sclerosis–a review. Eur J Neurol 26:27–40

Doss MX, Sachinidis A (2019) Current challenges of iPSC-based disease modeling and therapeutic implications. Cell 8:403

Dugger BN, Dickson DW (2017) Pathology of neurodegenerative diseases. Cold Spring Harb Perspect Biol 9:a028035
Eiraku M, Takata N, Ishibashi H et al (2011) Self-organizing optic-cup morphogenesis in three-dimensional culture. Nature 472:51–56
Eiraku M, Watanabe K, Matsuo-Takasaki M et al (2008) Self-organized formation of polarized cortical tissues from ESCs and its active manipulation by extrinsic signals. Cell Stem Cell 3: 519–532
Fang R, Liu K, Zhao Y et al (2014) Generation of naive induced pluripotent stem cells from rhesus monkey fibroblasts. Cell Stem Cell 15:488–497
Fuller HR, Mandefro B, Shirran SL et al (2016) Spinal muscular atrophy patient iPSC-derived motor neurons have reduced expression of proteins important in neuronal development. Front Cell Neurosci 9:506
Gage FH (2000) Mammalian neural stem cells. Science 287:1433–1438
Garreta E, Sanchez S, Lajara J et al (2018) Roadblocks in the path of iPSC to the clinic. Curr Transplant Rep 5:14–18
Grinand L, Matsuzaki F (2012) A focus on mammalian OSVZ progenitors
Haridhasapavalan KK, Raina K, Dey C et al (2020) An insight into reprogramming barriers to iPSC generation. Stem Cell Rev Rep 16:56–81
Hernández-Sapiéns MA, Reza-Zaldívar EE, Cevallos RR et al (2020) A three-dimensional Alzheimer's disease cell culture model using iPSC-derived neurons carrying A246E mutation in PSEN1. Front Cell Neurosci 14:151
Hickman S, Izzy S, Sen P et al (2018) Microglia in neurodegeneration. Nat Neurosci 21:1359–1369
Huang L, Dong H, Zheng J et al (2019) Highly efficient single base editing in aspergillus Niger with CRISPR/Cas9 cytidine deaminase fusion. Microbiol Res 223:44–50
Iwatani S, Yoshida M, Yamana K et al (2019) Isolation and characterization of human umbilical cord-derived mesenchymal stem cells from preterm and term infants. JoVE (J Visual Exp) e58806
Jennekens FG (2014) A short history of the notion of neurodegenerative disease. J Hist Neurosci 23: 85–94
Jeon I, Choi C, Lee N et al (2014) In vivo roles of a patient-derived induced pluripotent stem cell line (HD72-iPSC) in the YAC128 model of Huntington's disease. Int J Stem Cells 7:43–47
Koehler KR, Mikosz AM, Molosh AI et al (2013) Generation of inner ear sensory epithelia from pluripotent stem cells in 3D culture. Nature 500:217–221
Koehler KR, Nie J, Longworth-Mills E et al (2017) Generation of inner ear organoids containing functional hair cells from human pluripotent stem cells. Nat Biotechnol 35:583–589
Krause MN, Sancho-Martinez I, Belmonte JCI (2016) Understanding the molecular mechanisms of reprogramming. Biochem Biophys Res Commun 473:693–697
Kuhn S, Gritti L, Crooks D, Dombrowski Y (2019) Oligodendrocytes in development, myelin generation and beyond. Cell 8:1424
Kuijlaars J, Oyelami T, Diels A et al (2016) Sustained synchronized neuronal network activity in a human astrocyte co-culture system. Sci Rep 6:1–14
Lancaster MA, Renner M, Martin C-A et al (2013) Cerebral organoids model human brain development and microcephaly. Nature 501:373–379
Laperle AH, Sances S, Yucer N et al (2020) iPSC modeling of young-onset Parkinson's disease reveals a molecular signature of disease and novel therapeutic candidates. Nat Med 26:289–299
Leal AF, Alméciga-Díaz CJ (2022) Efficient CRISPR/Cas9 nickase-mediated genome editing in an in vitro model of mucopolysaccharidosis IVA. Gene Therapy:1–8
Lei Y, Schaffer DV (2013) A fully defined and scalable 3D culture system for human pluripotent stem cell expansion and differentiation. Proc Natl Acad Sci 110:E5039–E5048
Lepousez G, Nissant A, Lledo P-M (2015) Adult neurogenesis and the future of the rejuvenating brain circuits. Neuron 86:387–401
Li K, Li J, Zheng J, Qin S (2019) Reactive astrocytes in neurodegenerative diseases. Aging Dis 10: 664

Li PP, Margolis RL (2021) Use of single guided Cas9 nickase to facilitate precise and efficient genome editing in human iPSCs. Sci Rep 11:1–10

Li Y, Chan L, Nguyen HV, Tsang SH (2016) Personalized medicine: cell and gene therapy based on patient-specific iPSC-derived retinal pigment epithelium cells. Retinal Degenerative Diseases:549–555

Liang G, Zhang Y (2013) Embryonic stem cell and induced pluripotent stem cell: an epigenetic perspective. Cell Res 23:49–69

Liao J, Cui C, Chen S et al (2009) Generation of induced pluripotent stem cell lines from adult rat cells. Cell Stem Cell 4:11–15

Loh Y-H, Agarwal S, Park I-H et al (2009) Generation of induced pluripotent stem cells from human blood. Blood 113:5476–5479

Maherali N, Sridharan R, Xie W et al (2007) Directly reprogrammed fibroblasts show global epigenetic remodeling and widespread tissue contribution. Cell Stem Cell 1:55–70

Malik N, Rao MS (2013) A review of the methods for human iPSC derivation. Pluripotent Stem Cells:23–33

Masrori P, Van Damme P (2020) Amyotrophic lateral sclerosis: a clinical review. Eur J Neurol 27: 1918–1929

McGinley MP, Goldschmidt CH, Rae-Grant AD (2021) Diagnosis and treatment of multiple sclerosis: a review. JAMA 325:765–779

Mebarki M, Abadie C, Larghero J, Cras A (2021) Human umbilical cord-derived mesenchymal stem/stromal cells: a promising candidate for the development of advanced therapy medicinal products. Stem Cell Res Ther 12:1–10

Moore KA, Lemischka IR (2006) Stem cells and their niches. Science 311:1880–1885

Mossahebi-Mohammadi M, Quan M, Zhang J-S, Li X (2020) FGF signaling pathway: a key regulator of stem cell pluripotency. Front Cell Develop Biol 8:79

Myszczynska M, Ferraiuolo L (2016) New in vitro models to study amyotrophic lateral sclerosis. Brain Pathol 26:258–265

Nakano C, Kitabatake Y, Takeyari S et al (2019) Genetic correction of induced pluripotent stem cells mediated by transcription activator-like effector nucleases targeting ALPL recovers enzyme activity and calcification in vitro. Mol Genet Metab 127:158–165

Ohnuki M, Takahashi K, Yamanaka S (2009) Generation and characterization of human induced pluripotent stem cells. Curr Protoc Stem Cell Biol 9:4A

Omole AE, Fakoya AOJ (2018) Ten years of progress and promise of induced pluripotent stem cells: historical origins, characteristics, mechanisms, limitations, and potential applications. PeerJ 6:e4370

Paschon DE, Lussier S, Wangzor T et al (2019) Diversifying the structure of zinc finger nucleases for high-precision genome editing. Nat Commun 10:1–12

Payer B, Lee JT, Namekawa SH (2011) X-inactivation and X-reactivation: epigenetic hallmarks of mammalian reproduction and pluripotent stem cells. Hum Genet 130:265–280

Puri MC, Nagy A (2012) Concise review: embryonic stem cells versus induced pluripotent stem cells: the game is on. Stem Cells 30:10–14

Raja WK, Mungenast AE, Lin Y-T et al (2016) Self-organizing 3D human neural tissue derived from induced pluripotent stem cells recapitulate Alzheimer's disease phenotypes. PLoS One 11: e0161969

Rakic P (1995) Corticogenesis in human and nonhuman primates

Robinton DA, Daley GQ (2012) The promise of induced pluripotent stem cells in research and therapy. Nature 481:295–305

Rowe RG, Daley GQ (2019) Induced pluripotent stem cells in disease modelling and drug discovery. Nat Rev Genet 20:377–388

Santos AJ, Lo Y-H, Mah AT, Kuo CJ (2018) The intestinal stem cell niche: homeostasis and adaptations. Trends Cell Biol 28:1062–1078

Sasai Y (2013) Next-generation regenerative medicine: organogenesis from stem cells in 3D culture. Cell Stem Cell 12:520–530

Schaap-Oziemlak AM, Kühn PT, van Kooten TG, van Rijn P (2014) Biomaterial–stem cell interactions and their impact on stem cell response. RSC Adv 4:53307–53320
Sharma R, Khristov V, Rising A et al (2019) Clinical-grade stem cell–derived retinal pigment epithelium patch rescues retinal degeneration in rodents and pigs. Sci Transl Med 11:eaat5580
Sim X, Cardenas-Diaz FL, French DL, Gadue P (2014) A doxycycline-inducible system for genetic correction of iPSC disease models. In: Patient-specific induced pluripotent stem cell models. Springer, pp 13–23
Spyrou J, Gardner DK, Harvey AJ (2019) Metabolism is a key regulator of induced pluripotent stem cell reprogramming. Stem Cells Int 2019:1
Stadtfeld M, Maherali N, Breault DT, Hochedlinger K (2008) Defining molecular cornerstones during fibroblast to iPS cell reprogramming in mouse. Cell Stem Cell 2:230–240
Suga H, Kadoshima T, Minaguchi M et al (2011) Self-formation of functional adenohypophysis in three-dimensional culture. Nature 480:57–62
Takahashi K, Tanabe K, Ohnuki M et al (2007) Induction of pluripotent stem cells from adult human fibroblasts by defined factors. Cell 131:861–872
Tharmapalan P, Mahendralingam M, Berman HK, Khokha R (2019) Mammary stem cells and progenitors: targeting the roots of breast cancer for prevention. EMBO J 38:e100852
Vázquez-Vélez GE, Zoghbi HY (2021) Parkinson's disease genetics and pathophysiology. Annu Rev Neurosci 44:87–108
Volarevic V, Markovic BS, Gazdic M et al (2018) Ethical and safety issues of stem cell-based therapy. Int J Med Sci 15:36
Wang Q, Liu J, Janssen JM et al (2021) Precise and broad scope genome editing based on high-specificity Cas9 nickases. Nucleic Acids Res 49:1173–1198
Weller J, Budson A (2018) Current understanding of Alzheimer's disease diagnosis and treatment. F1000Research 7
Wu SM, Hochedlinger K (2011) Harnessing the potential of induced pluripotent stem cells for regenerative medicine. Nat Cell Biol 13:497–505
Wu Y-C, Sonninen T-M, Peltonen S et al (2021) Blood–Brain Barrier and Neurodegenerative Diseases—Modeling with iPSC-Derived Brain Cells. Int J Mol Sci 22:7710
Yamanaka S (2012) Induced pluripotent stem cells: past, present, and future. Cell Stem Cell 10: 678–684
Yamanaka S (2020) Pluripotent stem cell-based cell therapy—promise and challenges. Cell Stem Cell 27:523–531
Yamashiro C, Sasaki K, Yabuta Y et al (2018) Generation of human oogonia from induced pluripotent stem cells in vitro. Science 362:356–360
Yip BH (2020) Recent advances in CRISPR/Cas9 delivery strategies. Biomol Ther 10:839
Yu J, Vodyanik MA, Smuga-Otto K et al (2007) Induced pluripotent stem cell lines derived from human somatic cells. Science 318:1917–1920
Zakrzewski W, Dobrzyński M, Szymonowicz M, Rybak Z (2019) Stem cells: past, present, and future. Stem Cell Res Ther 10:1–22
Zhao N, Sheng M, Wang X et al (2021) Differentiation of human induced pluripotent stem cells into male germ cells. Curr Stem Cell Res Ther 16:622–629

Isolation, Characterization, and Detailed History of Exosomes Derived from Stem Cells and their Epigenetic Biology

7

Sakshi Chaudhary, Ayush Madan, Bibhas Bhunia, and Deepika Arora

Abstract

Exosomes, the nanosized (40–160 nanometers) extracellular bio-vesicles, are gaining attention worldwide owing to their regulatory role in several pathogenesis of various diseases as well as cell-to-cell communication. Mechanistically, it has now been recognized that these exosomes that were considered earlier as cellular trash bags actually help in transmitting some important biomolecules (e.g., DNA, RNA, miRNA, proteins, metabolites, enzymes, lipids, etc.) among cells. Notably, their biocompatible, immunomodulation capability and less toxic susceptibility towards cells are their imperative assets; thus, their potential in clinical applications (diagnosis as well as therapeutic carriers) has been endorsed. However, the precise information about their characterization, labeling, and biological role in exchange of genetic materials and alteration of cellular functions is still under scrutiny and remains to be established. The future prospects of exosomes as innovative therapeutic approach agents in the treatment of diseases are also explored.

Keywords

Extracellular vesicles · Exosomes · Stem cells · Epigenetics · Therapeutics

S. Chaudhary · A. Madan · B. Bhunia
Department of Biotechnology, School of Engineering and Technology, Shobhit Institute of Engineering & Technology (Deemed-to-be-University), Meerut, India

D. Arora (✉)
Department of Biotechnology, School of Engineering and Technology, Shobhit Institute of Engineering & Technology (Deemed-to-be-University), Meerut, India

Research & Development, Regrow Biosciences Private Limited., Lonavala, Maval, Pune, India

S. Jahan, A. J. Siddiqui (eds.), *Applications of Stem Cells and Derived Exosomes in Neurodegenerative Disorders*, https://doi.org/10.1007/978-981-99-3848-3_7

7.1 Introduction

Extracellular vesicles (EVs) refer to be the lipid bilayer membrane-bound heterogeneous vesicles actively released by many different types of living cells into the extracellular space (Doyle and Wang 2019). EVs have the potential to spot unknown cellular and molecular mechanisms in intercellular communication and also in organ physiological state and sickness. All types of cells including prokaryotes and eukaryotes normally secrete EVs as the natural element of their physiology through nonheritable anomalies. EVs are generally categorized into three major subtypes including microvesicles (MVs), apoptotic bodies, and exosomes based upon their morphology, size, content, release pathways, surface marker, and activities. Moreover, EVs are often divided into two classes: ectosomes and exosomes. Ectosomes are the vesicles that pinch off the cytoplasm membrane via outward budding consisting of microvesicles and microparticles with size ranging from ~50 nm to 1000 nm in diameter, while exosomes are comparatively smaller in size ranging from 40 nm to 160 nm with an average size of approximately 100 nm in diameter (Kalluri and LeBleu 2020). The presence of exosomes in extracellular vesicles was known as early as the late 1980s. The term "exosomes" ought to be outlining the secreted membrane vesicles that originated within the intracellular multivesicular compartments discharging upon union of those sections with the cytoplasmic membrane. Because of this intracellular basis, exosomes are internal vesicles with a diameter of around 100 nm, similar to those of multivesicular compartments (Théry 2011). Depending on cell of origin, exosome can contain various types of cellular constituents including genetic materials (like DNA and RNA), proteins (e.g., soluble proteins, peripheral membrane proteins, transmembrane proteins, and lipid-anchored membrane proteins of the exosome lumen), lipids (e.g., phosphatidylethanolamine (PE), phosphatidylcholine (PC), phosphatidylserine (PS), and others), several kinds of enzymes (e.g., glycosyltransferases, proteases, RNA editing enzymes, lipases, glycosidases, and metabolic enzymes), and metabolites (Fig. 7.1). Generating exosomes from cells is largely unknown, and it was initially thought that exosomes carried metabolic wastes just to maintain the cellular homeostasis. However, recent investigations suggest that exosomes are very useful vehicles carrying the cargos loaded with bioactive molecules including nucleic acids, lipids, and proteins, delivering to the target cells they encounter. The delivered molecules ultimately reprogram the recipient cells apart from their originating cells and regulate the intracellular communication (Zhang et al. 2019).

Over the past few decades, regenerative medication has drawn a considerable attention to use human stem cells to repair impaired tissues. The embryonic stem cells (ESCs), mesenchymal stem cells (MSCs), and induced pluripotent stem cells (iPSCs) have shown their potential in differentiation and proliferation, thus being capable of regenerating human tissues. Stem cells generally release a variety of regulatory bioactive molecules (in the form of exosomes) in an exceedingly paracrine fashion with relevant effects. Thus, exosomes play the necessary roles in regulations and cell-to-cell communication.

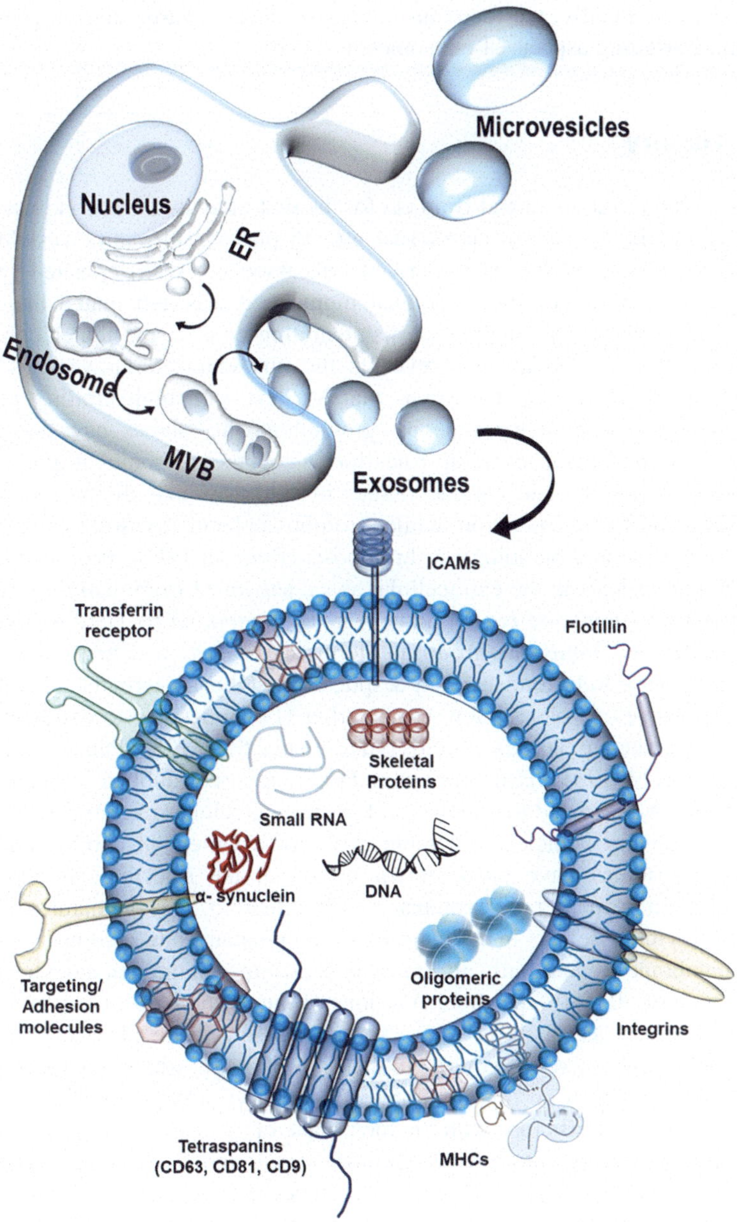

Fig. 7.1 Schematic representation of exosome biogenesis and its composition

With these perspectives, in this chapter, we aim to present an overview of exosomes and their history derived from stem cells and isolation and characterization. Further, this chapter deals with mode of action of exosomes isolated from various sources. Finally, we focus on future prospects towards disease prevention and health intriguing aspects of exosomes.

7.2 History

In the late 1960s, Anderson and Bonucci for the first time used electron microscopy to identify matrix vesicles in epiphyseal growth plates, and matrix vesicles were defined as membrane-enclosed cell-derived units associated with deposited hydroxyapatite particles (Bonucci 1967). At that moment, it was anticipated that matrix vesicles were formed by budding or disintegrating cells in the upper epiphyseal plate. Similarly, Wolf et al. also revealed the tiny extracellular vesicles (EVs), produced by platelets, also known as platelet dust with high clotting potential (Wolf 1967). These contemporaneous, groundbreaking studies have since been expanded and confirmed, revealing roles for osteogenic exosomes in pathological calcification of arteries and cardiac tissues in cardiovascular disease, as well as normal bone and tooth deposition, while thrombus-inducing exosomes can contribute to both normal and pathological thrombosis (Bonucci 1967). Post discovery of small ~50 nm vesicles in the extracellular space jettisoned from maturing mammalian reticulocyte (immature red blood cell) in 1983 reported by Harding and Stahl group and Pan and Johnstone group, the term "exosomes" was first coined a few years later by Rose Johnstone in 1987, despite the fact that the term had already been used as "exosome complex" a few years earlier (1981) denoted to extramembrane fragments generated in biological fluids (Harding and Stahl 1983; Pan and Johnstone 1983; Trams et al. 1981). Exosomes were largely ignored (dubbed "cellular dust") until studies have revealed that they could act as intercellular messengers, potential medication delivery vehicles, and biomarkers for a variety of chronic and acute disorders. Exosomes have been shown to aid in the selective elimination of diversified plasma membrane proteins, as the reticulocyte later matures into red blood cell (erythrocyte). As in the majority of mammalian cells, a component of the plasma membrane is recurrently attributed as endosomes in the reticulocyte with the recycling rate of 50–180% per hour. Portions of the membranes of few endosomes are then absorbed as small vesicles. These endosomes are also referred to as multivesicular bodies (MVBs) owing to their morphology, which embraces numerous tiny vesicles (ILVs or "intraluminal endosomal vesicles") engulfed in greater body. When these MVBs fuse with the membrane of the cell, consequently internal vesicles are ejected out into the extracellular space, and ILVs are converted into exosomes. New directions in exosome research began to emerge in the early 1980s. The notable contribution of Trams et al. on ectoenzymes was among them, as was the discovery of exosomes from the prostate and epididymis in seminal fluid and exosome complexes that not only are essential for sperm maturation but also transport proteins and lipids from the prostate to and into the sperm membrane (Trams et al. 1981). The promise of these early discoveries has been recognized by a nearly explosive growth in the field of exosome biology over the intervening three decades, resulting in the formation of various societies including the American

Society for Exosomes and Microvesicles and International Society for Extracellular Vesicles and even a dedicated journal i.e., *Journal of Extracellular Vesicles*. Moreover, many international meetings have been conducted and thousands of research articles on exosomes have already been published to date.

7.3 Isolation of Exosomes and Challenges

Over the past 10 years, exosome separation techniques have advanced dramatically, and there has been promising progress in solving the exosome mystery. It is challenging to accurately and quickly identify exosomes due to the complexity of biological samples, interference from other extracellular vesicles brought on by a significant physicochemical and biochemical overlapped properties, and also the heterogeneous nature of exosomes themselves. For example, differential ultracentrifugation, which is currently a known gold standard for exosome isolation, is labor-intensive and typically yields exosomes with proteins and lipoproteins as contaminants. Exosomes have been isolated from a variety of bodily fluids including blood, plasma, cerebrospinal fluid, saliva, urine, epididymal fluid, synovial fluid, amniotic fluid, and breast milk. The techniques used to isolate EVs are crucial to the success of isolation procedures, and efforts have been made to improve and standardize these techniques in order to properly understand the mechanisms of action and their use in biomedical research (Li et al. 2019). Because of their small size, isolation of exosomes remains difficult. However, using some high-throughput techniques, viz., (1) ultracentrifugation, (2) ultrafiltration, (3) affinity capture on antibody-coupled magnetic beads, (4) chromatography, and (5) polymer-based precipitation, a number of laboratories have been successful in isolating exosomes. Moreover, the technique used for isolation and scaling up is also dependent on the type of samples used as exosome source. Integrating different isolation methods, such as ultracentrifugation and immunoaffinity capture, can produce benefits by combining the benefits of the physical and biological worlds, but one must consider the additional labor and cost. Post isolation, extracted vesicles are generally gone through a series of testing including proteomic, immunoblotting, or protein staining techniques for their purity and functionality. Different isolation techniques and associated advantages and limitations are discussed over here (Kurian et al. 2021).

7.3.1 Ultracentrifugation

Low- and high-density vesicles from yeast were typically isolated by ultracentrifugation and density gradient ultracentrifugation, respectively. However, when compared to ultracentrifugation or precipitation-based approaches, density gradient ultracentrifugation has been shown to produce the purest form of exosome population (Van Deun et al. 2014). It is worth noting that exosome separation appears to be affected due to the density of the cargo present in the vesicles. Clinical samples with small sample quantities are ineligible for ultracentrifugation, whereas large sample volume is suitable for this technique. The ultracentrifugation process necessitates high centrifugal forces of up to 1,000,000 g and this process is subdivided into two categories: analytical and preparative ultracentrifugation. Analytical ultracentrifugation is used to look into the physicochemical characteristics of particulate materials as well as the molecular interactions of polymeric materials, whereas preparative ultracentrifugation is utilized to separate biological components, e.g., viruses, bacteria, subcellular organelles, and EVs (Li et al. 2017). To overcome the limitations of these approaches, relatively less complex isolation techniques that can be used with small sample volumes have also been developed. Although exosome isolation from clinical samples requires numerous overnight centrifugation procedures, which are labor-intensive and time-consuming, ultracentrifugation is a commonly used method for pelleting extravesicular protein complexes, aggregates, and other pollutants (Witwer et al. 2013).

7.3.2 Differential Centrifugation

Differential centrifugation is the most popular method for isolating exosomes (Théry et al. 2006). The main goal of this technique is to pellet down the exosomes, shedding vesicles, and apoptotic bodies and cell debris sequentially. However, because of their similar sedimentation characteristics, differential centrifugation may produce disappointing results like low yield with insufficient purity of the exosome population (Witwer et al. 2013; Lane et al. 2015). Furthermore, similar centrifugation techniques are frequently used with different rotors excluding the variations in viscosity that finally cause data dissimilarity between research groups. Notably, the documented variations in exosomes and shedding vesicle proportions are most likely the result of this heterogeneity. Due to these flawed practices using a standard centrifugation process with different rotors, this method frequently produces inconsistent and inappropriate results (Tauro et al. 2012; Cvjetkovic et al. 2014).

7.3.3 Size-Exclusion Chromatography (SEC)

Size-exclusion chromatography (SEC) with Sepharose 2B- or CL-4B-packed columns is often utilized to isolate exosomes from biofluids. Few commercially available membrane filters with pores 50–450 nm in size, such as polyvinylidene difluoride or polycarbonate, can also be used to separate cells and large EVs from biological materials. After sieving cells and large EVs from biofluid, ultracentrifugation is generally followed in conjunction with filtering techniques, where membranes are used to separate exosomes from proteins (Momen-Heravi et al. 2012; Sabapatha et al. 2006). Exosomes can be separated from proteins utilizing SEC; however, it can be very challenging to do so when they are mixed to certain other macroparticles e.g., MVs, protein aggregates, lipoparticles, or particulate debris. How EVs are typically separated depends on how differently they can pass through the physical barriers, like chromatography columns or filters. Using column chromatography, different sized EV fractions can be eluted sequentially from a single column. Filter techniques alone, on the other hand, are ineffective for exosome enrichment. Therefore, in order to enrich exosomes with higher yields than either procedure alone, SEC has also been employed in combination with ultracentrifugation (Lai et al. 2010).

7.3.4 Immune Affinity Capture, Immuno-Affinity Purification of Exosomes

Exosome membranes are well-known to express a large amount of surface marker protein. Immunoaffinity techniques are effective for separating exosomes by utilizing the interactions between these surface marker proteins (antigens) and their respective antibodies, as well as the specific interactions between ligands and their receptors (Li et al. 2017). This technique is especially helpful when surface markers produced on the exosome membranes lack of their soluble counterparts. Immunoaffinity purification (IP) approaches have been utilized to selectively capture particular exosomes from a complex population based on unique surface markers, overcoming the impurity of exosome preparations. This method is considered to be quick and simple and can also be performed using standard laboratory tools. This technique separates specific set of exosomes through using streptavidin-coated magnetic beads that have high affinity to any biotinylated antibody, such as CD81, CD9, or CD63. For instance, according to Tauro et al., immunoaffinity capture technique was more efficient than density gradient or ultracentrifugation techniques for isolating exosomes from colon cancer cells. This technique is promising as it has the capability to identify specific exosomes expressing certain exosomal markers, such as CD63 or proteins connected to cancer (Tauro et al. 2012; Grasso et al. 2015). ELISA (enzyme-linked immunosorbent assay) technique can also be utilized for extracting and measuring exosomes from urine plasma and serum because it uses a variety of targeted antibodies. In terms of specificity, exosomes isolated using immunoaffinity techniques have high yields comparable to those obtained using centrifugation processes. Immunoaffinity has been improved using submicron-sized magnetic particles, which have been reported to produce yields that are 10 to 15 times greater than those of ultracentrifugation processes (Zarovni et al. 2015). Exosomes isolated using magnetic microbeads coated with anti-CD34 antibodies have typical size, shape, biological activity, and molecular profiles (Hong et al. 2014). Additionally, due to their larger surface area and higher surface tension, immunoaffinity techniques using magnetic beads offer a higher capture efficiency and sensitivity than other microplate-based techniques. Furthermore, there are no volume restrictions with these methods (Li et al. 2017). The immunoaffinity technique is also used to separate specific exosomes based on the expression of surface proteins such as tetraspanins, CD9, CD63, and CD81. Exosomes can also be isolated using immunoaffinity techniques by incubating the source samples with gold-loaded porous ferric oxide nano-cubes coated with antibodies against specific surface marker proteins (Greening et al. 2015). Some affinity techniques also use very specific molecules such as heat shock protein and heparin or parent cell markers including chondroitin sulfate proteoglycan 4 (CSPG4) and epithelial cellular adhesion molecule (EPCAM) that can bind to the exosomes (Tauro et al. 2012).

The key drawback of these techniques is that the users must choose a group of marker-specific vesicles that may not accurately reflect in all exosomes. Only exosomes recognized by an antibody are collected, resulting in a lower exosomal yield, but separated exosomes have a higher degree of purity. Furthermore, the integrity of the exosomes may be compromised if the antibodies cannot be simply removed from the vesicles following precipitation (Taylor and Gercel-Taylor 2008). Additionally, the quality and specificity of the antibody are other shortcomings of these approaches, as most commercially available antibodies for immunoprecipitation are nonspecific in nature. Immunoaffinity technique is one of the most expensive techniques for isolating exosomes from a large sample volume because it requires a large quantity of antibody-conjugated beads, which may limit its application. Any possible therapeutic application could thus be hampered because it might only be suitable for research purposes with a small sample size.

7.3.5 Microfluidic-Based Isolation Methods

Advanced methods are required to address the subject of high-purity exosomes especially for clinical settings. Traditional processes have several disadvantages, including poor yield and impurity, time commitment, costly, and complications in standardizing. Microfluidics is an advanced technology that deals with physical and biological characteristics of exosomes which has recently emerged as an effective tool that has the capability to isolate, identify, and analyze exosomes at a microscale. It involves not only traditional sorting (such as electrophoresis, electromagnetic operations, and nanowire-based traps (NTs)) but also novel separation determinants (such as viscoelastic flow, size, density, immunoaffinity, and nano-sized deterministic lateral displacement). Importantly, these procedures also require less volume of samples and reagents (Lee et al. 2015; Davies et al. 2012; Contreras-Naranjo et al. 2017). Furthermore, antibody detection apprehension on a microfluidic device improves its specificity and subtyping capabilities (Chen et al. 2010). For instance, Wang et al. designed a porous silicon nanowire-on-micropillar edifices to separate exosomes from the rest part of the cells (Wang et al. 2013). This microfluidic system is capable of selectively trapping the exosomes (size from 40 to 100 nm) and removing proteins, other EVs, and cellular debris. In microfluidic-based immunoaffinity capture (Mf-IAC), “capture antibodies” or “capture beads” target specific surface indicators of EV subpopulations. For instance, Kanwar et al. developed “ExoChip” device that effectively collected EVs from circulation (Kanwar et al. 2014). Similarly, EVs have also been successfully extracted from the plasma membrane via mica surface-coated immunoaffinity method (Ashcroft et al. 2012). In this reference, using magnetic capture beads and a magnet-separating Mf-IAC system, Shao et al. not only isolated the exocytic vesicles but also obtained a high yield (Shao et al. 2016).

7.3.6 Polyethylene Glycol (PEG)-Based Precipitation

Polyethylene glycol (PEG) has been used as a solvent to separate exosomes from biofluids. Polyethylene glycol (PEG)-based precipitation has been useful to separate exosomes from proteins, but not from other MVs and lipoparticles. To make exosomes easier to precipitate from biological fluids, PEG that excludes water is used to change their solubility and dispersibility. This method facilitates the formation of exosome aggregates by encasing them in an aqueous PEG solution, which can then be precipitated by centrifugation at low speed (i.e.,1500 g) (Weng et al. 2016). Although the isolated exosomal size range is consistent with other techniques, e.g., differential ultracentrifugation, the purity and specificity have to be compromised as the soluble non-exosomal proteins, immune complexes, and other contaminants. These limitations can be overcome by employing exosome-specific markers (e.g., CD9 or other tetraspanins). However, this method typically results in a "biased" isolation, such as isolating the CD9+ exosome population only while excluding the CD9- one (Hurwitz and Meckes 2017). Because of its nonspecific mechanism, this method typically yields low-quality exosome separation with a high yield. When combined with an immunoprecipitation assay, it can produce pure exosomal fractions based on immunological markers. To avoid the drawbacks of using PEG alone, PEG-based isolation can be combined with alternative exosomal enrichment methods that provide numerous advantages including use of multiple samples at one go, being cost-effective and not harming the exosomes (Gallart-Palau et al. 2015). However, the contamination of the final exosome pellet is one of the major disadvantages that prevents exosomes from being examined further using omics-based tests. Regardless, this super-hydrophilic polymer performs well in clinical research settings and, along with its other associated advantages, becomes a desirable tool for rapid exosome extraction and examination (Rider et al. 2016).

7.3.7 Size-Based Filtration and Ultrafiltration (UF)

This method basically employs standard membrane filtration, with the components primarily separated based on their molecular weight, and exosomes can be retrieved via membranes within the limits of their size exclusion. However, in comparison, ultracentrifugation is relatively simpler and faster, but the use of specialized nanomembranes for exosome isolation and diagnostic purposes may be advantageous (Konoshenko et al. 2018). When exosomes are isolated from cell-free sources, e.g., urine, and cell culture medium, commercially available syringe filter-based isolation kits are preferred. These fast fractionators based on syringe filters can be used in both research and clinical settings. In the syringe filter cartridges, two membranes are tandemly positioned next to each other in such a way that the lower membrane collects small vesicles such as exosomes, while the larger components remain at the top membrane (Watson et al. 2018).

Another technique for size-based exosome separation is ultrafiltration (UF) wherein the isolation process is performed again based on the size of exosomes.

Exosomes can be isolated using size exclusion limits, i.e., by using membrane filters of fixed molecular weight. UF is relatively less expensive than costly equipped centrifugation process. Exosome preparations with high purity can be created using this technique and can also be combined with size-exclusion chromatography (SEC) to achieve the best performance (Zeringer et al. 2015). Despite producing pure vesicles, ultrafiltration has the disadvantage of having difficulty in eliminating contaminating proteins.

7.3.8 Polymer Precipitation

Exosomes between 50 and 150 nm in size can be captured and collected in "polymer nets" using low-speed centrifugation (1500 g). This technique is rapid and simple as it does not require any specialized equipment (Zeringer et al. 2015). By utilizing already existing technologies, this strategy enables simple incorporation into clinical practice and is also saleable for higher sample quantities. Notably, it is important that the polymers used to isolate exosomes should be made from harmless, neutral substances that do not elicit immunological reactions in vivo or in vitro (Ibrahim et al. 2014). However, the drawbacks of this method includes mixture of exosomes of various sizes and the chances of inclusion of non-exosomal contents, such as protein aggregates (Peterson et al. 2015).

7.3.9 Isolation by Sieving

Exosomes can be separated from biological liquids by sieving them through a membrane using pressure or electrophoresis (Davies et al. 2012). The technique can produce pure exosomes but requires less separation time (Sparks and Phillips 1992). However, the disadvantages include non-selectivity to a specific type and limited recovery of isolated exosomes.

7.4 Characterization of Exosomes

As per the International Society of Extracellular Vesicles (ISEV) suggestions, at least two different technologies for individual extracellular vesicle (EV) characterization are necessary. It is essential as there are only minor variances in the physical states of various EV subpopulations. Initially, the only criterion used to distinguish separated extracellular vesicles was their protein contents. The higher protein concentration in isolated EVs is obtained as it is usually observed overvalued due to the presence of impurities of other factors (Zaborowski et al. 2015). Therefore, several sophisticated approaches are being used to evaluate EVs and their applications in different biomedical fields. Physical analysis and chemical/biochemical/compositional studies are two common forms of investigation that are now conducted on the separated vesicles. Physical analysis is performed to measure EV particle size and its concentration characteristics by using electron microscopy, atomic force microscopy (AFM), dynamic light scattering (DLS), nanoparticle tracking analysis (NTA), and tunable resistive pulse sensing (TRPS), whereas the chemical, biological, and compositional studies are performed using staining, immunoblotting, or proteomic analysis to reveal information about the isolated vesicles' contents (Yáñez-Mó et al. 2015) (Table 7.1). However, the most difficult part is that the proteome profiles of exosomes change when they are extracted from the same cell line using various techniques.

7.4.1 Nanoparticle Tracking Analysis (NTA)

The physicochemical characteristics (e.g., size, surface charge, density, shape, and absorbency) of exosomes must be evaluated in order to understand their interactions with biological entities. Exosomes have been regularly identified and characterized in a number of techniques like electron microscopy, resistive pulse sensing, resistive flow cytometry, DLS, NTA, and AFM. But there are some limitations to each of these approaches that must be taken care of. Exosomes have been examined using a variety of methods, including the recent one (microfluidics platform). Biophysical approaches are commonly used to characterize the range of exosome size distribution. One such biophysical technique that can measure the exosome concentration and size distribution (10 nm to 2 μm) is referred to as optical particle tracking or more specifically, it can be called as NTA (Dragovic et al. 2011). This method enables atomic-level estimation of the Brownian motion of individual nanoparticles in a suspension. The movement of each exosome can be tracked by NTA using image analysis (de Necochea-Campion et al. 2018). Particle size, distribution, concentration, and other phenotypic characteristics can easily be determined using this method. Additionally, the technique employs fluorescently tagged antibodies to identify the presence of antigens on EVs. One of the major advantages of NTA is that it can measure tiny particles (approx. 30 nm). Moreover, advantages such as the ease of sample preparation and measurement and quick analysis are the reasons this method is used frequently. However, the success of NTA is significantly influenced by the proper dilution factor during sample preparation. Because samples may be returned to their original form after the tests, this procedure is even more desirable (Szatanek et al. 2017).

7.4.2 Dynamic Light Scattering

The exosome size can also be determined using photon correlation spectroscopy (also known as dynamic light scattering (DLS)). In DLS, a monochromatic coherent laser beam is passed through a suspension of particles (Szatanek et al. 2017). It is noticed that in samples, a time-dependent variation in interference is produced by the relative Brownian movements of the particles. By using this technology, it is possible to measure particles having a size from 1 nm to 6 nm; however, it functions

Table 7.1 Methods for the characterization of exosomes

S. no	Method	Advantages	Disadvantages	Ref.
1.	Electron microscopy	• Determination of size of vesicles (nm resolution). • Discrimination between EVs and other small particles (proteins, lipoproteins). • Visualize biogenesis.	• Preparation can result in loss of vesicle shape (cup-shaped). • Specialized equipment. • Time intensive. • Expensive.	(Tauro et al. 2013; Colombo et al. 2014)
2.	Western blot	• Cost Effective. • Well-established workflow.	• No information about co-expression of markers. • Large sample volume required.	(Khodakov et al. 2016; Cheng et al. 2014)
3.	Simple Western	• Automated. • Well-established workflow. • Small sample volume.	• No information about co-expression of markers.	(Yokoyama et al. 2017)
4.	ELISA	• Highly customizable. • When combined with exosome standards, can be used for quantification. • Small sample volume.	• Limited in scope to maximum expression of two protein markers (capture and detection antibodies). • Analysis of bulk populations of exosomes.	(Filipe et al. 2010)
5.	Flow cytometry	• Single-vesicle resolution. • Co-expression of multiple protein markers.	• Cytometers must be specifically calibrated for EVs. • Careful filtration of all reagents to minimize debris.	(Erdbrügger and Lannigan 2016; Galindo-Hernandez et al. 2013; Headland et al. 2014)
6.	Atomic force microscopy	• No need of sample fixation and staining. • Less sample quantity.	• It can only obtain surface information from samples. • Single scan image size.	(Wang et al. 2013; Coumans et al. 2017; Parisse et al. 2017)
7.	Resistive pulse sensing	• Rapid. • Easy sample preparation. • Capable of detecting non-vesicular material.	• For unknown size distribution. • Insufficient for detection of all particle sizes >70 nm.	(Maas et al. 2015; Van der Pol et al. 2014)

(continued)

Table 7.1 (continued)

S. no	Method	Advantages	Disadvantages	Ref.
8.	Dynamic light scattering	• Accurate for monodisperse sample (lower size <30 nm). • Rapid preparation of sample. • Downstream analysis.	• Unable to measure complex and large-sized exosome. • Can't distinguish contaminated protein for the source.	(Sokolova et al. 2011)
9.	Nanoparticle tracking analysis	• Accurate for both monodisperse and polydisperse samples. • Calibration particle standards. • Rapid. • Applicable to particle aggregates and heterogeneous samples.	• Complex and sometimes it is hard to identify contaminated protein from exosome. • Quantification of exosome concentration is not possible due to low-throughput camera and detection system.	(Aatonen et al. 2014; Dragovic et al. 2015; Filipe et al. 2010)
10.	Transmission electron microscopy	• To observe morphological structure of exosomes. • Internal structure information. • Assess the particular size distribution.	• Complex processing of TEM. • Large sample volume. • Applicable to limited samples. • SEM resolution is advanced than TEM.	(Théry et al. 2018; Skotland et al. 2017)

best when the sample is a monodispersed suspension (Bryant and Thomas 1995; Hoo et al. 2008). The efficiency of this technique has been confirmed with the volume and distribution of EVs in red blood cells and with ovarian cancer cell EVs (Lawrie et al. 2009). Although this method can offer information regarding a wide range of vesicle diameters under examination, its major drawback is the nonavailability of the biochemical composition and cellular origin of EVs (Gercel-Taylor et al. 2012).

7.4.3 Resistive Pulse Sensing (TRPS)

The in situ single-particle characterization and exosome content measuring are the key aspects of this technique. It has the ability to characterize things objectively, particle by particle. Both magnetic beads and a number of different biomolecules have been measured effectively using TRPS on a variety of nanoparticle solutions (Anderson et al. 2015). This method's drawback is that TRPS measurements are susceptible to system stability issues, where the pores could be blocked by the particles, and sensitivity where the particle sizes are too small. The authors have shown that by optimizing system parameters including system noise and cutoff limits of sensors, the system's drawback, i.e., susceptibility to sensitivity and stability, may be solved. Vogel et al. confirmed the reliability and adaptability of TRPS via sensing the multimodal mixtures of carboxylated and bare polystyrene particles and mixed anionic and cationic liposomes and exosomes. Moreover, they also performed an in situ time-course analysis of DNA attachment onto magnetic nanoparticles (Vogel et al. 2017). TRPS was also employed by Patko et al. to scrutinize leukemia-derived EVs binding to the extracellular matrix, with a size range between 200 and 300 nm (Patko et al. 2013). TRPS has also been used in many reports, where the size distributions of EVs designed to deliver enzymes to combat Alzheimer's disease (150–200 nm) and anticancer miRNAs to tumor cells were measured and characterized (Shimbo et al. 2014; Katsuda et al. 2013).

7.4.4 Atomic Force Microscopy

For examining exosomes, AFM offers a distinctive substitute to optical and electron diffraction methods. It appears to be a reliable method which identifies and records contacts between a probing tip and the sample surface. This method's competence to analyze samples in their natural environments, with limited requirements with preparation of samples and without using any harmful modes of operation, is a key component (Yuana et al. 2010). Abundance, shape, biomechanics, and biomolecular makeup of exosomes are all characterized at the nanoscale using AFM. This technique has aided the single-vesicle and sub-vesicular levels of exosomes. With heterogenic populations, e.g., tumor samples, AFM can be used not only to quantify the abundance of EVs but also reveal the structure, biomechanics, and biomolecular contents of exosomes (Sharma et al. 2017). With sub-nanometer accuracy, AFM enables the dimensional measurement of the nano-objects. The drawback of this approach is that the sample's characterization was done using outside analyses, which meant that it was done under various experimental settings, such as temperature or altering scan speed. AFM has been effectively used in several investigations to describe EVs made from blood, saliva, and synovial fluid (Hardij et al. 2013; Sharma et al. 2010; György et al. 2011). The isolation processes, detection techniques, screening of membrane components, mechanical properties, and depiction of the morphology have also been described in these works.

7.4.5 Transmission Electron Microscopy

TEM is a method that is frequently used to describe the size and structural morphology of diverse biological entities. TEM works by producing pictorial descriptions such as an electron beam traveling through a sample, generating secondary electrons in the process. Special lenses helps to gather and amplify these electrons. TEM and cryo-electron microscopy are two forms of EM that are frequently employed in biological sample research (cryo-EM). Specimens must be dehydrated and preserved with glutaraldehyde before being examined under a TEM. It is necessary to take TEM images in a vacuum. The only way to view EVs is with TEM, and the pictures that are produced can be utilized to calculate the vesicular diameter. The intensive, multistep sample preparation required for TEM, which could affect the EVs' shape, is a crucial factor to take into account. Spherical exosomes and EVs have various shapes, according to Colombo et al.'s observation using TEM (Colombo et al. 2014). Additionally, the electron beam may occasionally harm biological samples. Exosomes that have been extracted and analyzed by TEM often have a cup-shaped structure, whereas exosomes that have been frozen and examined by cryo-TEM have circular structures (Raposo and Stoorvogel 2013).

Cryo-EM is used for EV investigation because it uses a unique way of sample preparation and processing which prevents damaging the sample from the electron beam. Due to the fact that here the samples are kept in −196 °C and the cells are preserved under these circumstances, cryo-EM is free from the issues of sample dehydration and its fixation with no ultrastructural variations. The best technique for viewing proteins and nanoparticles without dehydration artifacts is cryo-TEM. Cryo-TEM images of exosomes involves the membrane architecture and lumens. The identification of individual proteins inside the exosomes is crucial to understanding how exosomes function biologically. Exosomal proteins are typically labeled and seen using certain fluorescent dyes. Exaggerated fluorescence signals, however, can occasionally make it impossible to discriminate among exosomes, depending on their size and shape (Zaborowski et al. 2015). In order to properly define the function of these proteins, an alternate way is the imaging of exosomes with particular immunogold EM antibody binding.

7.4.6 Flow Cytometry

Flow cytometry is a molecular method for identifying the surface proteins of exosomes along with the depiction of their size and structure (Pospichalova et al. 2015). It is one of the commonly used methods for EV analysis since it can predict the important information of exosomes, i.e., biological source. When employing this technique to separate and illustrate exosomes, the preliminary sample size (with its volume) is crucial. Parallelly, ultracentrifugation, Western blotting, and electron microscopy are some of the methods that are still the most dependable (Dragovic et al. 2011). However, none of these methods offer hope for use in clinical or diagnostic research. Contrarily, flow cytometry technology is well appropriate to the consistent examination even in clinical settings, enabling the investigation of biological characteristic features of cells and particles in suspension and permitting analysis of the size and characteristics of exosomes. Based on forward-scattered light, conventional flow cytometers can quantify particles larger than 300 nm but are unable to identify smaller particles (FSC). As a result, these tools prevent the direct identification of exosomes (Suárez et al. 2017). By passing a laser beam of a particular wavelength through a stream of fluid containing suspended particles, the presence of particles in the samples can be assessed. Additionally, this method measures particles that have been fluorescently dyed. Although this technique is capable of finding relative size and granulation of particles, because of size detection restrictions, a sizable portion of particles evade through detection (Suárez et al. 2017). In order to identify labeled exosomes from background pollutants, a high-end (with higher-sensitivity forward scatter detection) flow cytometer with advance fluorescence amplification system has recently been created (Erdbrügger et al. 2014). A better particle resolution (with different angles) is provided by the new generation of flow cytometers (Chandler et al. 2011). The benefits of this method also include (i) the rapid measurement of suspended exosomes, (2) the high-throughput identification of EVs (even smaller than 300 nm), and (3) the exact quantification data of exosomes (on the basis of antigen expression levels) (Orozco and Lewis 2010).

There are also a number of additional molecular methods for characterizing exosomes. One of these is Raman spectroscopy, which uses a laser to illuminate materials. This method reveals details on the chemical composition of exosomes (Smith et al. 2015). Exosomes binding to certain antibodies on microfluidic channels have been studied using a microfluidic-based approach, which also elutes the associated vesicles (He et al. 2014). Exosomes can also be recognized with the presence of cargo molecules, like RNA which can be estimated advanced methods of RNA quantifications like microarray analysis, next-generation sequencing, etc. (Ramirez et al. 2018).

7.4.7 Electron Microscopy

When combined with immunogold labeling, where colloidal gold-conjugated antibodies are used for staining, electron microscopy is a valuable device for assessing the size and outer morphology of EVs and identifying the protein present in EVs (Chiriacò et al. 2018). Transmission electron microscopy (TEM) and scanning electron microscopy (SEM) are the two popular methods of electron microscopy used to evaluate the outer and inner structures of exosomes. Using an electron beam, the SEM and TEM create high-quality photographs of submicron particles where electrons are found to be present in the two methods is the distinction. Simply expressed, the electrons that scatter in the SEM and pass through the sample in the TEM are detected (Skoog et al. 2017).

More specifically, in SEM, when the electrons interact with the sample's particles, they scatter. This particle image is created by first detecting and then capturing the scattered electrons. TEM, on the other hand, uses a fluorescent screen to detect the electrons without interacting with the particles and pass through the material. The sample's particles cast shadows or dark patches on the fluorescent screen, which results in the creation of an image. Both TEM and SEM in the case of exosomes show comparable particle size distributions but slightly distinct morphologies (Wu et al. 2015). Thus, the exosomes in TEM and SEM often have a divot in their center. This is probably because of the drying procedure involved in getting the samples ready for TEM and SEM.

7.4.8 Western Blot

Immunoblotting, also known as Western blotting, is based on the idea of the affinity of an antigen (target proteins) to its antibody that has been particularly designed to identify the antigen. Contrary to flow cytometry, which allows for the viewing of intact vesicles, Western blotting lyses the vesicles and causes the proteins to be decreased and denatured (Gallagher et al. 2011). SDS-PAGE is used to separate the proteins after denaturation, and the separated proteins are then transferred to polyvinylidene fluoride (PVDF) or nitrocellulose membranes. The membrane's remaining open holes are filled with protein (derived from nonfat milk) and/or detergent before being exposed to an antibody against a target antigen. In an ideal world, the antigen on the membrane's surface is recognized specifically by the antibody. The membrane is then targeted to a secondary antibody that has reactivity toward its initial (primary) antibody. With its fluorescent tag or the conventional horseradish peroxidase/alkaline phosphatase group attached to the secondary antibody, the detection of secondary antibody can be done easily. Due to its simplicity, accessibility, and capacity to identify both exosomal surface proteins and interior proteins, Western blotting is one of the most often used analysis method for exosome analysis. The main drawback is that it is not adequately multiplexed, and the effectiveness of the utilized antibodies limits its ability to be specific and reproducible. Exosomal protein must be employed in huge quantities in order to obtain a small amount of information since multiplicity is lacking. More multiplexed analytical techniques would be beneficial because the separation of exosomes is frequently a time-consuming and low-yield operation (Bordeaux et al. 2010).

7.4.9 Simple Western

Proteins that are connected to or co-isolated with EVs can be found via Western blotting, which can also reveal important details regarding the impurities content during EV preparation. It is significant that it can also verify the target protein's molecular weight. The lack of housekeeping internal reference proteins to utilize for normalization in immunoblotting investigations in EV samples as compared to cell lysates is a drawback (Takov et al. 2019).

Therefore, similar protein amounts, EV separation volumes, or particle counts are frequently utilized. To make accurate estimation of the enrichment of proteins in the EV isolate, the inclusion of the original raw sample, the EV-depleted sample, and the technical controls is essential. Western blotting can be difficult because it needs a lot of EVs in order to be sensitive enough. Dot blotting and capillary electrophoresis immunoassays are two alternatives that can offer significantly improved sensitivity (Nelson et al. 2017). Although there is disagreement over which proteins should be looked into as potential pollutants, MISEV offers the best direction. It may be helpful to confirm the removal of extra endoplasmic or plasma membrane proteins as well as lipoproteins and serum albumin, depending on the source of the EVs (Théry et al. 2018).

7.4.10 ELISA

ELISA is a popular immunolabeling technique that uses antibody recognition to quantify peptides and proteins (Wang et al. 2017). Exosome profiling and diagnostics use ELISA, which enables the identification of protein markers and quantification of tumor antigens and exosome-specific antigens. For some protein markers, ELISA is an affordable option, and mass spectrometry enables protein quantitation in a complicated biological sample (Jeppesen et al. 2014). Mass spectrometry has significant technical requirements, which prevents it from being widely used in clinical research and makes it difficult to access (Lai et al. 2022). An established method called ELISA is capable of highly sensitive detection in multi-well formats.

Sandwich ELISA is also another type of antibody detection method which is extensively used due to its microplate setup and higher throughput than immunoblotting. However, this method also consists of some limitations such as risk of false-positive signal with low-specificity antibodies. Using enzyme-linked or fluorescent detection is likely necessary by merging distinct capture and detection antibodies. Another high-throughput and sensitive immunoassay variant associated with DELFIA (dissociation-enhanced lanthanide fluorescence immunoassay) has been proficiently used for detecting EVs and their associated molecules (Bordeaux et al. 2010; Welton et al. 2015). Immunoassays, which are similar to dot blots in that they offer good sensitivity for small sample sizes, call for well-confirmed antibodies and do give information about the molecular weight.

7.5 Epigenetics Biology of Exosomes

Exosomes, which are derived from endosomal vesicles with a typical size range of 30 to 100 nm, contribute in a number of biological processes, such as facilitating intercellular communication by carrying the genetic materials. Numerous tumor cells have also been found to release exosomes which may bind to the specific receptors of the target cells and release their contents into the recipient cells causing functional modifications (Hannafon and Ding 2013; Shah and Calin 2013). Nowadays, it is believed that bioactive materials, such as proteins, RNAs, DNA, and microRNAs, have been horizontally transferred and play significant roles in a number of tumor-relevant processes including incursion, metastasis, development, inflammation, angiogenesis, and renewal of stemness in cells and its expansion. These exosomes provide signaling cues for tumor cell stimulation, initiation, propagation, and differentiation. mRNAs, microRNAs (miRNA), and proteins found in exosomes have the potential to be transported to target cells and cause genetic and epigenetic alterations (Behbahani et al. 2016).

In this section, we have emphasized on epigenetic molecules associated with the exosomes.

7.5.1 Exosomal Noncoding RNA

Nonprotein coding transcripts, also known as long noncoding RNAs (lncRNAs) or short noncoding RNAs, are commonly referred to as the noncoding RNA. miRNA is considered to be the most extensively studied noncoding RNA. The key finding in this field is the detection of circulating tumor-associated miRNAs in the blood serum of a cancer patient. Circulating noncoding RNAs could also be used as noninvasive indicators for cancer diagnosis (Lawrie et al. 2008). According to a recent study, tumor cells have the ability to produce extracellular vesicles containing noncoding RNAs which are quite stable in nature and can be easily detected in the blood serum while they are in the circulating state (Ma et al. 2012). Moreover, these circulating noncoding miRNAs are also found to be protected from endogenous RNases (Mitchell et al. 2008). Although the noncoding RNAs in extracellular vesicles have the potential to depict cancer status and their involvement in cellular talk and change the cancer microenvironment has yet to be determined.

7.5.2 Exosomal Long Noncoding RNA

Two hundred nucleotide-long nonprotein coding transcripts are designated as long noncoding RNAs (lncRNAs), and they are well-known for their record of post- or pre-transcriptional regulation of the epigenetic process (Perkel 2013). Despite receiving less attention than miRNAs, lncRNAs were shown to be the most predominant in exosomal RNA species, accounting for 3.36% of all mappable reads, in the sequence analysis of 14 size-selected sequencing libraries that were previously described (Huang et al. 2013). The lncRNA composition of exosomes differs in different donors, as previously shown for miRNAs, indicating preferential secretion of lncRNAs (Grammatikakis et al. 2014). In a previous study, six lncRNAs (lincRNA-p21, MALAT1, GAS5, HOTAIR, TUG1, and CCND1-ncRNA) were compared between donor cells and exosomes in MCF-7 and HeLa cells, to determine how they differed from one another under DNA damage stress (Gezer et al. 2014). Although MALAT1 was found to be common, its level in exosomes was quite low; in contrast, lincRNA-p21 was found to be highly expressed in exosomes.

7.5.3 Exosomal miRNAs

MicroRNAs (miRNAs), the small noncoding RNA molecules with 22–25 nucleotide sequences, are implicated in RNA silencing by posttranscriptional epigenetic modification (Ambros 2004; Bartel 2004). MiRNAs are considered to be the most predominant RNA species in exosomes, accounting for more than 76% of all mappable reads and 42% of all raw reads in 14 size-selected sequencing libraries. Certain groups of proteins are mainly involved in loading of miRNAs into the exosomes. For instance, argonaute-2 (AGO2) protein and GW182 localized in the P-bodies precisely and selectively load the miRNAs into exosomes (Pegtel et al. 2010; Zomer et al. 2010; Gibbings et al. 2009). Exosomal miRNAs have varied functions including cell proliferation, migration, cell-cell communication, and apoptosis. For instance, exosomes possessing MiR-126 isolated from chronic myelogenous leukemia cell are shown to be transported into the endothelial cells causing variation in motility and their adherence properties (Taverna et al. 2014). Brain metastatic cancer cells that release extracellular vesicles carrying miRNA-181c can disrupt the blood-brain barrier (Tominaga et al. 2015). Tight junctions are shown to be damaged by exosome-mediated miR-105 released by cancer cells (Snow et al. 1989). Breast cancer cells can be entered into dormant state in a metastatic niche by the exosomal transport of miRNA-23b isolated from bone marrow cells (Ono et al. 2014). MiR-210, which is secreted by metastatic cancer cells, travels to endothelial cells and control the spread of cancer cells (Kosaka et al. 2013). Currently, extracellular vesicular miRNAs have been proposed as potential tools for cancer diagnosis, prognostic evaluation, and tumor control (Rabinowits et al. 2009; Ohno et al. 2013).

7.5.4 DNA Methylation

Epigenetic DNA changes of oncogenes or anti-oncogenes are crucial for the onset, progression, and dissemination of many malignancies (Chik et al. 2012). DNA methylation is one of the most universal factors responsible for the generation of oncogenes and anti-oncogenes. Several DNA methyltransferases (DNMT) including DNMT1, DNMT3a, and 3b DNMT3b typically cause gene silencing when they add methyl groups to the cytosines in the CpG islands of the regulatory sequences (Robertson et al. 1999). The degree of methylation in the promoter region determines the level of oncogene or anti-oncogene transcription, which eventually affects the tumor growth. Unmethylated portions of mammalian genomes are shielded from de novo methylation by demethylating enzymes such as thymine DNA glycosylase (TDG) and activation-induced cytidine deaminase (AICDA) (Shen et al. 2013; Popp et al. 2010). Extracellular vesicles have a complicated bioactive cargo that can turn normal cells into cancerous ones. Extracellular vesicles' protein, DNA, or RNA content may influence recipient cells' genome methylation state, leading to epigenetic alterations. Microvesicles are the extracellular vesicles that have a diameter of 100–1000 nm and are generally produced by the plasma membrane of tumor cells, whereas the diameter of exosomes, which are often produced by endosomes, lies in the range of 30–100 nm (Raposo and Stoorvogel 2013; Ratajczak et al. 2006). It is shown that macrovesicles generated by leukemia cells enhance the level of overall DNA methylation in the recipient cells. In line of this segment in an elegant study, the hypermethylation of promoter regions of few genes, e.g., *P53* and *RIZ1*, was noticed when cells were treated with leukemia-derived microvesicles. Moreover, an upregulation of DNMT3a and DNMT3b mRNA expression and an enhanced level of activation-induced cytidine deaminase (AICDA) were also recorded in the treated groups. All these findings indicate that genomic instability may have facilitated leukemic transformation of the recipient cells. The amount of DNMT3a, DNMT3b, and AICDA dropped when microvesicles were treated with RNase, indicating that leukemia-derived macrovesicles might have the ability to affect the methylation state of recipient cells via communicating through microvesicular RNA. Additionally, breakpoint cluster region-Abelson leukemia gene human homolog 1 (BCR-ABL1) has been discovered to be the prevalent onco-mRNA in the microvesicles produced by the leukemia cell line K562 (Zhu et al. 2014).

7.5.5 Histone Acetylation

Two important enzymes named histone acetyltransferase (HAT) and histone deacetylase (HDAC) are mainly responsible for histone modification. The latter reduces positive charges of histone proteins, loosening the compacted chromatin structure and enhancing gene transcription (Kuo 1998; Grunstein 1997). Thus, the cancer microenvironment may affect DNA and histone modification-based epigenetic regulation of the expression of genes relevant to the cancer. Though the functions of extracellular vesicles in histone modification are currently debatable, genes important for transgenerational epigenetic inheritance, as well as the components of exosomes secreted by various cells including cancer cells, have been found to be strikingly overlapping as shown in various reports in the bioinformatics studies (Grunstein 1997). By looking at the GO biological processes connected to these overlapping mRNAs and proteins, it becomes clear that the exosome content is concentrated on a small network of processes, including several processes involved in epigenetic modification, rather than having a broad impact on cellular activities. A remarkably high percentage of the genes identified as being important for transgenerational epigenetic inheritance include those related to histone acetylation or deacetylation, histone ubiquitination, other histone modifications, and even chromatin remodeling (Sharma 2014).

These results suggest that exosomal mRNAs and proteins may have a direct or indirect role in epigenetic alteration, particularly histone modification, as well as the response to environmental exposure. A similar correlation between environmental exposure and histone modification has also been found by other investigators working in the exosomal miRNA field (Sharma 2014). These extracellular vesicle-related components may participate in cancer development, growth, and metastasis in the cancer microenvironment. For instance, G26/24 oligodendroglioma cell line produces extracellular vesicles purportedly including the differentiation-specific linker histone H1, which is not produced by the healthy astrocytes (Schiera et al. 2013).

Notably, the H1 histone family is considered to be the most diverse of all the histone families, and each subtype or variant of the H1 protein is linked to a particular function or distribution (Izzo et al. 2008; Marzluff et al. 2002; Kowalski and Pałyga 2012; Happel and Doenecke 2009). Usually, H1 is connected to the terminal differentiation of cells. A promising key molecular diagnostic approach for oligodendroglioma diagnosis could be the enhancement of histone H1 in cancer cell-derived extracellular vesicles, while the precise pathophysiological implications of this event are yet unclear (Zlatanova and Doenecke 1994; Gabrilovich et al. 2002). Additionally, recipient cells in the tumor microenvironment may be subsequently affected by extracellular vesicles carrying histone H1.

7.6 Future Prospects

Exosomes' ability to be employed clinically as soon as possible will primarily depend on how well existing exosome issues can be optimized for. The number one objective is to figure out how to increase the content and purity of exosomes, which has always been the bottleneck preventing wider application of its transformation. Recent studies have demonstrated that the proper combination of multiple exosome extraction and purification methods can significantly reduce the aforementioned issues, but more research is still needed to determine the optimal ways to combine these methods. Moreover, it is still unknown how exosomes fuse and secrete their contents. It is necessary to learn more about how exosome heterogeneity affects medication loading effectiveness. It's also important to optimize and enhance exosome loading capacity and targeting strategies. Conduction of comprehensive and multidisciplinary studies to examine exosomes biological processes pharmacokinetics, toxicological, and clinical testing research) will aid a better understanding of the body's condition as well as the diagnosis and treatment of disorders.

Such developments will hasten our grasp of the role that exosomes play in health and disease and enable researchers to use this information to exosome-based therapeutics and diagnostics. Finally, we would want to emphasize the logical fallacy in attempting to identify exosome subclasses based on their purported place of budding, which is further made worse by the quickly coming horizon of single-exosome investigations. Conceptually, it is challenging to determine the exact origin of secreted tiny vesicles from the cells, and there is almost currently no proper molecular marker or purification procedure that can unmistakably distinguish or identify exosomes that are formed by budding via endosomes or the plasma membrane.

Oncogenes and compounds susceptible to epigenetic reprogramming are among the several types of biomolecules found in extracellular vesicles produced by cancer cells. They are expelled into the cancer microenvironment and could hasten the development of the disease. A significant part of this process appears to be played by epigenetic regulation. Extracellular vesicles include a large number of mRNAs and proteins that are associated with GO (gene ontology) biological processes that are involved in epigenetic control. Hepatocellular carcinoma cells' extracellular vesicles contain the lncRNA TUC339, which is involved in cell cycle progression, adhesion, and tumor formation. Extracellular vesicles might be used as targets for therapeutic intervention due to their unique messenger capability, i.e., mediating intercellular communication. In fact, extracellular vesicles themselves, whether isolated naturally or engineered artificially, provide a promising new approach of drug delivery. Moreover, a noninvasive method of cancer diagnosis is exemplified by the identification of the biomarkers inherent in these extracellular vesicles.

7.7 Conclusion

Under many normal and pathological circumstances, EVs play important roles. Exosomes are crucial in the communication between cancer enviornment and the cells they are targeting. Exosome release is necessary for processes like gene manipulation and the targeting of medicinal interventions to particular cells. Exosome production, composition, and yield are influenced by the health of the cells they come from. Existing techniques for exosome segregation and characterization are routinely employed for prognostic and diagnostic applications, despite the fact that various novel technologies are being developed. Here, we've included a summary of exosomes' development, isolation, characterization, biological functions, importance, and prospective therapeutic uses for a range of illnesses. The ideas, benefits, and drawbacks of various exosome isolation and detection approaches have also been covered.

Exosome properties like number and size can be analyzed using DLS and NTA techniques. The greatest method for examining exosome structural details is electron microscopy; however, the ideal methodology would be able to examine both structural and biological traits with a single piece of equipment. Therefore, the separation and characterization of exosomes require more advanced techniques. It is crucial to choose the appropriate methods for both the extraction and characterization of exosomes in order to improve the attribution of the isolated exosomes and the reliability of the results from their utilization. The ability to distinguish exosomes originating from normal cells from those formed from cells with diseased states is the main technological barrier in exosome detection in therapeutic applications. To distinguish exosome subtypes in diverse samples, a combination of different quantitative approaches may be required. This will create new opportunities for measuring and detecting exosomes. The ability to separate various subpopulations of vesicles, to which an origin and function may be attributed, should be a key component of emerging methodologies. Once these challenges are overcome, the origin of exosomes and their isolation procedures, composition, and interactions with specific receptors will help us better understand their role and make it easier to create new treatment approaches.

Exosomes have attracted a lot of attention from scientists both domestically and internationally as current research hotspots. Exosomal components can be employed as indicators for disease diagnosis and prognosis; however, the detection methods for these purposes are still being developed.

References

Aatonen MT, Öhman T, Nyman TA, Laitinen S, Grönholm M, Siljander PR-M (2014) Isolation and characterization of platelet-derived extracellular vesicles. J Extracell Vesicle 3:24692

Ambros V (2004) The functions of animal microRNAs. Nature 431:350–355

Anderson W, Lane R, Korbie D, Trau M (2015) Observations of tunable resistive pulse sensing for exosome analysis: improving system sensitivity and stability. Langmuir 31:6577–6587

Ashcroft BA, de Sonneville J, Yuana Y, Osanto S, Bertina R, Kuil ME, Oosterkamp TH (2012) Determination of the size distribution of blood microparticles directly in plasma using atomic force microscopy and microfluidics. Biomed Microdevices 14:641–649

Bartel DP (2004) MicroRNAs: genomics, biogenesis, mechanism, and function. Cell 116:281–297

Behbahani GD, Khani S, Hosseini HM, Abbaszadeh-Goudarzi K, Nazeri S (2016) The role of exosomes contents on genetic and epigenetic alterations of recipient cancer cells. Iran J Basic Med Sci 19:1031

Bonucci E (1967) Fine structure of early cartilage calcification. J Ultrastruct Res 20:33–50

Bordeaux J, Welsh AW, Agarwal S, Killiam E, Baquero MT, Hanna JA, Anagnostou VK, Rimm DL (2010) Antibody validation. BioTechniques 48:197–209

Bryant G, Thomas JC (1995) Improved particle size distribution measurements using multiangle dynamic light scattering. Langmuir 11:2480–2485

Chandler WL, Yeung W, Tait JF (2011) A new microparticle size calibration standard for use in measuring smaller microparticles using a new flow cytometer. J Thromb Haemost 9:1216–1224

Chen C, Skog J, Hsu C-H, Lessard RT, Balaj L, Wurdinger T, Carter BS, Breakefield XO, Toner M, Irimia D (2010) Microfluidic isolation and transcriptome analysis of serum microvesicles. Lab Chip 10:505–511

Cheng L, Sharples RA, Scicluna BJ, Hill AF (2014) Exosomes provide a protective and enriched source of miRNA for biomarker profiling compared to intracellular and cell-free blood. J Extracell Vesicle 3:23743

Chik F, Szyf M, Rabbani SA (2012) Role of epigenetics in cancer initiation and progression, Human Cell Transformation: Role of Stem Cells and the Microenvironment. Adv Exp Med Biol 720:91–104

Chiriacò MS, Bianco M, Nigro A, Primiceri E, Ferrara F, Romano A, Quattrini A, Furlan R, Arima V, Maruccio G (2018) Lab-on-chip for exosomes and microvesicles detection and characterization. Sensors 18:3175

Colombo M, Raposo G, Théry C (2014) Biogenesis, secretion, and intercellular interactions of exosomes and other extracellular vesicles. Annu Rev Cell Dev Biol 30:255–289

Contreras-Naranjo JC, Hung-Jen W, Ugaz VM (2017) Microfluidics for exosome isolation and analysis: enabling liquid biopsy for personalized medicine. Lab Chip 17:3558–3577

Coumans FAW, Brisson AR, Buzas EI, Dignat-George F, Drees EEE, El-Andaloussi S, Emanueli C, Gasecka A, Hendrix A, Hill AF (2017) Methodological guidelines to study extracellular vesicles. Circ Res 120:1632–1648

Cvjetkovic A, Lötvall J, Lässer C (2014) The influence of rotor type and centrifugation time on the yield and purity of extracellular vesicles. J Extracell Vesicle 3:23111

Davies RT, Junho Kim S, Jang C, Choi E-J, Gho YS, Park J (2012) Microfluidic filtration system to isolate extracellular vesicles from blood. Lab Chip 12:5202–5210

de Necochea-Campion R, Gonda A, Kabagwira J, Mirshahidi S, Cao H, Reeves ME, Wall NR (2018) A practical approach to extracellular vesicle characterization among similar biological samples. Biomed Phys Engg Exp 4:065013

Doyle LM, Wang MZ (2019) Overview of extracellular vesicles, their origin, composition, purpose, and methods for exosome isolation and analysis. Cell 8:727

Dragovic RA, Collett GP, Hole P, Ferguson DJP, Redman CW, Sargent IL, Tannetta DS (2015) Isolation of syncytiotrophoblast microvesicles and exosomes and their characterisation by multicolour flow cytometry and fluorescence nanoparticle tracking analysis. Methods 87:64–74

Dragovic RA, Gardiner C, Brooks AS, Tannetta DS, Ferguson DJP, Hole P, Carr B, Redman CWG, Harris AL, Dobson PJ (2011) Sizing and phenotyping of cellular vesicles using nanoparticle tracking Analysis. Nanomedicine 7:780–788

Erdbrügger U, Lannigan J (2016) Analytical challenges of extracellular vesicle detection: a comparison of different techniques. Cytometry A 89:123–134

Erdbrügger U, Rudy CK, Etter ME, Dryden KA, Yeager M, Klibanov AL, Lannigan J (2014) Imaging flow cytometry elucidates limitations of microparticle analysis by conventional flow cytometry. Cytometry A 85:756–770

Filipe V, Hawe A, Jiskoot W (2010) Critical evaluation of nanoparticle tracking analysis (NTA) by NanoSight for the measurement of nanoparticles and protein aggregates. Pharm Res 27:796–810

Gabrilovich DI, Cheng P, Fan Y, Bin Y, Nikitina E, Sirotkin A, Shurin M, Oyama T, Adachi Y, Nadaf S (2002) H1° histone and differentiation of dendritic cells. A molecular target for tumor-derived factors. J Leukocyte Biol 72:285–296

Galindo-Hernandez O, Villegas-Comonfort S, Candanedo F, Gonzalez-Vazquez M-C, Chavez-Ocana S, Jimenez-Villanueva X, Sierra-Martinez M, Salazar EP (2013) Elevated concentration of microvesicles isolated from peripheral blood in breast cancer patients. Arch Med Res 44:208–214

Gallagher S, Winston SE, Fuller SA, Hurrell JGR (2011) Immunoblotting and immunodetection. Curr Protoc Cell Biol 52:6.2. 1–6.2. 28

Gallart-Palau X, Serra A, Wong ASW, Sandin S, Lai MKP, Chen CP, Kon OL, Sze SK (2015) Extracellular vesicles are rapidly purified from human plasma by PRotein organic solvent PRecipitation (PROSPR). Sci Rep 5:1–12

Gercel-Taylor C, Atay S, Tullis RH, Kesimer M, Taylor DD (2012) Nanoparticle analysis of circulating cell-derived vesicles in ovarian cancer patients. Anal Biochem 428:44–53

Gezer U, Özgür E, Cetinkaya M, Isin M, Dalay N (2014) Long non-coding RNAs with low expression levels in cells are enriched in secreted exosomes. Cell Biol Int 38:1076–1079

Gibbings DJ, Ciaudo C, Erhardt M, Voinnet O (2009) Multivesicular bodies associate with components of miRNA effector complexes and modulate miRNA activity. Nat Cell Biol 11: 1143–1149

Grammatikakis I, Panda AC, Abdelmohsen K, Gorospe M (2014) Long noncoding RNAs (lncRNAs) and the molecular hallmarks of aging. Aging (Albany NY) 6:992

Grasso L, Wyss R, Weidenauer L, Thampi A, Demurtas D, Prudent M, Lion N, Vogel H (2015) Molecular screening of cancer-derived exosomes by surface plasmon resonance spectroscopy. Anal Bioanal Chem 407:5425–5432

Greening DW, Rong X, Ji H, Tauro BJ, Simpson RJ (2015) A protocol for exosome isolation and characterization: evaluation of ultracentrifugation, density-gradient separation, and immunoaffinity capture methods. Methods Mol Biol 1295:179–209

Grunstein M (1997) Histone acetylation in chromatin structure and transcription. Nature 389:349–352

György B, Módos K, Pállinger E, Pálóczi K, Pásztói M, Misják P, Deli MA, Sipos A, Szalai A, Voszka I (2011) Detection and isolation of cell-derived microparticles are compromised by protein complexes resulting from shared biophysical parameters. Blood 117:e39–e48

Hannafon BN, Ding W-Q (2013) Intercellular communication by exosome-derived microRNAs in cancer. Int J Mol Sci 14:14240–14269

Happel N, Doenecke D (2009) Histone H1 and its isoforms: contribution to chromatin structure and function. Gene 431:1–12

Hardij J, Cecchet F, Berquand A, Gheldof D, Chatelain C, Mullier F, Chatelain B, Dogné J-M (2013) Characterisation of tissue factor-bearing extracellular vesicles with AFM: comparison of air-tapping-mode AFM and liquid peak force AFM. J Extracell Vesicle 2:21045

Harding C, Stahl P (1983) Transferrin recycling in reticulocytes: pH and iron are important determinants of ligand binding and processing. Biochem Biophys Res Commun 113:650–658

He M, Crow J, Roth M, Zeng Y, Godwin AK (2014) Integrated immunoisolation and protein analysis of circulating exosomes using microfluidic technology. Lab Chip 14:3773–3780

Headland SE, Jones HR, D'Sa ASV, Perretti M, Norling LV (2014) Cutting-edge analysis of extracellular microparticles using imagestreamx imaging flow cytometry. Sci Rep 4:5237

Hong CS, Muller L, Boyiadzis M, Whiteside TL (2014) Isolation and characterization of CD34+ blast-derived exosomes in acute myeloid leukemia. PLoS One 9:e103310

Hoo CM, Starostin N, West P, Mecartney ML (2008) A comparison of atomic force microscopy (AFM) and dynamic light scattering (DLS) methods to characterize nanoparticle size distributions. J Nanopart Res 10:89–96

Huang X, Yuan T, Tschannen M, Sun Z, Jacob H, Meijun D, Liang M, Dittmar RL, Liu Y, Liang M (2013) Characterization of human plasma-derived exosomal RNAs by deep sequencing. BMC Genomics 14:1–14

Hurwitz SN, Meckes DG (2017) An adaptable polyethylene glycol-based workflow for proteomic analysis of extracellular vesicles. Methods Mol Biol 1660:303–317

Ibrahim AG-E, Cheng K, Marbán E (2014) Exosomes as critical agents of cardiac regeneration triggered by cell therapy. Stem Cell Rep 2:606–619

Izzo A, Kamieniarz K, Schneider R (2008) The histone H1 family: specific members, specific functions? vol 389, p 333

Jeppesen DK, Hvam ML, Primdahl-Bengtson B, Boysen AT, Whitehead B, Dyrskjøt L, Ørntoft TF, Howard KA, Ostenfeld MS (2014) Comparative analysis of discrete exosome fractions obtained by differential centrifugation. J Extracell Vesicle 3:25011

Kalluri R, LeBleu VS (2020) The biology, function, and biomedical applications of exosomes. Science 367:eaau6977

Kanwar SS, Dunlay CJ, Simeone DM, Nagrath S (2014) Microfluidic device (ExoChip) for on-chip isolation, quantification and characterization of circulating exosomes. Lab Chip 14:1891–1900

Katsuda T, Kosaka N, Takeshita F, Ochiya T (2013) The therapeutic potential of mesenchymal stem cell-derived extracellular vesicles. Proteomics 13:1637–1653

Khodakov D, Wang C, Zhang DY (2016) Diagnostics based on nucleic acid sequence variant profiling: PCR, hybridization, and NGS approaches. Adv Drug Deliv Rev 105:3–19

Konoshenko MY, Lekchnov EA, Vlassov AV, Laktionov PP (2018) Isolation of extracellular vesicles: general methodologies and latest trends. Biomed Res Int 2018:8545347

Kosaka N, Iguchi H, Hagiwara K, Yoshioka Y, Takeshita F, Ochiya T (2013) Neutral sphingomyelinase 2 (nSMase2)-dependent exosomal transfer of angiogenic microRNAs regulate cancer cell metastasis. J Biol Chem 288:10849–10859

Kowalski A, Pałyga J (2012) Linker histone subtypes and their allelic variants. Cell Biol Int 36: 981–996

Kuo M-H, Allis CD (1998) Roles of histone acetyltransferases and deacetylases in gene regulation. BioEssays 20:615–626

Kurian TK, Banik S, Gopal D, Chakrabarti S, Mazumder N (2021) Elucidating methods for isolation and quantification of exosomes: a review. Mol Biotechnol 63:249–266

Lai JJ, Chau ZL, Chen S-Y, Hill JJ, Korpany KV, Liang N-W, Lin L-H, Lin Y-H, Liu JK, Liu Y-C (2022) Exosome processing and characterization approaches for research and technology development. Adv Sci 9:2103222

Lai RC, Arslan F, Lee MM, Sze NSK, Choo A, Chen TS, Salto-Tellez M, Timmers L, Lee CN, El Oakley RM (2010) Exosome secreted by MSC reduces myocardial ischemia/reperfusion injury. Stem Cell Res 4:214–222

Lane RE, Korbie D, Anderson W, Vaidyanathan R, Trau M (2015) Analysis of exosome purification methods using a model liposome system and tunable-resistive pulse sensing. Sci Rep 5: 7639

Lawrie AS, Albanyan A, Cardigan RA, Mackie IJ, Harrison P (2009) Microparticle sizing by dynamic light scattering in fresh-frozen plasma. Vox Sang 96:206–212

Lawrie CH, Gal S, Dunlop HM, Pushkaran B, Liggins AP, Pulford K, Banham AH, Pezzella F, Boultwood J, Wainscoat JS (2008) Detection of elevated levels of tumour-associated microRNAs in serum of patients with diffuse large B-cell lymphoma. Br J Haematol 141: 672–675

Lee K, Shao H, Weissleder R, Lee H (2015) Acoustic purification of extracellular microvesicles. ACS Nano 9:2321–2327

Li L-M, Liu H, Liu X-H, Hong-Bin H, Liu S-M (2019) Clinical significance of exosomal miRNAs and proteins in three human cancers with high mortality in China. Oncol Lett 17:11–22

Li P, Kaslan M, Lee SH, Yao J, Gao Z (2017) Progress in exosome isolation techniques. Theranostics 7:789

Ma R, Jiang T, Kang X (2012) Circulating microRNAs in cancer: origin, function and application. J Exp Clin Cancer Res 31:1–9

Maas SLN, De Vrij J, Van Der Vlist EJ, Geragousian B, Van Bloois L, Mastrobattista E, Schiffelers RM, Wauben MHM, Broekman MLD, Nolte ENM (2015) Possibilities and limitations of current technologies for quantification of biological extracellular vesicles and synthetic mimics. J Control Release 200:87–96

Marzluff WF, Gongidi P, Woods KR, Jin J, Maltais LJ (2002) The human and mouse replication-dependent histone genes. Genomics 80:487–498

Mitchell PS, Parkin RK, Kroh EM, Fritz BR, Wyman SK, Pogosova-Agadjanyan EL, Peterson A, Noteboom J, O'Briant KC, Allen A (2008) Circulating microRNAs as stable blood-based markers for cancer detection. Proc Natl Acad Sci 105:10513–10518

Momen-Heravi F, Balaj L, Alian S, Trachtenberg AJ, Hochberg FH, Skog J, Kuo WP (2012) Impact of biofluid viscosity on size and sedimentation efficiency of the isolated microvesicles. Front Physiol 3:162

Nelson GM, Guynn JM, Chorley BN (2017) 'Procedure and key optimization strategies for an automated capillary electrophoretic-based immunoassay method. J Vis Exp 2017(127):55911

Ohno S-i, Takanashi M, Sudo K, Ueda S, Ishikawa A, Matsuyama N, Fujita K, Mizutani T, Ohgi T, Ochiya T (2013) Systemically injected exosomes targeted to EGFR deliver antitumor microRNA to breast cancer cells. Mol Ther 21:185–191

Ono M, Kosaka N, Tominaga N, Yoshioka Y, Takeshita F, Takahashi R-u, Yoshida M, Tsuda H, Tamura K, Ochiya T (2014) Exosomes from bone marrow mesenchymal stem cells contain a microRNA that promotes dormancy in metastatic breast cancer cells. Sci Signal 7:ra63-ra63

Orozco AF, Lewis DE (2010) Flow cytometric analysis of circulating microparticles in plasma. Cytometry A 77:502–514

Pan BT, Johnstone RM (1983) Fate of the transferrin receptor during maturation of sheep reticulocytes in vitro: selective externalization of the receptor. Cell 33:967

Parisse P, Rago I, Ulloa Severino L, Perissinotto F, Ambrosetti E, Paoletti P, Ricci M, Beltrami AP, Cesselli D, Casalis L (2017) Atomic force microscopy analysis of extracellular vesicles. Eur Biophys J 46:813–820

Patko D, Gyorgy B, Nemeth A, Szabó-Taylor KE, Kittel A, Buzás EI, Horvath R (2013) Label-free optical monitoring of surface adhesion of extracellular vesicles by grating coupled interferometry. Sensors Actuators B Chem 188:697–701

Pegtel DM, Cosmopoulos K, Thorley-Lawson DA, van Eijndhoven MAJ, Hopmans ES, Lindenberg JL, de Gruijl TD, Würdinger T, Middeldorp JM (2010) Functional delivery of viral miRNAs via exosomes. Proc Natl Acad Sci 107:6328–6333

Perkel JM (2013) Visiting "noncodarnia". In: Future Science, vol 54, p 301

Peterson MF, Otoc N, Sethi JK, Gupta A, Antes TJ (2015) Integrated systems for exosome investigation. Methods 87:31–45

Popp C, Dean W, Feng S, Cokus SJ, Andrews S, Pellegrini M, Jacobsen SE, Reik W (2010) Genome-wide erasure of DNA methylation in mouse primordial germ cells is affected by AID deficiency. Nature 463:1101–1105

Pospichalova V, Svoboda J, Dave Z, Kotrbova A, Kaiser K, Klemova D, Ilkovics L, Hampl A, Crha I, Jandakova E (2015) Simplified protocol for flow cytometry analysis of fluorescently labeled exosomes and microvesicles using dedicated flow cytometer. J Extracell Vesicle 4: 25530

Rabinowits G, Gerçel-Taylor C, Day JM, Taylor DD, Goetz H (2009) Exosomal microRNA: a diagnostic marker for lung cancer. Clin Lung Cancer 10:42–46

Ramirez SH, Andrews AM, Paul D, Pachter JS (2018) Extracellular vesicles: mediators and biomarkers of pathology along CNS barriers. Fluids Barriers CNS 15:1–21

Raposo G, Stoorvogel W (2013) Extracellular vesicles: exosomes, microvesicles, and friends. J Cell Biol 200:373–383

Ratajczak J, Wysoczynski M, Hayek F, Janowska-Wieczorek A, Ratajczak MZ (2006) Membrane-derived microvesicles: important and underappreciated mediators of cell-to-cell communication. Leukemia 20:1487–1495

Rider MA, Hurwitz SN, Meckes DG (2016) ExtraPEG: a polyethylene glycol-based method for enrichment of extracellular vesicles. Sci Rep 6:1–14

Robertson KD, Uzvolgyi E, Liang G, Talmadge C, Sumegi J, Gonzales FA, Jones PA (1999) The human DNA methyltransferases (DNMTs) 1, 3a and 3b: coordinate mRNA expression in normal tissues and overexpression in tumors. Nucleic Acids Res 27:2291–2298

Sabapatha A, Gercel-Taylor C, Taylor DD (2006) Specific isolation of placenta-derived exosomes from the circulation of pregnant women and their Immunoregulatory consequences 1. Am J Reprod Immunol 56:345–355

Schiera G, Di Liegro CM, Saladino P, Pitti R, Savettieri G, Proia P, Di Liegro I (2013) Oligodendroglioma cells synthesize the differentiation-specific linker histone H1 and release it into the extracellular environment through shed vesicles. Int J Oncol 43:1771–1776

Shah MY, Calin GA (2013) The mix of two worlds: non-coding RNAs and hormones. Nucleic Acid Therapeutics 23:2–8

Shao H, Chung J, Issadore D (2016) Diagnostic technologies for circulating tumour cells and exosomes. Biosci Rep 36:e00292

Sharma A (2014) Bioinformatic analysis revealing association of exosomal mRNAs and proteins in epigenetic inheritance. J Theor Biol 357:143–149

Sharma S, Zuñiga F, Rice GE, Perrin LC, Hooper JD, Salomon C (2017) Tumor-derived exosomes in ovarian cancer–liquid biopsies for early detection and real-time monitoring of cancer progression. Oncotarget 8:104687

Sharma S, Rasool HI, Palanisamy V, Mathisen C, Schmidt M, Wong DT, Gimzewski JK (2010) Structural-mechanical characterization of nanoparticle exosomes in human saliva, using correlative AFM, FESEM, and force spectroscopy. ACS Nano 4:1921–1926

Shen L, Hao W, Diep D, Yamaguchi S, D'Alessio AC, Fung H-L, Zhang K, Zhang Y (2013) Genome-wide analysis reveals TET-and TDG-dependent 5-methylcytosine oxidation dynamics. Cell 153:692–706

Shimbo K, Miyaki S, Ishitobi H, Kato Y, Kubo T, Shimose S, Ochi M (2014) Exosome-formed synthetic microRNA-143 is transferred to osteosarcoma cells and inhibits their migration. Biochem Biophys Res Commun 445:381–387

Skoog DA, Holler FJ, Crouch SR (2017) Principles of instrumental analysis. Cengage Learning

Skotland T, Sandvig K, Llorente A (2017) Lipids in exosomes: current knowledge and the way forward. Prog Lipid Res 66:30–41

Smith ZJ, Lee C, Rojalin T, Carney RP, Hazari S, Knudson A, Lam K, Saari H, Ibañez EL, Viitala T (2015) Single exosome study reveals subpopulations distributed among cell lines with variability related to membrane content. J Extracell Vesicle 4:28533

Snow AD, Lara S, Nochlin D, Wight TN (1989) Cationic dyes reveal proteoglycans structurally integrated within the characteristic lesions of Alzheimer's disease. Acta Neuropathol 78:113–123

Sokolova V, Ludwig A, Hornung S, Rotan O, Horn PA, Epple M, Giebel B (2011) Colloids and surfaces B: biointerfaces characterisation of exosomes derived from human cells by nanoparticle tracking analysis and scanning electron microscopy. Colloids Surf B Biointerfaces 87:146–150

Sparks DL, Phillips MC (1992) Quantitative measurement of lipoprotein surface charge by agarose gel electrophoresis. J Lipid Res 33:123–130

Suárez H, Gámez-Valero A, Reyes R, López-Martín S, Rodríguez MJ, Carrascosa JL, Cabañas C, Borràs FE, Yáñez-Mó M (2017) A bead-assisted flow cytometry method for the semi-quantitative analysis of extracellular vesicles. Sci Rep 7:11271

Szatanek R, Baj-Krzyworzeka M, Zimoch J, Lekka M, Siedlar M, Baran J (2017) The methods of choice for extracellular vesicles (EVs) characterization. Int J Mol Sci 18:1153

Takov K, Yellon DM, Davidson SM (2019) Comparison of small extracellular vesicles isolated from plasma by ultracentrifugation or size-exclusion chromatography: yield, purity and functional potential. J Extracell Vesicle 8:1560809

Tauro BJ, Greening DW, Mathias RA, Ji H, Mathivanan S, Scott AM, Simpson RJ (2012) Comparison of ultracentrifugation, density gradient separation, and immunoaffinity capture methods for isolating human colon cancer cell line LIM1863-derived exosomes. Methods 56: 293–304

Tauro BJ, Mathias RA, Greening DW, Gopal SK, Ji H, Kapp EA, Coleman BM, Hill AF, Kusebauch U, Hallows JL (2013) Oncogenic H-ras reprograms Madin-Darby canine kidney (MDCK) cell-derived exosomal proteins following epithelial-mesenchymal transition. Mol Cell Proteomics 12:2148–2159

Taverna S, Amodeo V, Saieva L, Russo A, Giallombardo M, De Leo G, Alessandro R (2014) Exosomal shuttling of miR-126 in endothelial cells modulates adhesive and migratory abilities of chronic myelogenous leukemia cells. Mol Cancer 13:1–15

Taylor DD, Gercel-Taylor C (2008) MicroRNA signatures of tumor-derived exosomes as diagnostic biomarkers of ovarian cancer. Gynecol Oncol 110:13–21

Théry C (2011) Exosomes: secreted vesicles and intercellular communications. F1000 Bio Rep 3:15

Théry C, Amigorena S, Raposo G, Clayton A (2006) Isolation and characterization of exosomes from cell culture supernatants and biological fluids. Curr Protoc Cell Biol 30:3.22.1–3.22.29

Théry C, Witwer KW, Aikawa E, Alcaraz MJ, Anderson JD, Andriantsitohaina R, Antoniou A, Arab T, Archer F, Atkin-Smith GK (2018) Minimal information for studies of extracellular vesicles 2018 (MISEV2018): a position statement of the International Society for Extracellular Vesicles and update of the MISEV2014 guidelines. J Extracell Vesicle 7:1535750

Tominaga N, Kosaka N, Ono M, Katsuda T, Yoshioka Y, Tamura K, Lötvall J, Nakagama H, Ochiya T (2015) Brain metastatic cancer cells release microRNA-181c-containing extracellular vesicles capable of destructing blood–brain barrier. Nat Commun 6:6716

Trams EG, Lauter CJ, Salem JN, Heine U (1981) Exfoliation of membrane ecto-enzymes in the form of micro-vesicles. Biochim Biophys Acta 645:63–70

Van der Pol E, Coumans FAW, Grootemaat AE, Gardiner C, Sargent IL, Harrison P, Sturk A, Van Leeuwen TG, Nieuwland R (2014) Particle size distribution of exosomes and microvesicles determined by transmission electron microscopy, flow cytometry, nanoparticle tracking analysis, and resistive pulse sensing. J Thromb Haemost 12:1182–1192

Deun V, Jan PM, Sormunen R, Cocquyt V, Vermaelen K, Vandesompele J, Bracke M, De Wever O, Hendrix A (2014) The impact of disparate isolation methods for extracellular vesicles on downstream RNA profiling. J Extracell Vesicle 3:24858

Vogel R, Pal AK, Jambhrunkar S, Patel P, Thakur SS, Reátegui E, Parekh HS, Saá P, Stassinopoulos A, Broom MF (2017) High-resolution single particle zeta potential characterisation of biological nanoparticles using tunable resistive pulse sensing. Sci Rep 7: 17479

Wang L, Skotland T, Berge V, Sandvig K, Llorente A (2017) Exosomal proteins as prostate cancer biomarkers in urine: from mass spectrometry discovery to immunoassay-based validation. Eur J Pharm Sci 98:80–85

Wang Z, Hung-jen W, Fine D, Schmulen J, Ye H, Godin B, Zhang JXJ, Liu X (2013) Ciliated micropillars for the microfluidic-based isolation of nanoscale lipid vesicles. Lab Chip 13:2879–2882

Watson DC, Yung BC, Bergamaschi C, Chowdhury B, Bear J, Stellas D, Morales-Kastresana A, Jones JC, Felber BK, Chen X (2018) Scalable, cGMP-compatible purification of extracellular vesicles carrying bioactive human heterodimeric IL-15/lactadherin complexes. J Extracell Vesicle 7:1442088

Welton JL, Webber JP, Botos L-A, Jones M, Clayton A (2015) Ready-made chromatography columns for extracellular vesicle isolation from plasma. J Extracell Vesicle 4:27269

Weng Y, Sui Z, Shan Y, Yechen H, Chen Y, Zhang L, Zhang Y (2016) Effective isolation of exosomes with polyethylene glycol from cell culture supernatant for in-depth proteome profiling. Analyst 141:4640–4646

Witwer KW, Buzás EI, Bemis LT, Bora A, Lässer C, Lötvall J, Esther N, Nolte-'t Hoen, Melissa G Piper, Sarada Sivaraman, and Johan Skog (2013) Standardization of sample collection, isolation and analysis methods in extracellular vesicle research. J Extracell Vesicle 2:20360

Wolf P (1967) The nature and significance of platelet products in human plasma. Br J Haematol 13: 269–288

Wu Y, Deng W, Klinke II DJ (2015) Exosomes: improved methods to characterize their morphology, RNA content, and surface protein biomarkers. Analyst 140:6631–6642

Yáñez-Mó M, Siljander PR-M, Andreu Z, Zavec AB, Borràs FE, Buzas EI, Buzas K, Casal E, Cappello F, Carvalho J (2015) Biological properties of extracellular vesicles and their physiological functions. J Extracell Vesicle 4:27066

Yokoyama S, Takeuchi A, Yamaguchi S, Mitani Y, Watanabe T, Matsuda K, Hotta T, Shively JE, Yamaue H (2017) Clinical implications of carcinoembryonic antigen distribution in serum exosomal fraction—measurement by ELISA. PLoS One 12:e0183337

Yuana Y, Oosterkamp TH, Bahatyrova S, Brian Ashcroft P, Rodriguez G, Bertina RM, Osanto S (2010) Atomic force microscopy: a novel approach to the detection of nanosized blood microparticles. J Thromb Haemost 8:315–323

Zaborowski MŁP, Balaj L, Breakefield XO, Lai CP (2015) Extracellular vesicles: composition, biological relevance, and methods of study. Bioscience 65:783–797

Zarovni N, Corrado A, Guazzi P, Zocco D, Lari E, Radano G, Muhhina J, Fondelli C, Gavrilova J, Chiesi A (2015) Integrated isolation and quantitative analysis of exosome shuttled proteins and nucleic acids using immunocapture approaches. Methods 87:46–58

Zeringer E, Barta T, Li M, Vlassov AV (2015) Strategies for isolation of exosomes. Cold Spring Harb Protoc 2015:pdb. top074476

Zhang Y, Liu Y, Liu H, Tang WH (2019) Exosomes: biogenesis, biologic function and clinical potential. Cell Biosci 9:1–18

Zhu X, You Y, Li Q, Zeng C, Fu F, Guo A, Zhang H, Zou P, Zhong Z, Wang H (2014) BCR-ABL1–positive microvesicles transform normal hematopoietic transplants through genomic instability: implications for donor cell leukemia. Leukemia 28:1666–1675

Zlatanova J, Doenecke D (1994) Historie H1°: a major player in cell differentiation? FASEB J 8: 1260–1268

Zomer A, Vendrig T, Hopmans ES, van Eijndhoven M, Middeldorp JM, Michiel D, Pegtel. (2010) Exosomes: fit to deliver small RNA. Commun Integr Biol 3:447–450

8

Stem Cells Vs Exosomes: Promising Therapeutic Approach and Biomarkers Agent against Neurodegenerative Disorders

Johra Khan and I. Irem Tatli

Abstract

Neurodegenerative disorders in older population are one of biggest cause of death and related disabilities causing huge burden on health systems. Dose-dependent and short-term pharmacological treatment produces many side effects forcing the medical community to develop effective treatment. Some researches proposed a novel therapeutic approach using stem cells to treat different neurodegenerative diseases like Parkinson's disease, stroke, and Alzheimer's disease. Stem cells and their secretions are found effective as a therapeutic atmosphere for treating neurodegenerative diseases. Extracellular vesicles like exosomes released from stem cells were found to be effective in targeting specific cell types and to modify certain proteins, lipids, and nucleic acids. In this chapter we discuss different therapeutic applications of stem cells and exosomes and related biomarkers against neurodegenerative disorders.

Keywords

Stem cells · Exosomes · Alzheimer's disease · Parkinson's disease · Blood-brain barriers · Neurodegenerative disease

J. Khan (✉)
Department of Medical Laboratory Sciences, College of Applied Medical Sciences, Majmaah University, Al-Majmaah, Saudi Arabia
e-mail: j.khan@mu.edu.sa

I. I. Tatli
Department of Pharmaceutical Botany, Faculty of Pharmacy, Hacettepe University, Ankara, Turkey

S. Jahan, A. J. Siddiqui (eds.), *Applications of Stem Cells and Derived Exosomes in Neurodegenerative Disorders*, https://doi.org/10.1007/978-981-99-3848-3_8

8.1 Introduction

In the twentieth century, the rapid increase in the elderly population due to the developments in medicine and the increase in the average age of the individuals who take an active role in society have made dementia an important health problem (Loy et al. 2014). Dementia is a diagnosis made in the case of the clinical integrity of symptoms that develop due to damage to the central nervous system without acute unconsciousness or delirium, affect the daily life of the patient, and progress progressively with destruction in at least two cognitive areas.

Vascular Dementia (VD) It is the second most known type of dementia next to Alzheimer's disease (AD), accounting for 20% of dementia cases. VD is a progressive disease that results from decreased blood flow to the brain. VDs include a heterogeneous group of dementias and can be classified according to the characteristics of the affected vessel, the location of the lesion, the number of lesions, the temporal relationship with stroke, the type of damage, the course of the disease, and whether it is accompanied by other degenerative pathologies (Scheltens et al. 2016). Disease factors such as hypertension (HT), diabetes mellitus (DM), and hyperlipidemia increase the development of VD as in cerebrovascular diseases.

Parkinson's Disease Dementia Idiopathic Parkinson's disease (PD) is an advanced neurodegenerative disease, manifested by latent tremor, rigidity, bradykinesia, and postural impairment (Beitz 2014). Deterioration of neurons in the compacta part of the substantia nigra and the presence of Lewy bodies in their cytoplasm constitute the classical pathological findings of the disease. During the yearslong course of PD, a picture of dementia may emerge and this picture may be severe and restrictive, overshadowing the movement disorder of the disease (Tysnes and Storstein 2017). Dementia is defined as the occurrence of losses in more than one cognitive area such as attention, memory, language, executive functions, praxis, and visuospatial functions, these losses express a significant decrease from the previous level, and this decrease is severe enough to affect daily, professional, and social life. PD dementia is mild or moderate dementia that starts insidiously, progresses slowly, and affects some areas of cognitive functions, especially executive functions, and psychosis often develops during its course.

Huntington's Disease Dementia (HHD) Huntington's disease (HD) dementia is an autosomal dominant, progressive, and fatal neurodegenerative disease. The disease gene was isolated in 1993. The gene localized to the short arm of chromosome 4 is referred to as IT15 and encodes the Huntington protein (Roos 2010). In HD dementia, there is an increase in the "cytosine–adenine–guanine (CAG)" repeat sequence in this protein more than normal. Although there is an inverse correlation between the increase in CAG and the age of onset of the disease, there is no correlation between clinical findings and the number of CAG repeats (Ross and Tabrizi 2011). Choreic movements, cognitive disorders causing dementia, and psychiatric findings are the main features of the clinical picture. In addition, for sleep and circadian rhythm disorders, signs of autonomic involvement can also be seen (Frank 2014).

1. Dementia caused by Down syndrome: Down syndrome is the most common genetic cause of mental retardation. Many people with Down syndrome experience health problems associated with dementia in middle and old age.
2. Mixed dementia.
3. Dementia with Lewy bodies.
4. Creutzfeldt–Jakob disease.
5. Frontotemporal dementia.
6. Normal-pressure hydrocephalus.
7. Korsakoff syndrome.
8. Posterior cortical atrophy.

8.2 Alzheimer's Disorder

Alzheimer's disorder is a very complex disease categorized by severe synaptic losses and neuronal death, especially in regions with cognitive functions such as the cerebral cortex, hippocampus, entorhinal cortex, and ventral striatum. Generally, in an average of 10 years, the stage of minor cognitive diminishing is passed to the advanced stage of AD, and the patient is lost in a completely helpless state at the end of this period. Due to this long duration of the disease and the fact that it affects the vital structures that determine who we are, it creates a great emotional and financial load on patients' relatives and people (Ramirez-Bermudez 2012). The German psychiatrist and neuropathologist Alois Alzheimer (June 14, 1864, to December 19, 1915) first published the case of "presenile dementia," and later this disease was defined as "Alzheimer's disease" by the famous psychiatrist Emil Kraepelin. During his studies in Frankfurt, Alzheimer encountered 51-year-old Auguste Deter, who had strange behavior and short-term memory loss. When Deter died in April 1906, he had his file and his brain brought to Kraepelin's laboratory in Munich for autopsy, where new dyeing techniques were used in the patient's brain, and then he detected "amyloid plaques" and "neurofibrillary tangles (NFTs)." In November 1906, he presented the pathology and clinical symptoms of the disease for the first time in congress. Since the pioneering work of Alois Alzheimer in 1907, neuropathologists have recognized amyloid plaques and NFTs in brains of patients in autopsy examinations and stated that these pathologies causes disease (Anand et al. 2014). Amyloid plaques have been found to be extracellular deposits related to amyloid-beta (Aβ) found inside the brain parenchyma and cerebral blood vessels (Golde et al. 2011). It has been determined that the NFTs observed in the cell are composed of hyper-phosphorylated tau protein associated with microtubules and clustered in helical filaments (Palmer 2011). Additional pathological data for amyloid plaques and NFTs can be listed as intracellular granulovacuolar degeneration, decrease in the number of synapses, cholinergic cell losses in Meynert's basal nucleus, and astroglial activation (Lou et al. 2014) (Fig. 8.1).

8.2.1 Alzheimer's Disease: Genetics, Epigenetics, and Epidemiology

Studies conducted to understand AD indicate that the disease arises as a result of complex interactions of many genetic, epigenetic, and environmental factors. The Aβ peptide, which plays an important role in the pathogenesis of the disease, is formed as a result of the cleavage of amyloid precursor protein (APP) by secretases (Bakulski et al. 2012). APP is a protein with a single transmembrane domain expressed in almost every cell. Cleavage of this protein is via α-secretase-soluble APP; successive cleavage via β- and gamma-secretase leads to the formation of Aβ peptides. Alterations in the genes responsible for proteins including presenilin-1 (PS1) and presenilin-2 (PS2), which are portions of the APP and gamma-secretase enzyme complex, cause early-onset autosomal dominant AD, which constitutes 1% of the disease (Bertram et al. 2010).

These mutations cause a change in the type and rate of amyloid peptides produced by affecting molecular processes such as cleavage of APP from the region related to secretases (Bertram et al. 2010; Bekris et al. 2011). The fact that early-onset dementia and pathological markers of AD are observed in Down syndrome patients who carry a spare copy of chromosome number 21 with the APP gene and that replication of the APP gene also causes early-onset AD supports the hypotheses regarding the role of APP overexpression in the disease (Sanchez et al. 2012; Chouraki and Seshadri 2014). In addition, extensive genetic studies performed in different populations show that carriers of the apolipoprotein E (ApoE) 4 allele are associated with late-onset AD cases and reduce the age at which the disease is diagnosed (Crean et al. 2011; Holtzman et al. 2012).

The epigenetic method also performs a part in the pathogenesis of AD (Day and Sweatt 2011). Researches in postmortem brain tissues, peripheral leukocytes, and transgenic animals in humans have revealed the presence of epigenetic changes such as atypical DNA methylation and histone modification in AD (Chouliaras et al. 2010). However, it is uncertain whether these observed epigenetic variations are a cause of the disease or a result of developing pathological processes. Studies performed on twins indicate that epigenetic devices modulate disease risk (Chouliaras et al. 2010; Gouras et al. 2010). Inhibition of DNA methylation in the hippocampus immediately after a learning test impaired memory consolidation in healthy mice, while potentiation of histone acetylation promoted learning in Alzheimer's transgenic mice (Gouras et al. 2010; Peleg et al. 2010). The results point to the character of epigenetic alterations in normal learning and memory methods as well as in disease. Increasing age is the most important epidemiological hazardous factor in delayed-onset AD. Although other probable risk factors have been identified as head trauma, low education level, hyperlipidemia, HT, DM, homocysteinemia, and obesity, it should be noted that there are conflicting results

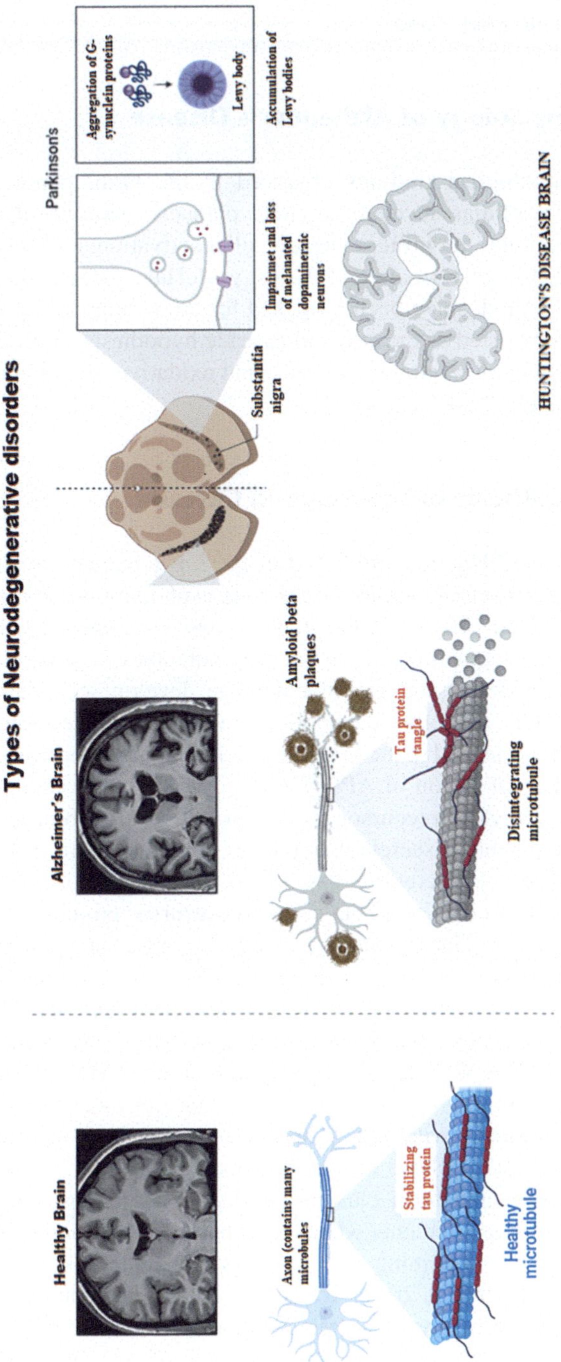

Fig. 8.1 Different types of neurodegenerative disorders (Biorender)

about some of them (Daviglus et al. 2010; Sharp and Gatz 2011; Sivanandam and Thakur 2012; Douaud et al. 2013).

8.2.2 Pathophysiology of Alzheimer's Disease

The main histopathological findings observed in the brain parenchyma of the patients have extracellular located amyloid plaques, neurofibrillary structures consisting of intracellular tau protein clusters, glial activation, and traces of inflammation (Bloom 2014; Silverman and Krinsky-McHale 2021). Based on these symptoms, many methods have been proposed for the pathogenesis of the disease. The main ones can be listed as the amyloid cascade hypothesis, cholinergic damage hypothesis, neuronal cytoskeleton hypothesis, and oxidative stress hypothesis (Jiang et al. 2012; Sharma et al. 2019).

8.2.3 The Hypothesis of the Amyloid Cascade

The study of Hardy and Higgins, published in 1992, is a pioneering study in which they put forward the amyloid cascade hypothesis explaining the neuropathological mechanism of AD (Reitz 2012). In the study, it has been suggested that the main neuropathological mechanism of AD is the accumulation of Aβ peptides, and as a result, NFTs, neuron death, and vascular damage develop secondary to amyloid deposition. Aβ peptides and other metabolites are formed by the proteolytic degradation of APP. APP is cleaved in the first step by either α- or β-secretase enzymes. α-Secretase affects the Aβ region of APP, forming the 83 amino acid sequence C83 and the α-secretory amyloid precursor protein (α-APPs) (Tayeb et al. 2013; Crary 2016). In a healthy person, α-secretase activity is dominant in the brain and α-APP release is higher. The beta-secretase enzyme, on the other hand, affects the amino terminus of APP to form C99 and beta-amyloid precursor protein (β-APPs) with a 99 amino acid sequence. Gamma secretase, which comes into play, later on, is a protease complex consisting of many subunits and has not been fully characterized, but it is known that it basically consists of four important proteins. These are PS1, PS2, nicastrin, and "anterior pharynx-defective-1 (APH-1)" proteins. This enzyme affects both C83 and C99, forming the intercellular domain APP with its peptide P3 from C83 and Aβ from C99. Aβ peptides can have 38, 40, and 42 amino acid sequences and accumulate with the "β-sheet" conformation (Kumar and Walter 2011; Ivanova et al. 2021). Since amyloid deposits are in the "β-sheet" structure, they polymerize easily and form clusters rapidly. These deposits have very low solubility and are seen as filaments when examined under a microscope and show cytotoxic properties by combining with other proteins such as proteoglycan and apolipoproteins (Fändrich et al. 2011). It has been determined that genetic mutations in PS proteins in the structure of secretases that control APP and Aβ processes increase the level of Aβ42. In the early occurrence of AD, autosomal dominant mutations in PS1, PS2, or APP must be found to affect APP metabolism and lead to

Aβ accumulation and aggregation (Atsmon-Raz and Miller 2016). Manifestation of mutant APP in humans alone or in combination with transformed PS1 in transgenic rats triggered AH-like molecular and cognitive changes. It is thought that the Aβ peptide plays a role in the regulation of neuronal and synaptic activity; therefore, its accumulation leads to synaptic depression with excessive neuronal activity (Bertram et al. 2010; Palop and Mucke 2010). It is known that functional impairments in inhibitory interneurons and excessive glutamatergic transmission causing excitotoxicity are important mechanisms that play a role in the pathogenic cascade. This excessive neuronal activity may subsequently lead to the production of more Aβ peptides and trigger a neurodegenerative cycle (Suzuki et al. 1994). For example, it regulates gamma-secretase traffic by binding to PS1. It has been determined that the activation of the Arc (activity-regulated cytoskeleton-associated protein) gene plays a role in the production of Aβ, which occurs depending on neuronal activity (Wu et al. 2011). It is also known that Aβ forms fibrils and activates local microglia and astrocytes, and molecules released from these cells create neurotoxic effects on neurons. Based on the evidence from experimental models, it is suggested that soluble, non-fibrillar Aβ, which forms dimers, trimers, and larger oligomers, is less pathogenic than monomeric Aβ and Aβ fibrils in amyloid plaques (Chowdhury et al. 2006). However, which amyloid deposits are more disease causing and the way their buildup triggers synaptic and neuronal dysfunction are questions that are still not fully answered by intensive studies and scientific discussions. It has been determined that these structures can effectively bond with proteins and lipids both inside and outside the cell. Many cell surface molecules, such as receptor tyrosine kinases and advanced glycation end product receptors, play a role in Aβ oligomer-induced toxicity (Cissé et al. 2011; Srikanth et al. 2011). The potential mechanisms triggered by the continuation of these interactions include the distribution of neurotransmitter receptors and related signal molecules, changes in their activities, disruption of Ca^{+2} homeostasis, axonal transport, and mitochondrial functions (Mruthinti et al. 2004).

8.2.4 Hypothesis of Neuronal Cell Skeleton

Alzheimer's is a disease in which many proteins show pathogenic conformation. In addition to abnormal amyloid plaque accumulation, intracellular neurofibrillary clusters (neurofilament tangles, NFT) are observed in AH. It has been found that the main component of these clusters is the tau protein, which shows irregular posttranslational alterations such as enlarged acetylation and phosphorylation (Cohen et al. 2011). In the disease-associated neuronal cytoskeleton hypothesis, it is suggested that these tau protein-related changes and subsequent NFTs have a causal role in inducing AD pathology. Tau protein is a protein found mostly in microtubules in neurons and is involved in the controlling of polymerization, depolymerization, and stabilization processes of microtubules (Cohen et al. 2011). It has been shown that Aβ peptides activate intracellular kinases such as glycogen synthase kinase and cause hyper-phosphorylation of tau proteins in microtubules, triggering NFT accumulations, and this process causes structural–functional changes in neuron soma and dendrites. Recent studies point out that tau proteins have roles in various cellular mechanisms apart from the stabilization of microtubules, which is their main function, and it is suggested that tau aggregation may contribute to the pathogenesis of the disease by disrupting these mechanisms (Morris et al. 2011). As a result of examinations in cell culture and mice, it was found that tau protein facilitates excitatory transmission by regulating the distribution of signal molecules that are effective in synaptic activity. However, it is thought that tau with abnormal conformation may accumulate in dendritic spins and disrupt normal nerve conduction (Hoover et al. 2010). It has been shown that Aβ oligomers interact with the MARK enzyme family to increase the accumulation of tau in the postsynaptic membrane, and in parallel, it has been found that reducing the level of tau in cells or in transgenic animals reduces the neuronal dysfunction caused by Aβ (Zempel et al. 2010). It has also been shown that genetic mutations induced in APP and PS lead to NFT accumulation after amyloid plaque formation. While these findings indicate that Aβ is at the top of the pathological steps, they reveal that tau protein plays a role in its neuronal effects (Zempel and Mandelkow 2012).

8.2.5 Alzheimer's Disease Cholinergic Damage Hypothesis

In the cholinergic hypothesis, it is suggested that loss of cholinergic neurons and the resulting cholinergic conduction disorder are pathophysiological mechanisms leading to AD. Although acetylcholine (ACh), an important neurotransmitter in the central nervous system, is mostly found in interneurons, it forms an important part of the cholinergic system in neurons projecting from the basal forebrain (Meynert's basal nucleus) to the cerebral cortex and limbic structures (Chen and Mobley 2019). ACh synthesis in cholinergic neurons takes place at the axon terminal by acetylating free choline via choline acetyltransferase (ChAT). The ChAT enzyme is transported from the cholinergic cell body to the axonal end, with both ChAT and ACh concentrations reaching the highest amount at the axonal terminal. Since the ChAT enzyme is found only in cholinergic neurons, it is used in studies as a marker showing these neurons (Cummings et al. 1998). Released ACh is then degraded by acetylcholinesterase in the synaptic space to choline and acetate. ACh acts through nicotinic (nAChR) and G-protein-dependent muscarinic receptors (mAChR), which are excitatory cation channels in the central nervous system. Muscarinic receptors function depending on the Gq (M1, M3, M5) proteins that activate phospholipase C or the Gi/o (M2, M4) proteins that inhibit adenylate cyclase. In addition to the different effects of the receptors, the pre- or postsynaptic distribution in certain regions enables ACh to have different effects in the central nervous system. Although this neurotransmitter is primarily excitatory in the periphery, it is a neuromodulator that coordinates cholinergic transmission, neuronal excitability, presynaptic neurotransmitter release, and neuron firing in the central nervous system (Wonnacott 1997; Zhang and Sulzer 2004; Goutier et al. 2016). As an example of these heterogeneous effects, ACh plays an inhibitory autoreceptor role in presynaptic M2/M4 mAChR cholinergic terminals, decreasing glutamate release in corticostriatal synapses, and M1/M5 mAChR stimulates dopamine release in striatal synaptosomes and increases excitability in postsynaptic M1/M5 neurons (Douglas et al. 2002).

8.2.6 Alzheimer's Disease and Oxidative Stress Hypothesis

Although the physiological cause of aging is not fully known, the free radical theory states that increasing oxidative stress of aging and aging-related diseases plays a fundamental role in this process by causing cellular degeneration (Remigante and Morabito 2022). The increase in the number of free radicals observed in age-related neurodegenerative diseases and the fact that neurons are more sensitive to this damage has been determined as important characteristics. Therefore, it is thought that free radical production has an important role in the development and progression of AD (Higley et al. 2009). It has been stated that neurons are more susceptible to free radical damage for certain reasons.

These can be listed as follows:

1. Neurons are highly dependent on the oxidative phosphorylation reactions that take place in the mitochondria to provide the necessary energy.
2. High polyunsaturated fatty acids in their membranes can be used as substrates in lipid peroxidation reactions.
3. High levels of ionic iron catalyze free radical reactions.
4. The level of glutathione, an endogenous antioxidant, is lower than in other tissues. In addition to mitochondrial oxidative phosphorylation, enzymatic conversion of catecholamines (epinephrine, norepinephrine) and indoleamines (serotonin, melatonin) by monoamine oxidase, it is known that autoxidation of catecholamines and activities of lipoxygenase and cyclooxygenase enzymes also cause free radical production. Free radicals were found to be increased in AH, a radical that includes superoxide, hydrogen peroxide, hydroxyl radical, and nitric oxide. The antioxidant defense system in neurons is very important in preventing damage caused by reactive oxygen radicals (ROS) produced in many ways. This system comprises of enzymatic and nonenzymatic antioxidants that stabilize physiological ROS making with detoxification.

The antioxidant structure can be examined under three main headings:

1. Enzymatic antioxidant system: It is known that the enzyme superoxide dismutase has three different forms—mitochondria, cytosol, and extracellular. The superoxide dismutase enzyme functions by converting $O_2{\bullet}^-$ to H_2O_2. Selenium-dependent GSH-Px and CAT enzymes are other members of the enzymatic antioxidant system. All the enzymes provide detoxification by converting H_2O_2 to water.
2. Minor molecule antioxidant system: Vitamins like E and vitamin C reduce their activities by reacting with free radicals.

3. Chelator protein antioxidant system: Proteins containing small-molecular-weight thiol collections such as glutathione. These proteins prevent reactions that catalyze free radical production by bonding with metals.

After determining that ApoE 4 allele dominance is a hereditary risk factor for the progress of AD, many studies have been conducted to examine the structural features, neurobiological functions, and pathophysiological effects of the ApoE protein in the disease. It is a polymorphic protein with the three most collective isoforms in humans, ApoE, ApoE2, ApoE3, and ApoE4. ApoE has many important roles, such as initiating the formation of lipoprotein particles and directing lipid metabolism by interacting with specific cellular receptors and functioning in the reinnervation and restructuring of neuronal cells (Huang 2006; Castellano et al. 2011). It has been determined that these roles differ in isoform-specific and that ApoE proteins have different roles according to the cell type in which they are expressed.

Pathological effects of ApoE proteins in the pathology of AD can be examined in two groups, Aβ dependent and independent. With ApoE3 being faster and more stable, both ApoE3 and E4 isoforms can formulate stable bonds with Aβ peptides (Kim et al. 2009; Castellano et al. 2011). It was determined that astroglial ApoE3 can bind to Aβ peptides 20 times more than the ApoE4 isoform. Therefore, increasing the lipid content of ApoE3 in mice carrying the mutant human APP gene increased the clearance of the peptide and decreased the amyloid level (Aleshkov et al. 1997). The ability of ApoE2 and ApoE3 to scavenge higher amounts of Aβ peptide than the ApoE4 isoform indicates the importance of the isoform type in the astroglial degradation of the deposits. When microglial Aβ clearance was measured, ApoE4 was found to reduce this clearance by 40% (Frieden and Garai 2012). However, despite these findings, the contribution of these procedures to cognitive impairment in AD has not been clarified. The main reason for this is that there is no correlation between the number of amyloid plaque deposits and cognitive symptoms in patients.

Unlike ApoE4, the ApoE3 isoform was found to be protective in age-related or excitotoxicity-related neurodegeneration, and learning and memory impairment were found in transgenic mice in which ApoE4 was expressed alone in neurons (Castellano et al. 2011). When the reasons for the observed decrease in cognitive functions were examined, it was determined in in vivo and in vitro studies that ApoE4 reduced the density of dendritic spin by disrupting synaptogenesis (Dumanis et al. 2009). In addition, it was observed that the neurogenesis of neuronal stem cells decreased in mice with induced ApoE4 expression, and it was stated that this decrease may cause cognitive impairment (Li et al. 2009). When the effects of isoforms on neurite outgrowth were examined, it was observed that ApoE3 stimulated spin branching and growth, while ApoE4 inhibited it in dorsal root ganglion cells and neuronal cultures (Li et al. 2010). In addition, unlike ApoE4, astroglial ApoE3 was found to induce spin elongation in hippocampal neurons. Since Aβ accumulation was not observed in these models, it can be said that the determined changes are independent effects of ApoE4 from Aβ. In addition, it has been determined that neuronal ApoE expression is induced in brain damage and stress situations, and the increased ApoE4 is proteolytically degraded into toxic fragments. Unlike the ApoE3 isoform, ApoE4 produced in neurons undergoes proteolysis more easily and is broken down into toxic fragments. It was observed that the level of these fragments increased in patients (Hiekkanen et al. 2007). Mice expressing these fragments have spatial learning and memory impairment, as well as degeneration of the hippocampus. In addition, it was determined that ApoE4 fragments triggered tau phosphorylation and fibrillar aggregation. ApoE4-related dysfunction was also observed in neuronal mitochondria, while more dysfunction was found in carriers of this allele than in E3 carriers (Jones et al. 2011). In conclusion, it has been determined that the ApoE4 isoform affects neuronal functions through many different mechanisms, $A\beta$ dependent and independent, and it has been stated that further studies are needed to understand the contribution of these different effects on the development and progression of the disease (Chen et al. 2012).

8.3 Stem Cell Therapy and Neurodegenerative Diseases

Stem cells as we know are known for their capacity to renew themselves and are able to differentiate into different types of cell lines. These stem cells were first revealed in the 1960s and classified into four broad classes containing embryonic stem cells (ESCs), mesenchyme stem cells (MSCs), pluripotent stem cells (PSCs), and progenitor cells. On the basis of their isolation and differentiation, they are further classified as totipotent, multipotent, and pluripotent cells (Baek et al. 2019). Multipotent stem cells and progenitor cells can be isolated using different tissues and differentiate into many types of cells but belonging to only same family. Totipotent stem cells can be isolated from embryo at a four-stage cell and differentiate into all body cell types outside the embryo (Kumar et al. 2019). The development of technology has made stem cell therapies available for the treatment of different neurodegenerative disease. Different uses of stem cell therapy for the treatment of neurodegenerative disease including Alzheimer disease, Parkinson's disease, and many others are shown in Table 8.1.

8.3.1 Stem Cell and AD Therapy

Mesenchyme stem cells from an animal model are reported to be important for the treatment of AD; they are found to help in regulating immune system, reduce Aβ plaque load by internalization and breaking down of Aβ endosomal lysosomal trail oligomers and recreating potential (Evangelista et al. 2019). In this study, injection of green fluorescent protein–labelled bone marrow–derived MSCs in the hippocampus of an AD-induced animal model revealed a decrease in the mass of Aβ plaques in the brain and also helped in regulating the functioning of immunity (Bianco et al. 2008). An immunostained brain section from MSCs transplanted with antipolysialylated from a neural cell adhesion molecule showed increased

Table 8.1 Different types of stem cells and their applications in neurodegenerative disease

Stem cell types	Origin of cells	Functions	Application in neurodegenerative diseases	References
Mesenchyme stem cells	Obtained from the adipose bone marrow and peripheral blood	Have ability to differentiate into neural linkage cells	MSCs helps in regulating immune system and reduces Aβ plaque load through internalization and Aβ breakdown of endosome–lysosome pathway oligomers and regenerative potential	(Cheng et al. 2017)
Embryonic stem cells	Originate from blastocyst innermost cell corpus	Pluripotent and self-renewing capacity and differentiate into neural stem cells	ESC can ease plaque formation and cerebral dysfunction in the a5XFAD mouse model	(Cha et al. 2017, Zhao et al. 2020)
Pluripotent stem cells	Derived from adult somatic tissues	These are alternate cells which can function as embryonic stem cell	Human iPSC-obtained macrophage-like cells hereditarily modified to show neprilysin-2 or to alter tau Ex10 + 16; break down Aβ, separating into functional neurons; and decrease Aβ levels after xenograft management to the 5XFA	(Kim et al. 2021)
Neural stem cells	Derived only from neural cells especially from fetal brain cells	Potential to differentiate into glial cells and neurons	NSCs increase the existence and renewal of endogenous neurons by creating neurotrophic factors, vascular endothelial growth factor (VEGF), and vessel density in the cortex	(Zhu et al. 2011, Jinnou 2021)

neurogenesis. Another study used PKH26–111 injected into mice with induced AD via the tail vein to reach the brain. This study confirmed a significantly higher radioactivity of bone marrow–derived MSCs in comparison to controls (Fujita et al. 2018). Mammalian brain cells have the capacity to repair by neurogenesis, but with age, there is decrease in gliogenesis and brain cells lose their power to regenerate a sufficient amount of cells to regain normal functioning. Many recent studies used NSCs which express a phenotype analogous to brain cells and are able to provide better potential for the management of AD (Shi et al. 2012). NSCs were reported to increase the life span of endogenous neurons by the production of neurotrophic factors, VEGF (vascular endothelial growth factors), and the vessel mass of the cortex also increases their regeneration capacity and becomes a resilient candidate for the treatment of AD. A study based on human sources found that NSCs function significantly to diminish cerebral Aβ42 levels.

A study by Park et al. (2010) reported that the human NS cell line coding ChAT gene transferred into the APP (amyloid precursor protein) swe/PS1dE9 AD mice model with persuaded propagation of endogenous NSCs and manufacturing of growth and neurotrophic factors (Park et al. 2010). PSCs, after 10 years of their discovery, are now extensively used for the therapeutic purpose in AD for the regulation of endogenous neurogenesis, pathological changes, and neuronal loss. PSCs extracted from the skin fibroblasts of the mouse after giving them the protein extracts of ESCs have been found to ease plaque deposition and cerebral dysfunction in the 5XFAD transgenic animal model. PSC and PSC-derived cells from human sources were used to correct degenerative disorder.

8.3.2 Stem Cells and PD Therapy

Among other neurodegenerative disease, PD is very common, which is characterized by decreased motor function caused by loss of the dopamine neuron (DA) in the human midbrain (Ekstrand et al. 2007). Numerous studies were conducted on both motor and non-motor defects responsible for PD. In PD cognitive defect starts to develop at a very early stage and keeps on progressing till it becomes visible. Motor and non-motor symptoms are caused by a decline in the quality of life and the method of treatment of the patient that is used for the correction of disease (Chaudhuri et al. 2006; Sauerbier et al. 2016). Stem cell-like human MSC transplanted to 6-hydroxydopamine (6-OHDA)-persuaded injuries secured dopaminergic neurons and encouraged neurogenesis that results in a beneficial effect because of the soluble factor released like BDNF (brain-derived neurotrophic factor). MSCs were found to be effective in the regulation of DA neuron apoptosis and related oxidative stress and also improves motor functioning when applied at very-early-stage PD (Blandini et al. 2010).

Cai et al. (2010) performed a study defining the effect of stem cell homing by applying an intra-arterial mixture of MSCs in combination with passing the BBB disrupted due to treatment with mannitol. After a period of 28 days of 6-OHDA treatment, no change in progression of the damage was found but there was standardization of pathological receptiveness of the striatal neuron towards dopaminergic stimulus that were induced due to MSC infusion (Cai et al. 2010).

8.4 Exosomes

These are extracellular vesicles that originate in body fluids together with urine, cerebrospinal fluid (CSF), saliva, and blood with a diameter ranging from 50 to 100 nm (Rousseau et al. 2015). The content of exosomes depends on the parent cell and varies from disease to disease condition because of which they can be used as potential biomarkers for the diagnosis of different brain diseases (Pegtel and Gould 2019). Exosomes were first found in 1983, when they were first noted as small vesicles released from reticulocytes that carry the transferrin receptor into the extracellular space (Meldolesi 2018; Kalluri and LeBleu 2020). Exosomes are double-membrane saclike structures that appear spherical under electron microscopy. The double membrane and their small size make it difficult for them to be removed from the macrophage body due to which they are the choice for biotic treatment and therapy for different neurological disorders (Théry et al. 2002; Aryani and Denecke 2016). Different biodistribution studies on unmodified exosomes after the fourth stage of drug administration reported a fast collection of exosomes in the reticuloendothelial system of organs and some of them were found in the brain. This showed that with little modification in the target delivery of the exosome, they can be used successfully in brain therapy (Van Niel et al. 2006; Kowal et al. 2014).

Exosomes are made of a double membrane of phospholipids enclosing sucrose, cholesterol, lipid rafts, sphingomyelin, ceramide, and evolutionarily conserved biomarkers (Lin et al. 2015). These biomarkers help to distinguish them from microvesicles and apoptotic bodies. Some of the exclusive proteins initiated in exosomes are CD9, CD63, CD81, and CD82 (tetraspanins); Hsp60, Hsp70, and Hsp90 (heat shock proteins); MHC classes I and II (major histocompatibility component); lactadherin; Tsg101; lysosome-associated membrane glycoprotein 2; and Alix. Exosomes also contain protein messengers that are cell-specific, cytosol components including the endoplasmic reticulum, Golgi apparatus, and mitochondria (Cheng et al. 2017; Baek et al. 2019; Evangelista et al. 2019; Kumar et al. 2019). Exosomes are produced from the inward budding of endosomes that makes the intraluminal vesicle inside endosomal chambers that are known as multivesicular bodies (MVBs) (Bianco et al. 2008). When matured, these MVBs fuse with the plasma membrane to produce an exosome or break down their load by combining with lysosomes. No specific reason or triggering reason for the increase or decrease in production or release of exosome is clear till now. Some studies reported that in case of starvation, MVBs are broken and fused with autophagosomes that results in the decrease of exosome release (Fujita et al. 2018; Klymiuk et al. 2019). The MVBs that fuse with the plasma membrane cause exosome release into the extracellular space, interacting with the extracellular matrix affecting the cell microenvironment and enters the circulatory system via the blood or lymph nodes by paracrine signaling. Some secretion process results in the release of a large number of exosomes, as large as 300,0000 per microliter of blood serum (Karaöz et al. 2011). Fusion of MVBs to lysosomes causes degradation and recycling of exosome lipids, nucleotide, and proteins. Since the topology of exosome is the same as of a cell, these exosomes can communicate with the receptor cells due to which they can directly cooperate with related receptors found in the plasma membrane (Trajkovic et al. 2008; Juan and Fürthauer 2018). Exosomes can be excellent vehicles for the distribution of RNAs and many cytoplasmic proteins as they not only combine but also discharge the required content into the receiver cell by fusing straight with the plasma membrane or endoplasmic reticulum (ER) after endocytosis (Baietti et al. 2012) (Fig. 8.2).

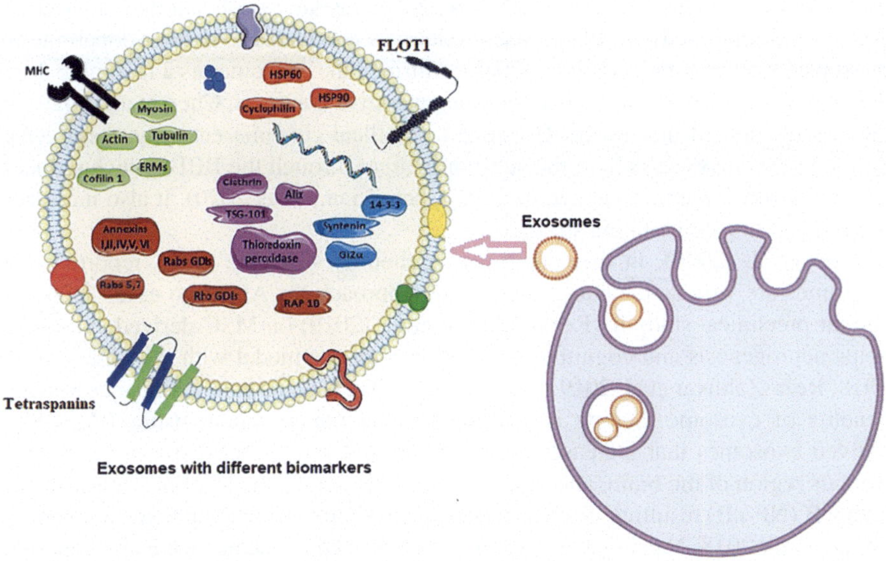

Fig. 8.2 Release of exosome and different types of biomarkers found in exosomes (Biorender)

8.5 Exosomes as Therapeutic Approach for Neurodegenerative Disease

As many studies found blood–brain barriers strictly limit the movement of large molecules to the brain, a large number of drugs that can be used for improving the therapeutic approach towards neurological disorders are not able to succeed in clinical trials or to reach clinics. To overcome these limitations, exosomes emerge as a promising approach. The nano-size structure of exosomes can easily pass through the blood–brain barrier and can deliver the bioactive molecules that are secreted by their cell origin, or as per requirement, they can be engineered to deliver drugs to targets (Huang et al. 2018).

Haney et al. (2015) studied and reported that exosomes can be made permeable using saponins, which function as secondary metabolite obtained from different plant species. In this in vivo study, the researchers confirmed the movement of exosomes via neuronal cells, resulting in a neuroprotective effect (Haney et al. 2015). The studies on Parkinson's disease (PD) by Kojima et al. (2018) developed a controlled device known as EXOtic (EXOsomal transfer into cell) that can deliver the catalase mRNA delivery without needing concentrated exosomes (Kojima et al. 2018; Qu et al. 2018). These devices were reported to increase the efficiency of exosome production especially in relation to specific mRNA and their delivery to the cytosol of target. Some studies focused on neuroinflammation and neurotoxicity using in vivo and in vitro models of Parkinson's disease and results confirmed the usefulness of EXOtic devices in RNA-based therapeutic applications (Chen et al. 2020). A similar study on PD mouse model was conducted to deliver dopamine to the substantia nigra and striatum, which confirmed the exosomes as a favorable drug delivery system for PD and other neurodegenerative therapies. Chen et al. (2020) in their study proved that hucMSC (human umbilical cord mesenchyme stem cell)-derived exosomes can deliver the substantia nigra through the BBB, which reduces apoptosis and dopaminergic neuron loss (Chen, Liang et al. 2020). It also increases the level of dopamine in the striatum.

Guo et al. (2020), in their study on Alzheimer's disease (AD), reported that exosomes are very suitable as a therapeutic approach for AD (Guo et al. 2020). A similar preclinical study by Reza-Zaldivar et al. (2019) on MSC-derived exosomes helps neurogenesis and cognitive function in a mouse model with AD (Ding et al. 2018; Reza-Zaldivar et al. 2019; Araldi et al. 2020). Nakano et al. (2020) showed the benefits of exosome therapy in treating an AD mouse model using BM-MSC-derived exosomes that increases the expression of microRNA-146a in the hippocampus region of the brain, and in astrocytes, it decreases the level of nuclear factor kappa B (NF-κB) resulting in synaptogenesis and correction of cognitive weakening (Wang et al. 2018; Nakano et al. 2020). MSC-derived exosomes were also reported to reduce the inducible nitric oxide synthesis (iNOS) in cultured neuron cells and improve neural impairment of CA1 synaptic transmission in AD. A study on the efficacy of intravenously delivered exosomes was done by Cui et al. (2019); they developed brain-targeting exosomes that were derived from MSCs by using CNS-specific rabies viral glycoprotein (RVG) peptide (Cui et al. 2019). RVG-modified exosomes researchers were able to improve the hippocampus and cortex and reduce the expression of inflammatory cytokines TNF-α (tumor necrosis factor alpha) and IL-6 (interleukin six), and it also improves the function of APP/PS1 in mouse models. In a recent study by Clark et al. (2019), it was proved that exosome-based therapy is very useful in treating multiple sclerosis. In this study they found that MSC-produced exosomes helped in myelin restoration in different mouse models in case of multiple sclerosis (Clark et al. 2019). A similar study by Li et al. (2019) supported the results that MSC-derived exosomes also decreased infiltration of CNS-related inflammations and reduced demyelination in autoimmune encephalomyelitis performed in a rat model (Li et al. 2019). In case of Theiler's murine encephalomyelitis virus (TMEV)-induced demyelination disease, similar results were also observed (Laso-García et al. 2018).

Bonafede et al. (2016) in their study on NSC-34 reported that murine adipose–derived exosomes can protect the overexpression of NSC-34 and SOD1 (G93A, G37R, A4W) from any oxidative damage which is responsible for causing amyotrophic lateral sclerosis (Bonafede et al. 2016). Lee et al. (2016) showed that adipose-derived exosomes can normalize the phosphor-cAMP reaction element binding protein CREB/CREB ratio and also modulate mitochondrial dysfunction and SOD-1 accumulation (Lee et al. 2016).

8.6 Conclusion

As the health services are increasing and with the development of technology, the life expectancy of people is increasing as per the report of the WHO (World Health Organization) by 2050 most people will live at the age of 60 years and above around the world causing the old-age people to grow in number and the burden of neurodegenerative diseases to also increase. The need to find the most effective therapeutic solution is required, and as per the studies conducted using stem cells and stem cell–derived exosomes, it can be said that exosomes are the future for the treatment of AZ, PD, and many other neurodegenerative disorders.

References

Aleshkov S, Abraham CR, Zannis VI (1997) Interaction of Nascent ApoE2, ApoE3, and ApoE4 Isoforms Expressed in Mammalian Cells with Amyloid Peptide β (1–40). Relevance to Alzheimer's Disease. Biochemistry 36(34):10571–10580

Anand R, Gill KD, Mahdi AA (2014) Therapeutics of Alzheimer's disease: past, present and future. Neuropharmacology 76:27–50

Araldi RP, D'Amelio F, Vigerelli H, de Melo TC, Kerkis I (2020) Stem cell-derived exosomes as therapeutic approach for neurodegenerative disorders: from biology to biotechnology. Cell 9(12):2663

Aryani A, Denecke B (2016) Exosomes as a nanodelivery system: a key to the future of neuromedicine? Mol Neurobiol 53(2):818–834

Atsmon-Raz Y, Miller Y (2016) Non-amyloid-β component of human α-synuclein oligomers induces formation of new Aβ oligomers: insight into the mechanisms that link Parkinson's and Alzheimer's diseases. ACS Chem Neurosci 7(1):46–55

Baek G, Choi H, Kim Y, Lee H-C, Choi C (2019) Mesenchymal stem cell-derived extracellular vesicles as therapeutics and as a drug delivery platform. Stem Cells Transl Med 8(9):880–886

Baietti MF, Zhang Z, Mortier E, Melchior A, Degeest G, Geeraerts A, Ivarsson Y, Depoortere F, Coomans C, Vermeiren E (2012) Syndecan–syntenin–ALIX regulates the biogenesis of exosomes. Nat Cell Biol 14(7):677–685

Beitz JM (2014) Parkinson's disease: a review. Front Biosci 6(1):65–74

Bekris LM, Yu C-E, Bird TD, Tsuang D (2011) The genetics of Alzheimer's disease and Parkinson's disease. Neurochem Mech Dis:695–755

Bertram L, Lill CM, Tanzi RE (2010) The genetics of Alzheimer disease: back to the future. Neuron 68(2):270–281

Bianco P, Robey PG, Simmons PJ (2008) Mesenchymal stem cells: revisiting history, concepts, and assays. Cell Stem Cell 2(4):313–319

Blandini F, Cova L, Armentero M-T, Zennaro E, Levandis G, Bossolasco P, Calzarossa C, Mellone M, Giuseppe B, Deliliers GL (2010) Transplantation of undifferentiated human mesenchymal stem cells protects against 6-hydroxydopamine neurotoxicity in the rat. Cell Transplant 19(2):203–218

Bloom GS (2014) Amyloid-β and tau: the trigger and bullet in Alzheimer disease pathogenesis. JAMA Neurol 71(4):505–508

Bonafede R, Scambi I, Peroni D, Potrich V, Boschi F, Benati D, Bonetti B, Mariotti R (2016) Exosome derived from murine adipose-derived stromal cells: neuroprotective effect on in vitro model of amyotrophic lateral sclerosis. Exp Cell Res 340(1):150–158

Cai J, Yang M, Poremsky E, Kidd S, Schneider JS, Iacovitti L (2010) Dopaminergic neurons derived from human induced pluripotent stem cells survive and integrate into 6-OHDA-lesioned rats. Stem Cells Dev 19(7):1017–1023

Castellano JM, Kim J, Stewart FR, Jiang H, DeMattos RB, Patterson BW, Fagan AM, Morris JC, Mawuenyega KG, Cruchaga C (2011) Human apoE isoforms differentially regulate brain amyloid-β peptide clearance. Sci Transl Med 3(89):89ra57-89ra57
Cha M-Y, Kwon Y-W, Ahn H-S, Jeong H, Lee YY, Moon M, Baik SH, Kim DK, Song H, Yi EC (2017) Protein-induced pluripotent stem cells ameliorate cognitive dysfunction and reduce Aβ deposition in a mouse model of Alzheimer's disease. Stem Cells Transl Med 6(1):293–305
Chaudhuri KR, Healy DG, Schapira AH (2006) Non-motor symptoms of Parkinson's disease: diagnosis and management. Lancet Neurol 5(3):235–245
Chen H-K, Liu Z, Meyer-Franke A, Brodbeck J, Miranda RD, McGuire JG, Pleiss MA, Ji Z-S, Balestra ME, Walker DW (2012) Small molecule structure correctors abolish detrimental effects of apolipoprotein E4 in cultured neurons. J Biol Chem 287(8):5253–5266
Chen H-X, Liang F-C, Gu P, Xu B-L, Xu H-J, Wang W-T, Hou J-Y, Xie D-X, Chai X-Q, An S-J (2020) Exosomes derived from mesenchymal stem cells repair a Parkinson's disease model by inducing autophagy. Cell Death Dis 11(4):1–17
Chen X-Q, Mobley WC (2019) Exploring the pathogenesis of Alzheimer disease in basal forebrain cholinergic neurons: converging insights from alternative hypotheses. Front Neurosci 13:446
Cheng L, Zhang K, Wu S, Cui M, Xu T (2017) Focus on mesenchymal stem cell-derived exosomes: opportunities and challenges in cell-free therapy. Stem Cells International 2017:6305295
Chouliaras L, Rutten BP, Kenis G, Peerbooms O, Visser PJ, Verhey F, van Os J, Steinbusch HW, van den Hove DL (2010) Epigenetic regulation in the pathophysiology of Alzheimer's disease. Prog Neurobiol 90(4):498–510
Chouraki V, Seshadri S (2014) Genetics of Alzheimer's disease. Adv Genet 87:245–294
Chowdhury S, Shepherd JD, Okuno H, Lyford G, Petralia RS, Plath N, Kuhl D, Huganir RL, Worley PF (2006) Arc/Arg3. 1 interacts with the endocytic machinery to regulate AMPA receptor trafficking. Neuron 52(3):445–459
Cissé M, Halabisky B, Harris J, Devidze N, Dubal DB, Sun B, Orr A, Lotz G, Kim DH, Hamto P (2011) Reversing EphB2 depletion rescues cognitive functions in Alzheimer model. Nature 469(7328):47–52
Clark K, Zhang S, Barthe S, Kumar P, Pivetti C, Kreutzberg N, Reed C, Wang Y, Paxton Z, Farmer D (2019) Placental mesenchymal stem cell-derived extracellular vesicles promote myelin regeneration in an animal model of multiple sclerosis. Cell 8(12):1497
Cohen TJ, Guo JL, Hurtado DE, Kwong LK, Mills IP, Trojanowski JQ, Lee VM (2011) The acetylation of tau inhibits its function and promotes pathological tau aggregation. Nat Commun 2(1):1–9
Crary JF (2016) Primary age-related tauopathy and the amyloid cascade hypothesis: the exception that proves the rule? J Neurol Neuromed 1(6):53
Crean S, Ward A, Mercaldi CJ, Collins JM, Cook MN, Baker NL, Arrighi HM (2011) Apolipoprotein E ε4 prevalence in Alzheimer's disease patients varies across global populations: a systematic literature review and meta-analysis. Dement Geriatr Cogn Disord 31(1):20–30
Cui G-H, Guo H-D, Li H, Zhai Y, Gong Z-B, Wu J, Liu J-S, Dong Y-R, Hou S-X, Liu J-R (2019) RVG-modified exosomes derived from mesenchymal stem cells rescue memory deficits by regulating inflammatory responses in a mouse model of Alzheimer's disease. Immun Ageing 16(1):1–12
Cummings JL, Vinters HV, Cole GM, Khachaturian ZS (1998) Alzheimer's disease: etiologies, pathophysiology, cognitive reserve, and treatment opportunities. Neurology 51(1 Suppl 1):S2–S17
Daviglus ML, Bell CC, Berrettini W, Bowen PE, Connolly ES Jr, Cox NJ, Dunbar-Jacob JM, Granieri EC, Hunt G, McGarry K (2010) NIH state-of-the-science conference statement: preventing Alzheimer's disease and cognitive decline. NIH Consens State Sci Statements 27(4):1–30
Day JJ, Sweatt JD (2011) Epigenetic mechanisms in cognition. Neuron 70(5):813–829
Ding M, Shen Y, Wang P, Xie Z, Xu S, Zhu Z, Wang Y, Lyu Y, Wang D, Xu L (2018) Exosomes isolated from human umbilical cord mesenchymal stem cells alleviate neuroinflammation and

reduce amyloid-beta deposition by modulating microglial activation in Alzheimer's disease. Neurochem Res 43(11):2165–2177

Douaud G, Refsum H, de Jager CA, Jacoby R, Nichols TE, Smith SM, Smith AD (2013) Preventing Alzheimer's disease-related gray matter atrophy by B-vitamin treatment. Proc Natl Acad Sci 110(23):9523–9528

Douglas CL, Baghdoyan HA, Lydic R (2002) Postsynaptic muscarinic M1 receptors activate prefrontal cortical EEG of C57BL/6J mouse. J Neurophysiol 88(6):3003–3009

Dumanis SB, Tesoriero JA, Babus LW, Nguyen MT, Trotter JH, Ladu MJ, Weeber EJ, Turner RS, Xu B, Rebeck GW (2009) ApoE4 decreases spine density and dendritic complexity in cortical neurons in vivo. J Neurosci 29(48):15317–15322

Ekstrand MI, Terzioglu M, Galter D, Zhu S, Hofstetter C, Lindqvist E, Thams S, Bergstrand A, Hansson FS, Trifunovic A (2007) Progressive parkinsonism in mice with respiratory-chain-deficient dopamine neurons. Proc Natl Acad Sci 104(4):1325–1330

Evangelista AF, Soares MBP, Villarreal CF (2019) Cell-free therapy: a neuroregenerative approach to sensory neuropathy? Neural Regen Res 14(8):1383

Fändrich M, Schmidt M, Grigorieff N (2011) Recent progress in understanding Alzheimer's β-amyloid structures. Trends Biochem Sci 36(6):338–345

Frank S (2014) Treatment of Huntington's disease. Neurotherapeutics 11(1):153–160

Frieden C, Garai K (2012) Structural differences between apoE3 and apoE4 may be useful in developing therapeutic agents for Alzheimer's disease. Proc Natl Acad Sci 109(23):8913–8918

Fujita Y, Kadota T, Araya J, Ochiya T, Kuwano K (2018) Clinical application of mesenchymal stem cell-derived extracellular vesicle-based therapeutics for inflammatory lung diseases. J Clin Med 7(10):355

Golde TE, Schneider LS, Koo EH (2011) Anti-Aβ therapeutics in Alzheimer's disease: the need for a paradigm shift. Neuron 69(2):203–213

Gouras GK, Tampellini D, Takahashi RH, Capetillo-Zarate E (2010) Intraneuronal β-amyloid accumulation and synapse pathology in Alzheimer's disease. Acta Neuropathol 119(5):523–541

Goutier W, Lowry JP, McCreary AC, O'Connor J (2016) Frequency-dependent modulation of dopamine release by nicotine and dopamine D1 receptor ligands: an in vitro fast cyclic voltammetry study in rat striatum. Neurochem Res 41(5):945–950

Guo M, Yin Z, Chen F, Lei P (2020) Mesenchymal stem cell-derived exosome: a promising alternative in the therapy of Alzheimer's disease. Alzheimers Res Ther 12(1):1–14

Haney MJ, Klyachko NL, Zhao Y, Gupta R, Plotnikova EG, He Z, Patel T, Piroyan A, Sokolsky M, Kabanov AV (2015) Exosomes as drug delivery vehicles for Parkinson's disease therapy. J Control Release 207:18–30

Hiekkanen H, Kurki T, Brandstack N, Kairisto V, Tenovuo O (2007) MRI changes and ApoE genotype, a prospective 1-year follow-up of traumatic brain injury: a pilot study. Brain Inj 21(12):1307–1314

Higley MJ, Soler-Llavina GJ, Sabatini BL (2009) Cholinergic modulation of multivesicular release regulates striatal synaptic potency and integration. Nat Neurosci 12(9):1121–1128

Holtzman DM, Herz J, Bu G (2012) Apolipoprotein E and apolipoprotein E receptors: normal biology and roles in Alzheimer disease. Cold Spring Harb Perspect Med 2(3):a006312

Hoover BR, Reed MN, Su J, Penrod RD, Kotilinek LA, Grant MK, Pitstick R, Carlson GA, Lanier LM, Yuan L-L (2010) Tau mislocalization to dendritic spines mediates synaptic dysfunction independently of neurodegeneration. Neuron 68(6):1067–1081

Huang S, Ge X, Yu J, Han Z, Yin Z, Li Y, Chen F, Wang H, Zhang J, Lei P (2018) Increased miR-124-3p in microglial exosomes following traumatic brain injury inhibits neuronal inflammation and contributes to neurite outgrowth via their transfer into neurons. FASEB J 32(1): 512–528

Huang Y (2006) Molecular and cellular mechanisms of apolipoprotein E4 neurotoxicity and potential therapeutic strategies. Curr Opin Drug Discov Devel 9(5):627–641

Ivanova MI, Lin Y, Lee Y-H, Zheng J, Ramamoorthy A (2021) Biophysical processes underlying cross-seeding in amyloid aggregation and implications in amyloid pathology. Biophys Chem 269:106507

Jiang T, Yu J-T, Tan L (2012) Novel disease-modifying therapies for Alzheimer's disease. J Alzheimers Dis 31(3):475–492

Jinnou H (2021) Regeneration using endogenous neural stem cells following neonatal brain injury. Pediatr Int 63(1):13–21

Jones PB, Adams KW, Rozkalne A, Spires-Jones TL, Hshieh TT, Hashimoto T, von Armin CA, Mielke M, Bacskai BJ, Hyman BT (2011) Apolipoprotein E: isoform specific differences in tertiary structure and interaction with amyloid-β in human Alzheimer brain. PLoS One 6(1): e14586

Juan T, Fürthauer M (2018) Biogenesis and function of ESCRT-dependent extracellular vesicles. Seminars in cell & developmental biology,, vol 74. Elsevier, p 66

Kalluri R, LeBleu VS (2020) The biology, function, and biomedical applications of exosomes. Science 367(6478):eaau6977

Karaöz E, Demircan PC, Sağlam Ö, Aksoy A, Kaymaz F, Duruksu G (2011) Human dental pulp stem cells demonstrate better neural and epithelial stem cell properties than bone marrow-derived mesenchymal stem cells. Histochem Cell Biol 136(4):455–473

Kim I-K, Park J-H, Kim B, Hwang K-C, Song B-W (2021) Recent advances in stem cell therapy for neurodegenerative disease: three dimensional tracing and its emerging use. World J Stem Cells 13(9):1215

Kim J, Basak JM, Holtzman DM (2009) The role of apolipoprotein E in Alzheimer's disease. Neuron 63(3):287–303

Klymiuk MC, Balz N, Elashry MI, Heimann M, Wenisch S, Arnhold S (2019) Exosomes isolation and identification from equine mesenchymal stem cells. BMC Vet Res 15(1):1–9

Kojima R, Bojar D, Rizzi G, Hamri GC-E, El-Baba MD, Saxena P, Ausländer S, Tan KR, Fussenegger M (2018) Designer exosomes produced by implanted cells intracerebrally deliver therapeutic cargo for Parkinson's disease treatment. Nat Commun 9(1):1–10

Kowal J, Tkach M, Théry C (2014) Biogenesis and secretion of exosomes. Curr Opin Cell Biol 29: 116–125

Kumar P, Kandoi S, Misra R, Vijayalakshmi S, Rajagopal K, Verma RS (2019) The mesenchymal stem cell secretome: a new paradigm towards cell-free therapeutic mode in regenerative medicine. Cytokine Growth Factor Rev 46:1–9

Kumar S, Walter J (2011) Phosphorylation of amyloid beta (Aβ) peptides–a trigger for formation of toxic aggregates in Alzheimer's disease. Aging (Albany NY) 3(8):803

Laso-García F, Ramos-Cejudo J, Carrillo-Salinas FJ, Otero-Ortega L, Feliú A, Gómez-de Frutos M, Mecha M, Díez-Tejedor E, Guaza C, Gutiérrez-Fernández M (2018) Therapeutic potential of extracellular vesicles derived from human mesenchymal stem cells in a model of progressive multiple sclerosis. PLoS One 13(9):e0202590

Lee M, Ban J-J, Kim KY, Jeon GS, Im W, Sung J-J, Kim M (2016) Adipose-derived stem cell exosomes alleviate pathology of amyotrophic lateral sclerosis in vitro. Biochem Biophys Res Commun 479(3):434–439

Li F-Q, Fowler KA, Neil JE, Colton CA, Vitek MP (2010) An apolipoprotein E-mimetic stimulates axonal regeneration and remyelination after peripheral nerve injury. J Pharmacol Exp Ther 334(1):106–115

Li G, Bien-Ly N, Andrews-Zwilling Y, Xu Q, Bernardo A, Ring K, Halabisky B, Deng C, Mahley RW, Huang Y (2009) GABAergic interneuron dysfunction impairs hippocampal neurogenesis in adult apolipoprotein E4 knockin mice. Cell Stem Cell 5(6):634–645

Li Z, Liu F, He X, Yang X, Shan F, Feng J (2019) Exosomes derived from mesenchymal stem cells attenuate inflammation and demyelination of the central nervous system in EAE rats by regulating the polarization of microglia. Int Immunopharmacol 67:268–280

Lin J, Li J, Huang B, Liu J, Chen X, Chen X-M, Xu Y-M, Huang L-F, Wang X-Z (2015) Exosomes: novel biomarkers for clinical diagnosis. Sci World J 2015:1

Lou K, Yao Y, Hoye AT, James MJ, Cornec A-S, Hyde E, Gay B, Lee VM-Y, Trojanowski JQ, Smith AB III (2014) Brain-penetrant, orally bioavailable microtubule-stabilizing small molecules are potential candidate therapeutics for Alzheimer's disease and related tauopathies. J Med Chem 57(14):6116–6127

Loy CT, Schofield PR, Turner AM, Kwok JB (2014) Genetics of dementia. Lancet 383(9919): 828–840

Bakulski M, Rozek LS, Dolinoy DC, Paulson HL, Hu H (2012) Alzheimer's disease and environmental exposure to lead: the epidemiologic evidence and potential role of epigenetics. Curr Alzheimer Res 9(5):563–573

Meldolesi J (2018) Exosomes and ectosomes in intercellular communication. Curr Biol 28(8): R435–R444

Morris M, Maeda S, Vossel K, Mucke L (2011) The many faces of tau. Neuron 70(3):410–426

Mruthinti S, Buccafusco JJ, Hill WD, Waller JL, Jackson TW, Zamrini EY, Schade RF (2004) Autoimmunity in Alzheimer's disease: increased levels of circulating IgGs binding Aβ and RAGE peptides. Neurobiol Aging 25(8):1023–1032

Nakano M, Kubota K, Kobayashi E, Chikenji TS, Saito Y, Konari N, Fujimiya M (2020) Bone marrow-derived mesenchymal stem cells improve cognitive impairment in an Alzheimer's disease model by increasing the expression of microRNA-146a in hippocampus. Sci Rep 10(1):1–15

Palmer AM (2011) Neuroprotective therapeutics for Alzheimer's disease: progress and prospects. Trends Pharmacol Sci 32(3):141–147

Palop JJ, Mucke L (2010) Amyloid-β–induced neuronal dysfunction in Alzheimer's disease: from synapses toward neural networks. Nat Neurosci 13(7):812–818

Park D-H, Eve DJ, Chung Y-G, Sanberg PR (2010) Regenerative medicine for neurological disorders. TheScientificWorldJOURNAL 10:470–489

Pegtel DM, Gould SJ (2019) Exosomes. Annu Rev Biochem 88:487–514

Peleg S, Sananbenesi F, Zovoilis A, Burkhardt S, Bahari-Javan S, Agis-Balboa RC, Cota P, Wittnam JL, Gogol-Doering A, Opitz L (2010) Altered histone acetylation is associated with age-dependent memory impairment in mice. Science 328(5979):753–756

Qu M, Lin Q, Huang L, Fu Y, Wang L, He S, Fu Y, Yang S, Zhang Z, Zhang L (2018) Dopamine-loaded blood exosomes targeted to brain for better treatment of Parkinson's disease. J Control Release 287:156–166

Ramirez-Bermudez J (2012) Alzheimer's disease: critical notes on the history of a medical concept. Arch Med Res 43(8):595–599

Reitz C (2012) Alzheimer's disease and the amyloid cascade hypothesis: a critical review. International Journal of Alzheimer's Disease 2012:369808

Remigante A, Morabito R (2022) Cellular and molecular mechanisms in oxidative stress-related diseases. Int J Mol Sci 23:8017

Reza-Zaldivar EE, Hernández-Sapiéns MA, Gutiérrez-Mercado YK, Sandoval-Ávila S, Gomez-Pinedo U, Márquez-Aguirre AL, Vázquez-Méndez E, Padilla-Camberos E, Canales-Aguirre AA (2019) Mesenchymal stem cell-derived exosomes promote neurogenesis and cognitive function recovery in a mouse model of Alzheimer's disease. Neural Regen Res 14(9):1626

Roos RA (2010) Huntington's disease: a clinical review. Orphanet J Rare Dis 5(1):1–8

Ross CA, Tabrizi SJ (2011) Huntington's disease: from molecular pathogenesis to clinical treatment. Lancet Neurol 10(1):83–98

Rousseau M, Belleannee C, Duchez A-C, Cloutier N, Levesque T, Jacques F, Perron J, Nigrovic PA, Dieude M, Hebert M-J (2015) Detection and quantification of microparticles from different cellular lineages using flow cytometry. Evaluation of the impact of secreted phospholipase A2 on microparticle assessment. PLoS One 10(1):e0116812

Sanchez MM, Heyn SN, Das D, Moghadam S, Martin KJ, Salehi A (2012) Neurobiological elements of cognitive dysfunction in down syndrome: exploring the role of APP. Biol Psychiatry 71(5):403–409

Sauerbier A, Jenner P, Todorova A, Chaudhuri KR (2016) Non motor subtypes and Parkinson's disease. Parkinsonism Relat Disord 22:S41–S46
Scheltens P, Blennow K, Breteler MM, De Strooper B, Frisoni GB, Salloway S, Van der Flier WM (2016) Alzheimer's disease. Lancet (London, England) 388(10043):505–517
Sharma P, Srivastava P, Seth A, Tripathi PN, Banerjee AG, Shrivastava SK (2019) Comprehensive review of mechanisms of pathogenesis involved in Alzheimer's disease and potential therapeutic strategies. Prog Neurobiol 174:53–89
Sharp ES, Gatz M (2011) The relationship between education and dementia an updated systematic review. Alzheimer Dis Assoc Disord 25(4):289
Shi Y, Kirwan P, Smith J, MacLean G, Orkin SH, Livesey FJ (2012) A human stem cell model of early Alzheimer's disease pathology in down syndrome. Sci Transl Med 4(124):124ra129-124ra129
Silverman W, Krinsky-McHale S (2021) Alzheimer's risk and quality of life: history of down syndrome as a case in point. Alzheimers Dement 13(1):e12171
Sivanandam TM, Thakur MK (2012) Traumatic brain injury: a risk factor for Alzheimer's disease. Neurosci Biobehav Rev 36(5):1376–1381
Srikanth V, Maczurek A, Phan T, Steele M, Westcott B, Juskiw D, Münch G (2011) Advanced glycation endproducts and their receptor RAGE in Alzheimer's disease. Neurobiol Aging 32(5): 763–777
Suzuki N, Cheung TT, Cai X-D, Odaka A, Otvos L Jr, Eckman C, Golde TE, Younkin SG (1994) An increased percentage of long amyloid β protein secreted by familial amyloid β protein precursor (βAPP717) mutants. Science 264(5163):1336–1340
Tayeb HO, Murray ED, Price BH, Tarazi FI (2013) Bapineuzumab and solanezumab for Alzheimer's disease: is the 'amyloid cascade hypothesis' still alive? Expert Opin Biol Ther 13(7):1075–1084
Théry C, Zitvogel L, Amigorena S (2002) Exosomes: composition, biogenesis and function. Nat Rev Immunol 2(8):569–579
Trajkovic K, Hsu C, Chiantia S, Rajendran L, Wenzel D, Wieland F, Schwille P, Brügger B, Simons M (2008) Ceramide triggers budding of exosome vesicles into multivesicular endosomes. Science 319(5867):1244–1247
Tysnes O-B, Storstein A (2017) Epidemiology of Parkinson's disease. J Neural Transm 124(8): 901–905
Van Niel G, Porto-Carreiro I, Simoes S, Raposo G (2006) Exosomes: a common pathway for a specialized function. J Biochem 140(1):13–21
Wang S-S, Jia J, Wang Z (2018) Mesenchymal stem cell-derived extracellular vesicles suppresses iNOS expression and ameliorates neural impairment in Alzheimer's disease mice. J Alzheimers Dis 61(3):1005–1013
Wonnacott S (1997) Presynaptic nicotinic ACh receptors. Trends Neurosci 20(2):92–98
Wu J, Petralia RS, Kurushima H, Patel H, Jung M-Y, Volk L, Chowdhury S, Shepherd JD, Dehoff M, Li Y (2011) Arc/Arg3. 1 regulates an endosomal pathway essential for activity-dependent β-amyloid generation. Cell 147(3):615–628
Zempel H, Mandelkow E-M (2012) Linking amyloid-β and tau: amyloid-β induced synaptic dysfunction via local wreckage of the neuronal cytoskeleton. Neurodegener Dis 10(1–4):64–72
Zempel H, Thies E, Mandelkow E, Mandelkow E-M (2010) Aβ oligomers cause localized Ca2+ elevation, missorting of endogenous tau into dendrites, tau phosphorylation, and destruction of microtubules and spines. J Neurosci 30(36):11938–11950
Zhang H, Sulzer D (2004) Frequency-dependent modulation of dopamine release by nicotine. Nat Neurosci 7(6):581–582

Zhao J, Su M, Lin Y, Liu H, He Z, Lai L (2020) Administration of amyloid precursor protein gene deleted mouse ESC-derived thymic epithelial progenitors attenuates Alzheimer's pathology. Front Immunol 11:1781

Zhu J, Zhao Y, Chen S, Zhang W, Lou L, Jin X (2011) Functional recovery after transplantation of neural stem cells modified by brain-derived neurotrophic factor in rats with cerebral ischaemia. J Int Med Res 39(2):488–498

Role of Stem Cells and Derived Exosomes as Novel Therapeutic Agents against Neuroinflammation and Stroke

9

Rabab Syeda Mirza, Nimisha Rawat, Deepanshi Thakur, Akanksha Bhardwaj, Shruti Gairola, and Tanisha Singh

Abstract

Stroke is a disorder resulting predominantly from the rupture or blockage of blood vessels and is the third primary cause of mortality and disability worldwide in the older population. Neuroinflammation, the key pathological event ensuing acute strokes, propagates brain damage via microglia and astrocyte stimulation, secretion of pro-inflammatory cytokines, distortion of the blood-brain barrier (BBB), and infiltration of white blood cells (WBCs) into the area of infarction. Regenerative medicine using stem cells is gaining popularity as a panacea for stroke consequences due to the properties of pluripotency and immunomodulation. Mesenchymal stem cell (MSC) transplantation rejuvenates the injured nervous tissue by converting microglia and monocytes into their anti-inflammatory phenotypes and stimulating neurogenesis, astrogenesis, and angiogenesis. Recent clinical evidence points towards treatment using stem cell-isolated exosomes as a promising remedial cell-free approach for strokes owing to unique features like low toxicity and immunogenicity, biodegradability, BBB penetrating capacity,

R. S. Mirza · D. Thakur
Department of Biotechnology, School of Chemical and Life Sciences, Jamia Hamdard University, Delhi, India

N. Rawat
Department of Biochemistry, Malla Reddy Institute of Medical Sciences, Hyderabad, Telangana, India

A. Bhardwaj
ICAR-Indian Agricultural Research Institute (ICAR-IARI), Delhi, India

S. Gairola
National Institute of Biologicals (NIB), Noida, Uttar Pradesh, India

T. Singh (✉)
School of Medicine, University of Pittsburgh, Pittsburgh, USA

S. Jahan, A. J. Siddiqui (eds.), *Applications of Stem Cells and Derived Exosomes in Neurodegenerative Disorders*, https://doi.org/10.1007/978-981-99-3848-3_9

and critical function in intercellular interaction. Exosomes from MSCs aid in neurorestoration and repair via induction of anti-apoptotic and anti-inflammatory actions as well as neurovascular remodeling. These extracellular nanovesicles (approximately 30–150 nm in diameter) can be tailored as targeted drug delivery vehicles for transporting extrinsic protein factors, chemical molecules, and foreign genes to host cells. This chapter largely focuses on the prospects of stem cells and their derived exosomes as novel therapeutic agents against neuroinflammation and strokes and the clinical advancements made regarding the same. Further studies are needed before either can be established as a treatment for attenuating and improving stroke consequences.

Keywords

Stroke · Neuroinflammation · Mesenchymal stem cells · Exosomes · Immunomodulation · Neurorestoration

9.1 Introduction

Stroke is a devastating neurological disorder, an outcome of blood vessel rupture or blockage in the brain, ranked the second-leading cause of death and the third-leading cause of combined death and disability in 2019 by the World Health Organization (WHO) (GBD 2017 DALYs and HALE Collaborators 2018; Krishnamurthi et al. 2020). From 1990 to 2019, the total number of incident stroke cases have risen by 70% and prevalent strokes elevated to 85%, whereas stroke-related deaths showed a 43% rise (Feigin et al. 2021). The prime source of hemorrhagic stroke is high blood pressure which results in the weakening of arteries in the brain, making them prone to rupture. Ischemic strokes are a consequence of blood clots that occur due to the narrowing or blockage of arteries with fatty deposits known as plaques, which obstruct the flow of blood and oxygen to the brain. Other factors that abet the process include smoking, obesity, high cholesterol levels, diabetes, excessive alcohol consumption, stress, and lack of exercise (National Health Service (NHS), UK).

Neuroinflammation is the prime pathological event engaged in ischemic stroke (Lian et al. 2021) that propagates brain damage via activation of astrocyte and microglia, pro-inflammatory cytokine production, disruption of the blood-brain barrier (BBB), and infiltration of leukocytes into the infarcted area (Jayaraj et al. 2019). The initiation and development of neuroinflammation occurs when dying cells release cellular components, including inflammatory factors, which activate microglia and infiltrate peripheral immune cells. This further generates various other inflammatory factors, chemokines, adhesion molecules, and tissue destruction enzymes and activates the complement system (Nakamura and Shichita 2019; Tsuyama et al. 2018; Ma et al. 2019) ultimately accelerating BBB damage and worsening brain injury (Tsuyama et al. 2018).

Brain injury due to stroke induces the differentiation of M1 microglia into its M2 phenotype, which plays a multiphasic role (Song et al. 2019). The M1 phenotype leads to the production of heightened pro-inflammatory cytokine levels, such as IL-6, IFN-γ, IL-1β, and TNF-α (Spellicy and Stice 2021), intensifies the degradation of the local inflammatory environment, exhibits nerve renewal after injury, and subsequently provides prolonged clinical remission (Zhang et al. 2021a, b). Similarly, the M2 phenotype is beneficial as it protects the neurons from hypoxia and ischemia, playing a significant role in endogenous repair and neuroprotective processes and additionally promoting long-term recovery after stroke.

The treatment strategy for ischemic stroke includes intravenous thrombolytic therapy and endovascular intracranial thrombectomy to accomplish recanalization; however, both techniques have a narrow therapeutic time window and poor functional outcome, limiting their clinical application (Li et al. 2018; Zhou et al. 2018). Contrarily, specific remedies for hemorrhagic stroke consist of minimally invasive surgery that removes the clot and intraventricular blood, leading to intracranial pressure management and effectively reducing mortality (Qureshi et al. 2009). Other than these approaches, numerous promising neuroprotective drug candidates that showed positive results in animal models failed in human clinical trials (Hasegawa et al. 2016; Alhadidi et al. 2016; Boltze et al. 2016).

Restorative and regenerative therapies are promising novel strategies as they enhance the process of endogenous tissue repair, by regulating neurogenesis, angiogenesis, and axonal outgrowth after stroke (Otero-Ortega et al. 2017a, b; Dong et al. 2020; Hira et al. 2018). Increasing evidence demonstrates the potential significance of exosome-mediated therapies from implanted cells, depicting positive results that can be applied as a treatment for the management of strokes (Zhang et al. 2019a, b, c).

Mesenchymal stem cells (MSCs) have immense neurorestorative potential. These multipotent stem cells are immunologically naive and can be easily extracted, maintained, and expanded in vitro without association with ethical concerns. These desirable properties make them suitable candidates for use in regenerative and restorative therapy against ischemic strokes (Baksh et al. 2007; Uccelli et al. 2008; Russell et al. 2018). The release of paracrine factors, mitochondrial transfer, and cell replacement are some of the action mechanisms for MSC-mediated neurorestoration. In addition, they release vascular endothelial growth factor (VEGF) and thereby induce angiogenesis (Li et al. 2000; Li et al. 2001; Chen et al. 2003; Shen et al. 2007). Regulation of MSC-released cytokines affects various pathways functioning in the regulation of immune cells and their responses, including reducing inflammation (Zhang et al. 2019a, b, c).

The therapeutic prospects of exosomes derived from MSCs were documented for the first time in rodent models of stroke and traumatic brain injury (TBI) in 2013 and 2015, respectively. The studies observed that rats subjected to focal cerebral ischemia or TBI, when administered with MSC-derived exosomes intravenously, considerably enhanced neurovascular remodeling with ameliorated neurological, behavioral, and cognitive effects during recovery (Zhang et al. 2015a, b; Xin et al. 2013a, b). Following numerous studies based on exosomal treatment in rodents, larger animal stroke and TBI models have been used for a superior comprehension of the beneficial outcomes of therapy. The main mechanism of exosome-based therapy involves transferring microRNAs (miRNAs) that target multiple genes and suppressing mRNA transcription, thereby regulating the differentiation, proliferation, apoptosis, and survival of target cells (Zhang et al. 2015a, b; Eulalio et al. 2008).

Multiple studies demonstrate that MSC-derived exosome therapy is a promising and valuable strategy against strokes in the new era. This chapter mainly focuses on the therapeutic application of stem cells and their derivative exosomes in targeting ischemic strokes along with their regulatory mechanism while modulating neuroinflammation.

9.2 Stroke: A Cerebrovascular Atack

9.2.1 Overview and Types

Stroke or cerebrovascular attack (CVA) is among the most medically severe conditions alongside metastatic tumors and cardiovascular diseases (Williams et al. 1999). Ischemic and hemorrhagic CVAs are the two most common forms of stroke. When 80% or more of the brain vessel is blocked by an obstruction, an ischemic stroke occurs. In contrast, a hemorrhagic stroke occurs after a vascular rupture (Ojaghihaghighi et al. 2017).

There are two types of hemorrhagic strokes, namely, intracranial hemorrhages (ICHs) and subarachnoid hemorrhages (SAHs). Myocardial infarctions, hypertension, and usage of thrombolytics are predisposing variables that dramatically raise the likelihood of suffering a hemorrhagic stroke (Rymer 2011). Hemorrhagic strokes can have a wide spectrum of clinical presentations that differ in different cases. Patients typically arrive with an intense headache, vomiting, and extremely elevated blood pressure. Focal neurological symptoms appear minutes after these clinical manifestations do. Although they can occasionally happen with other forms of stroke, these acute manifestations have been discovered to predominantly correspond to hemorrhagic strokes (An et al. 2017).

Meanwhile, thrombosis, hypoperfusion, and embolism are the three main causes of an ischemic stroke, of which thrombosis is presumed to be the most frequent cause. Clinical signs of ischemic strokes range in severity and advance more slowly than hemorrhagic strokes (over hours). An ischemic stroke can present clinically with paralysis, paresis, ataxia, vomiting, and ocular gaze. The particular location of the lesion determines the patient's symptoms and indicators (Ojaghihaghighi et al. 2017).

9.2.1.1 Ischemic Stroke

Stroke causes a high mortality and disability rate and is a debilitating neurological disease. Specifically, strokes are demarcated as either ischemic, resulting from blood vessel blockages, or hemorrhagic, originating from the ruptured blood vessels. The hallmark of an ischemic stroke is arterial blockage brought on by an embolus or thrombus (Catanese et al. 2017). The size of the ischemic area is determined by the size of the blocked artery, which in turn affects the metabolic and functional abnormalities arising throughout an ischemic stroke (Dugue et al. 2017). Approximately 25% of ischemic stroke is fatal; on the other hand, hemorrhagic stroke has a steep mortality of 50% (Andersen et al. 2009). Women are more likely to suffer from strokes than men, accounting for 60% of all stroke-related deaths after 60 years of age according to world statistics (Samai and Martin-Schild 2015). Stroke treatment accounts for 3–7% of healthcare budgets in developed countries (Chamorro et al. 2016).

During brain ischemia, nerve tissue damage is evident in two zones: "ischemic core," where blood circulation is <10 mL/100 g/min, and most cells die, and in the "ischemic penumbra," where blood circulation is 10–20 mL/100 g/min, and none of the neurons die but the structure of the tissue is altered (Dabrowska et al. 2019). Due to the lack of oxygen and glucose in the ischemic core, neuronal adenosine triphosphate (ATP) synthesis decreases, and cytoplasmic levels of Na^{+} and Ca^{2+} increase. An increased amount of Ca2+ enters cells as glutamate builds up, activating N-methyl-D-aspartate (NMDA) and amino-3-hydroxy-5-methyl-4-isoxazole propionic acid (AMPA) receptors (Roy-O'Reilly and McCullough 2014).

9.2.2 Pathophysiology of Stroke-Ensued Neuroinflammation

Brain tissue in the infarct core is deprived of oxygen and glucose, which causes ischemic brain damage. ATP deficit is caused by deprivation shortly after it starts (Prabhakaran and Naidech 2012). Energy synthesis deteriorates in the absence of ATP, and sodium-potassium pumps are unable to sustain the potential of cell membranes. Depolarization and a rise in the concentration of calcium ions inside the cell follow. This intracellular calcium increase, along with the oversecretion of reactive oxygen species (ROS), proteases, and phospholipases, results in cell death, inflammation, immune response activation, and BBB breakdown (Fig. 9.1) (Qureshi et al. 2009).

The infarct core and the penumbra are the two main zones that make up the ischemic stroke lesion region. There is a penumbra between the infarct core and the normal tissues surrounding the infarct. During the first few minutes, the infarct core's brain cells suffer irreparable damage, but the penumbra can be repaired. Restoring blood flow can slow the spread of the infarct into the penumbra region and reduce cell damage. The impairment from ischemia and the potential of inducing reperfusion damage both decrease with shorter times for blood supply restoration. New neuroprotective approaches to stroke appear to have a positive purpose in maintaining the penumbra (Lee et al. 2006).

9.2.3 Effects of Stroke and Associated Neuroinflammation

An organism's reaction to brain ischemia is characterized by local and systemic inflammation, when infectious pathogens are absent (Dabrowska et al. 2019). The

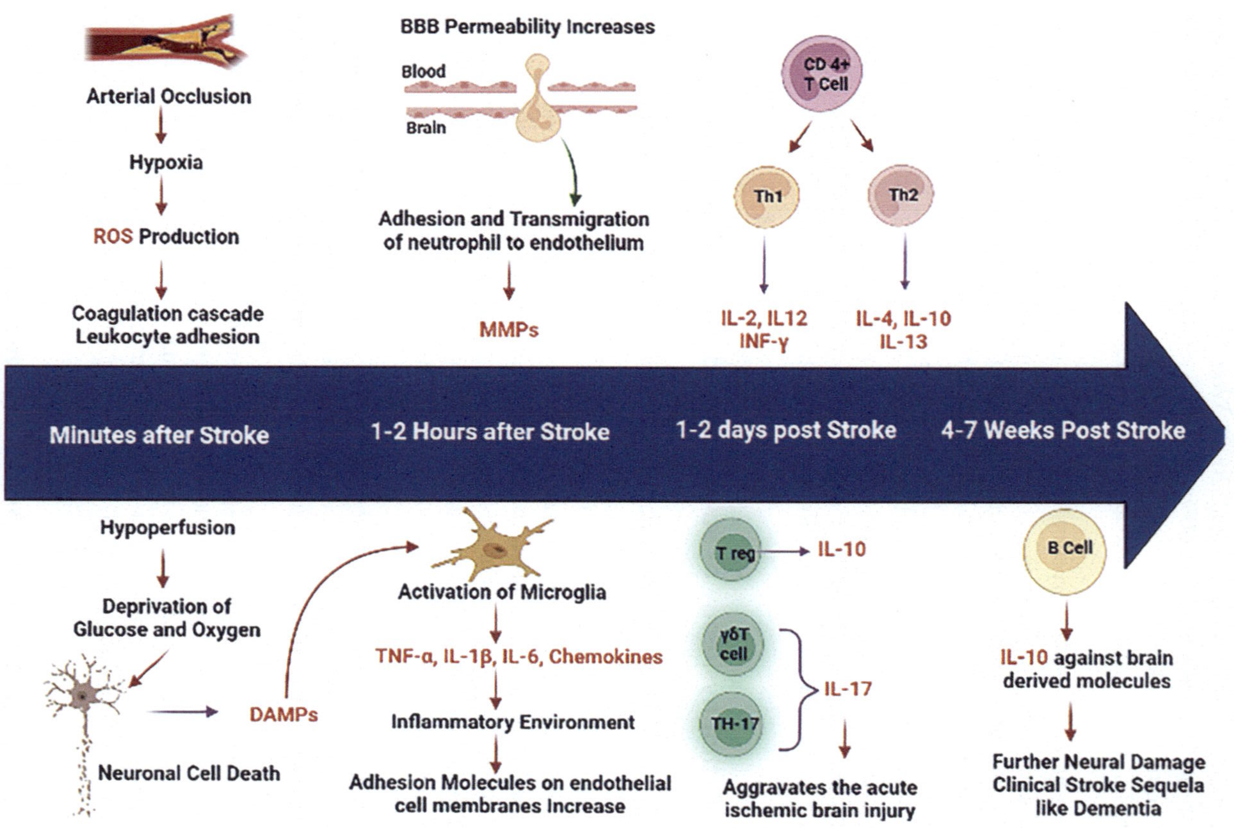

Fig. 9.1 Schematic figure representing the timeline of immunological response to ischemic stroke (taken from Jian et al. 2019, and modified using biorender.com). Commencement of ischemic stroke is marked by hypoxia of the blood vessels, resulting from arterial occlusion, producing ROS and triggering a

Fig. 9.1 (continued) coagulation cascade. The BBB permeability is enhanced by the release of MMPs from neutrophils. Additionally, cerebral hypoperfusion gives rise to a chronology of interlinked events, ultimately leading to neuronal death, which further sends DAMP signals causing microglial activation, that along with the activated T-cells stimulates pro-inflammatory cytokine and chemokine secretion resulting in neuroinflammation, thereby exacerbating the ischemic injury

damaged neurons and glial cells create DAMPs shortly after the ischemia event begins, which trigger astrocyte activation up to 28 days later. Astrocytes that have been activated can multiply quickly and alter their appearance and capabilities (Ahmad et al. 2014). They release metalloproteinases (MMPs), chemokines, and pro-inflammatory cytokines when activated (Dong and Benveniste 2001). The inflammatory response has both negative and positive effects on the development of cerebral ischemia.

Microglia, for example, produce MMP, IL-1, TNF, and reactive oxygen and nitrogen species, among other pro-inflammatory cytokines. Positive outcomes include reduced inflammation and atrophy of the brain caused by brain microglia/macrophages phagocytosing healthy, functional neurons (IL-10 and TGF secretion, phagocytic activity, and arginase synthesis). The release of growth factors like insulin-like growth factor 1 (IGF-1) and neurotrophic factors like brain-derived neurotrophic factor (BDNF), glial cell line-derived neurotrophic factor (GDNF), etc. by microglia negatively impacts the late ameliorative mechanisms, thus promoting neuronal formation and reorganization and scavenging and eliminating the necrotic remains.

Neuroinflammatory responses are closely associated with secondary cell death in acute and chronic strokes. Immediately after a stroke, the dead, dying, and decomposing cells and debris accumulate in the brain, resulting in acute neuroinflammation (Anthony et al. 2022). In an attempt to eliminate these damaged cells from the brain, microglia, which are resident macrophages, phagocytose injured brain cells (Cai et al. 2019). In an intracerebral hemorrhage (ICH), macrophages infiltrated with red blood cells are essential for the resolution of the hematoma. By promoting STAT6 activation, IL-4 improves microglia and macrophage phagocytosis and facilitates long-term erythrocyte clearance in mouse models of ICH (Xu et al. 2020). The acute neuroinflammatory response may enhance neuroprotection by creating IL-1 cytokines in the injured brain, which have then delegated surveillance duties to the cells in the injured brain. Neuroinflammation induced by microglia in the chronic phase of stroke exacerbates postischemic damage. The BBB is disrupted along with parenchymal and cerebral vascular infarction in the chronic inflammatory stage of stroke. When the BBB is compromised for a prolonged period of time, immune cells and serum proteins leak out, resulting in neuroinflammation, increased intracranial pressure, and macrophage death. This secondary inflammation-induced cell death, therefore, exacerbates the primary ischemic insult. Considering these observations, we can enhance neuroprotective acute inflammation and dampen deleterious chronic inflammation to improve stroke outcomes. A neurovascular impairment can trigger a variety of pathological symptoms throughout the whole system. The interactions between these cells may affect the entire system if one type of cell is missing, since physical and biochemical pathways allow them to communicate with one another.

9.2.4 Existing Stroke Therapies

Treating ischemic stroke is a "one in a million" opportunity. The therapeutic window for current stroke treatments is limited, lasting only up to 4.5 h for tissue plasminogen activator (tPA) given intravenously or 6–8 h for mechanical thrombectomy (MT) (Nogueira et al. 2018; Fugate and Rabinstein 2014). This time window is referred to as the "golden hours." Apart from time constraints, when the thrombus is massive or the stroke is severe, tPA is ineffective for patients with high-level arterial blockage or may cause hemorrhage (Bhaskar et al. 2018). Even in the case of MT, if the carotid arteries are mostly obstructed, not all patients can have surgery. Additionally, the surgery can disable people (Gervois et al. 2016). As a result, ischemic stroke protection is inherently difficult due to the rapid loss of brain cells, a small amount of spontaneous functional recovery, and a dearth of available treatments.

Patient recovery after an ischemic stroke is generally ineffective and often fails to restore lost functions. To protect neurons in the penumbra, prevent additional tissue destruction during reperfusion in the acute phase of the stroke, and replace dead cells, researchers have been searching for new treatment approaches (Gervois et al. 2016).

Research on stroke recovery has advanced in recent years, and it now demonstrates that stem cells and their exosomal derivatives modulate the synthesis of many biomolecules and protein factors ensuing a stroke to minimize infarct volume (Larpthaveesarp et al. 2021). Exosomes comprise various elements, while stem cells can replace necrotic cells by differentiating directly into glial cells; both have an impact on the recovery of motor function following a stroke (Han et al. 2021). As a result, current studies concentrate on both stopping pathologic processes that affect brain cells as well as regulating the local inflammatory response to ischemia, which is elaborately discussed in this chapter.

9.3 Stem Cells and their Types

Akin to blank canvases, stem cells are differentiated to have two fates: self-renewal and division to produce more stem cells or differentiation into specialized cells ranging from muscle to brain cells. They have mainly four sources of origin: adult tissues, fetal tissues, embryonic tissues, and genetically reprogrammed somatic cells (referred to as induced pluripotent stem cells or iPSCs). Discovered by biophysicist James Till and cellular biologist Ernest McCulloch in 1961, these cells have immense potential in cell therapies and tissue engineering.

Based on their division ability, cells can be differentiated into totipotent (have the highest differentiation potential that allows cells to form embryonic and extraembryonic structures), pluripotent (generation of germ layers), multipotent (cells of specialized lineage), oligopotent (differentiation into different cell types), and unipotent (narrowest differentiation potential, can divide repeatedly to form one type of cell). In line with their origin, there exist several types of stem cells. The most recognized types are adult stem cells, iPSCs, and embryonic stem cells (ESCs).

Adult stem cells are undifferentiated cells found in specific differentiated tissues that can renew or divide themselves to generate new cells that replenish dead or damaged tissue, with the main objective of maintaining tissue homeostasis. Also known as somatic stem cells, they are typically scarce in native tissues, which poses complications in their study and isolation for research purposes. Adult stem cells are present in numerous tissues such as the blood, muscle, brain, and heart (Bond et al. 2015; Bruyneel et al. 2016). The different stem cell types include hematopoietic stem cells, MSCs, neural stem cells (NSCs), and epithelial stem cells.

Originally derived from the inner cell mass of blastocysts, ESCs are capable of generating fully differentiated cells that form various tissues of the human body and are associated with tumorigenesis. Due to the destructive nature of their isolation, they have been mired in ethical conflicts despite their immense regenerative potential. However, the establishment of human ESC lines from single biopsied blastomeres of an in vitro fertilized embryo has been made possible with the use of LN-521 and E-cadherin in conjugation as the sole cell culture matrix in xeno-free conditions (Rodin et al. 2014). These pluripotent stem cells are only found during the initial stages of development.

Defined reprogramming factors were documented for the first time in 1987 by Davis et al., who recognized three genes that expressed predominantly in proliferative myoblasts, of which forced expression of myogenic differentiation 1 (*MYOD1*) single-handedly converted fibroblasts into myosin-expressing stable myoblasts (Davis et al. 1987). Advance to 2006, stem cells with properties similar to ESCs were generated from mouse fibroblasts by simultaneous introduction of four genes and were identified as iPSCs (Takahashi and Yamanaka 2006). After a year, utilizing the same process on human dermal fibroblasts combined with identical factors, i.e., Oct3/4, Sox2, Klf4, and c-Myc, human iPSCs were successfully generated (Takahashi et al. 2007). Owing to their property of pluripotency, iPSCs have found various applications in regenerative medicine and the generation of disease models. However, vehicles used for the transfer of genes, efficiency of stability of the iPSC line, and transcription of transgenes in iPSC-derived cells are issues that require further consideration.

9.3.1 Mesenchymal Stem Cells

MSCs are undifferentiated cells possessing the ability to self-renew and differentiate into numerous cell types, originating from tissues such as the bone marrow and adipose tissue. These stromal cells characteristically express the surface antigens like CD105, CD73, and CD90 while they lack the marker proteins on the cell surface-CD34, CD45, CD14, CD11, and major histocompatibility complex class II (MHC II) cell surface receptor (HLA-DR) (Jahan et al. 2017). MSCs express some markers, such as various integrins, mRNAs for cytokine and growth factors (macrophage colony-stimulating factor [M-CSF], IL-6, IL-11, and stem cell factor), and neuronal proteins (nestin and Tuj-1) (Majumdar et al. 2003; Majumdar et al. 2000; Tondreau et al. 2004). MSCs are capable of differentiating into cells of different lineages (ectodermal, endodermal, or mesodermal origin) under specific conditions, such as the influence of regulatory genes, growth factors, and transcription factors. They are highly migratory, with studies showing their migration to ischemic regions of the brain as a part of the host defense mechanism, exhibiting homing (Chen et al. 2001). In a way these stem cells, by upscaling growth factors, mimic molecular chaperones, preventing cell death due to injury and enhancing synaptic connectivity between the host and graft.

9.3.1.1 Immunomodulation of MSCs

MSCs exhibit powerful immunomodulatory effects targeting various cells of the immune system, including inhibition of T-lymphocyte stimulation, activation, and proliferation (Glennie et al. 2005; Keyser et al. 2007), resulting from the inhibition of cyclin D2. Studies suggest the role of prostaglandin E2, nitric oxide, histocompatibility locus antigen-G, and insulin-like growth factor-binding proteins, among other compounds, in influencing target cells (Siegel et al. 2009). Additionally, MSCs inhibit the proliferation of natural killer cells and B-lymphocytes.

9.3.1.2 Effect on Angiogenesis

In relation to angiogenesis, studies have shown that MSCs enhance wound healing, induce endothelial cell tube formation and cell migration, and improve blood vessel formation in the chorioallantoic membrane assay, exhibiting the potential to promote brain regeneration after an ischemic stroke by promoting revascularization of the regenerating tissue. ELISA has shown antibody arrays and immunohistochemistry of various angiogenic factors in the secretome of bone marrow-derived MSCs, including vascular endothelial growth factor (VEGF), fibroblast growth factor 2 (FGF-2), angiopoietin-1 (Ang-1), monocyte chemoattractant protein-1 (MCP-1), IL-6, and placental growth factor (PLGF) (Bronckaers et al. 2014). In mouse bone marrow-derived MSCs, the presence of cysteine-rich protein 61 (Cyr61; an angiogenic inducer) was confirmed by Estrada et al. (Estrada et al. 2009). This highlights the contribution of MSCs in promoting angiogenesis, performing a critical role in the rejuvenation and repair of injured tissues.

9.3.1.3 Role in Neurogenesis

MSCs have a distinct role in adult neurogenesis, which is the process of generation of functional neural cell types (such as neurons, astrocytes, and oligodendrocytes) from precursors, conventionally thought to take place during the embryonic and perinatal stages in mammals (Ming and Song 2005). The residence of neural progenitor stem cells in restricted regions of the brain, referred to as neurogenic niches, is well established. Two major niches have been identified in the postnatal brain: the subgranular zone of the hippocampus dentate gyrus and the subventricular zone lining the lateral ventricles (Ming and Song 2011). Other regions of the central nervous system (CNS) may exhibit extremely limited neurogenesis under conditions such as an injury (Gould 2007). Stem cells in the adult subventricular zone form transient amplifying progenitors which divide and give rise to neuroblasts. These neuroblasts then differentiate into subtypes of interneurons. In the subgranular zone, intermediate progenitor cells and neuroblasts originate from radial glia-like NSCs (type 1 cells) and finally differentiate into dentate granule neurons (Gurusamy et al. 2018).

The present understanding of multipotency and self-rejuvenation of adult NSCs is supported by (in vitro) evidence that in response to growth factors, precursor cells isolated from the adult CNS give rise to neurospheres or monolayer colonies, which can be induced to differentiate into various neural cell lineages following the removal of the growth factor (Palmer et al. 1999; Reynolds and Weiss 1992). However, in vivo evidence indicating the existence of endogenous adult NSCs with features of clonal-level self-revitalization and multilineage differentiation is unavailable. Various signaling pathways, such as Wnt and Sonic hedgehog (Shh) as well as those involving FGFs and BMPs, function as extrinsic factors and are responsible for regulating neural precursors in the adult brain. Meanwhile, intrinsic factors such as Pax6, Sox2, TLX, NeuroD, Mash1, PTEN, MLL1, and DISC1 regulate the proliferation and neuronal differentiation of adult neural precursors in vivo. However, predominantly due to complications in labeling and manipulating dormant neural precursors, the cellular and molecular mechanisms governing and controlling their behavior in the brain under basal physiological conditions are unknown (Bonaguidi et al. 2011).

MSCs express CXCR4, the receptor for SDF1α (CXCL12) which is a chemoattractant directing the migration of these cells to sites of inflammation; however, their neuroprotective attributes on neurological disorders are mainly an effect of the soluble factors secreted by them. In addition to their role in functional brain repair after trauma, neurogenesis helps maintain a flexible hippocampal neural network through the continuous addition of new immature neurons with distinct properties as well as induction of structural plasticity in adult neurons owing to the integration of naïve neurons (Christian et al. 2014).

The major MSC-mediated therapeutic mechanisms involved in facilitating functional neurological rehabilitation and repair, as elaborated above, have also been summarized in Table 9.1.

9.3.1.4 MSCs in Stroke Therapy

Activation of neurogenesis, astrogenesis, and oligodendrogenesis as well as neuroprotection and immunomodulation are some mechanisms by which the therapeutic action of MSCs is realized.

Brain neuron motifs, primary retinal ganglion cells (RGCs), or embryonic mouse spinal cord explant studies showed that in addition to promoting the survival of RGCs as well as cortical and dopaminergic neurons, MSCs induced neurite growth in the dorsal root ganglia (Crigler et al. 2006; Mead et al. 2014). The anti-apoptotic and pro-angiogenic effects of MSCs are possibly a result of increased levels of derived VEGF and MCP-1 (Dabrowska et al. 2019).

To verify the results of the experiments on the MSC-based curative effects in vitro, studies on numerous animal models were performed. One particular study showed the transfer of implanted bone marrow-derived mesenchymal stem cells (BMMSCs) to the infarcted region from the ischemic rats' cerebral cortex as well as

Table 9.1 Therapeutic effects and mechanism of the MSC proteins

Therapeutic effects	Proteins involved	Mechanism of action
Attenuate inflammation through immunomodulation (Glennie et al. 2005; Keyser et al. 2007; Siegel et al. 2009)	Interleukin-1, interferon-γ, tumor necrosis factor-α, monocyte chemoattractant protein-1 Interleukin-4, interleukin-10, tumor necrosis factor-β Prostaglandin E2 High mobility group box 1	Reduced pro-inflammatory cytokines to attenuate inflammation Elevated anti-inflammatory cytokines to reduce inflammation Regulate TNF-α and IFN-γ expression levels Delayed pro-inflammatory cytokine
Secrete trophic factors to modulate curative responses (Lakhal and Wood 2011; Zhao et al. 2020a, b)	BDNF GDNF NGF VEGF PDGF	Accelerated neurological healing and controlled MSC differentiation Decreased the volume of infarct Enhanced neuron growth and decreased neuron apoptosis New blood vessel generation Facilitates cell migration, encourages primary cortical neuron formation, reduces neuroinflammation, and stimulates new blood vessel formation and growth of axons
Induce angiogenesis (Bronckaers et al. 2014; Xiong et al. 2022)	Ang1 and tyrosine-protein kinase receptor Tie-2 receptor 2 (Flk1)	At the VEGF and VEGF location of vascular damage, these proteins were ↑ed. to ↑ blood vessel density.
Proliferate neuroblasts (Bonaguidi et al. 2011; Gurusamy et al. 2018; Zhang et al. 2022)	Axonal growth-associated and growth-inhibiting proteins Collagen IV and tight junction protein ZO-1 p53 protein	Rise in the former and decline in the latter to stimulate axonal growth Upregulation of these to inhibit BBB rupture and death of neurons Reduced protein activity to minimize neuronal apoptosis
Replace damaged cells (Bond et al. 2015; Bruyneel et al. 2016; Zhang et al. 2022)	Microtubule-associated protein 2 (MAP2) and NeuN GFAP and CNPase	Replaced damaged neurons with newly differentiated neurons Replaced by differentiated glial cells to repair damaged glial cells.

their survival in the host's brain and subsequent differentiation into adult nerve cells to assist the reinstitution of destroyed functions. This was validated by functional and behavioral examinations (Zhao et al. 2002). However, MSC-differentiated neural cells do not show the ability to migrate or further differentiate, giving rise to the conjecture that the paracrine activity of MSCs attributes to their clinical prospects. This has been demonstrated in studies by Leong et al. who noted that functional improvement in rodents with focal cerebral ischemia was likely due to the paracrine effects of transplanted dental pulp stem cells (Leong et al. 2012).

Clinically, due to the possibility of autologous transplantation, MSCs are ethically more favorable than ESCs. The main technical problem that poses a complication in the use of MSCs is the time interval essential for their harvesting, isolation, characterization, bacteriological testing, and multiplication, which ranges from 3 to 5 weeks. For autologous transplantation, this makes their administration in acute ischemic stroke difficult. Despite these complications, their favorable effects in preclinical studies using animal stroke models encouraged the translation to clinical trials.

Bang et al. supervised an early phase I/II clinical trial where BMMSCs were administered twice within 9 weeks to patients intravenously after ischemic stroke onset. During a one-year observation period, 5 study participants underwent MSC transplantation therapy while 25 were treated as control. In addition to improvement in the Barthel Index, the experimental group had no deaths following BMMSC transplantation, recurrence, or posttransplantation abnormalities (Bang et al. 2005).

In 2010, Lee et al. administered autologous BMMSCs in 85 patients through a long-term clinical investigation (5 years) centered on severe ischemic stroke, observing prolonged survival in patients administered with MSCs compared with the control group. No significant adverse events were reported after the administration of the therapy, and the incidence of comorbidities (such as recurrent stroke) was reported to be similar in the two groups. A rise in stromal cell-derived factor-1 (SDF-1) blood levels and a degree of subventricular region involvement along the lateral ventricle accompanied the improvement in the clinical condition of the participants undergoing MSC therapy (Lee et al. 2010).

Currently, over 62 clinical trials are enlisted for MSCs as therapy for strokes on clinicaltrials.gov, one of the largest clinical trial databases.

It has been emphasized that the therapeutic outcomes of MSCs are regulated by the secretion of soluble paracrine factors instead of cell proliferation and differentiation. Exosomes are the most crucial paracrine factors secreted by MSCs, and their function in relation to neuroinflammation and stroke remission has been discussed elaborately in the next section.

9.4 Exosomes

9.4.1 Biological Origin, Characteristics, and Applications

Exosomes are membrane-bound (bilayered) secretory nanovesicles with an estimated size of 30–150 nm, known to function as mediators of intercellular communication. They are one out of the three types of membrane-bound extracellular vesicles (EVs), the other two being membrane-shedding microvesicles and apoptotic bodies (Hong et al. 2019). All types of cells including immune, tumor, epithelial, glial, and neuronal cells as well as body fluids like plasma, blood, cerebrospinal fluid, urine, and saliva abundantly secrete exosomes, and their parent cell or fluid source is indicative of the functions that they perform, ranging from an efficient antigen to target cell association and stable conformational conditions for proteins to efficacy in distribution at the molecular level owing to the potential of these nanospheres and reaching distant organs via recirculation in body fluids

(Kowal et al. 2014; van Niel et al. 2018; Li et al. 2021a, b). Exosomes possess an endosomal origin, undergoing their biogenesis in mainly three stages, including (1) plasma membrane of the parent cell giving rise to endocytic vesicles, which mature from early to late endosomes, (2) budding of the late endosomal vesicular membrane inwards inducing the evolution of multivesicular bodies (MVBs) containing accumulated intraluminal vesicles (ILVs) within the endosomal compartment, and (3) MVBs fusing with the plasma membrane to release their vesicular content (ILVs) into the extracellular space, called exosomes (Li et al. 2021a, b; Jiang et al. 2022). These exosomes so formed and secreted are enriched with a unique cargo of bioactive metabolites like proteins, lipids (sphingolipids, cholesterol, phosphoglycerides), and nucleic acids (DNA, RNA, miRNA) derived from the parent cell. The proteins mainly comprise heat shock proteins (HSPs), transport proteins, tetraspanins, and MVB biogenesis-associated proteins. The accessory proteins in the MVB transport including Alix, TSG101, HSC70, and tetraspanins like CD63, CD9, and CD81 together are known as exosomal marker proteins (Yang et al. 2017a, b; Meldolesi 2018; van Niel et al. 2018). In general, the exosomal composition is crucial for its function as a biomarker of a particular disease and the determination of its biological role in various cellular processes. Exosomes act as novel shuttles, which, by virtue of their small and nontoxic nature, can traverse the BBB with ease without eliciting potent immune reactions. Thus, they offer an attractive complementary therapeutic platform for tissue regeneration post injury by tailoring themselves to serve as ideal drug delivery platforms including bioactive compounds and specialized immune modulators (Das et al. 2019; Li et al. 2021a, b).

9.4.2 Stem Cell-Derived Exosomes

Numerous studies have emphasized the excellent potential of cell-based therapies in ameliorating the neurological effects after a stroke, particularly those involving stem cells. The principal mechanism underlying the therapeutic activity of stem cells after stroke is their secreted paracrine factors and not cell replacement or differentiation of transplanted cells into brain cells (Qiu et al. 2018; Wang et al. 2018; Bang and Kim 2019; Cunningham et al. 2020). Stem cells have numerous paracrine effects and exosomes form an eminent part of the same (Fig. 9.2). They confer various advantages over whole cells, owing to their characteristic properties of low toxicity, immunogenicity, and tumorigenicity as well as high transport efficiency, inherent stability, and BBB-crossing capacity (Zhu et al. 2017; Gowen et al. 2020). They can be suitably modified based on their properties and the parent cells (stem cells) from which they are secreted to improve clinical efficacy, thereby offering promising treatment alternatives. Moreover, they can be generated in abundance from small cell quantities and possess long-term stability when stored at −80 °C. They can also enhance functional recovery after ischemic stroke, by demonstrating positive effects in terms of increasing brain plasticity (Xin et al. 2013a, b; Gao et al. 2020; Jiang et al. 2022). Distinct varieties of cells carry different miRNAs in their exosomes that are crucial for the repair of injured nerve cells (Deng et al. 2019). Stem cells are prolific

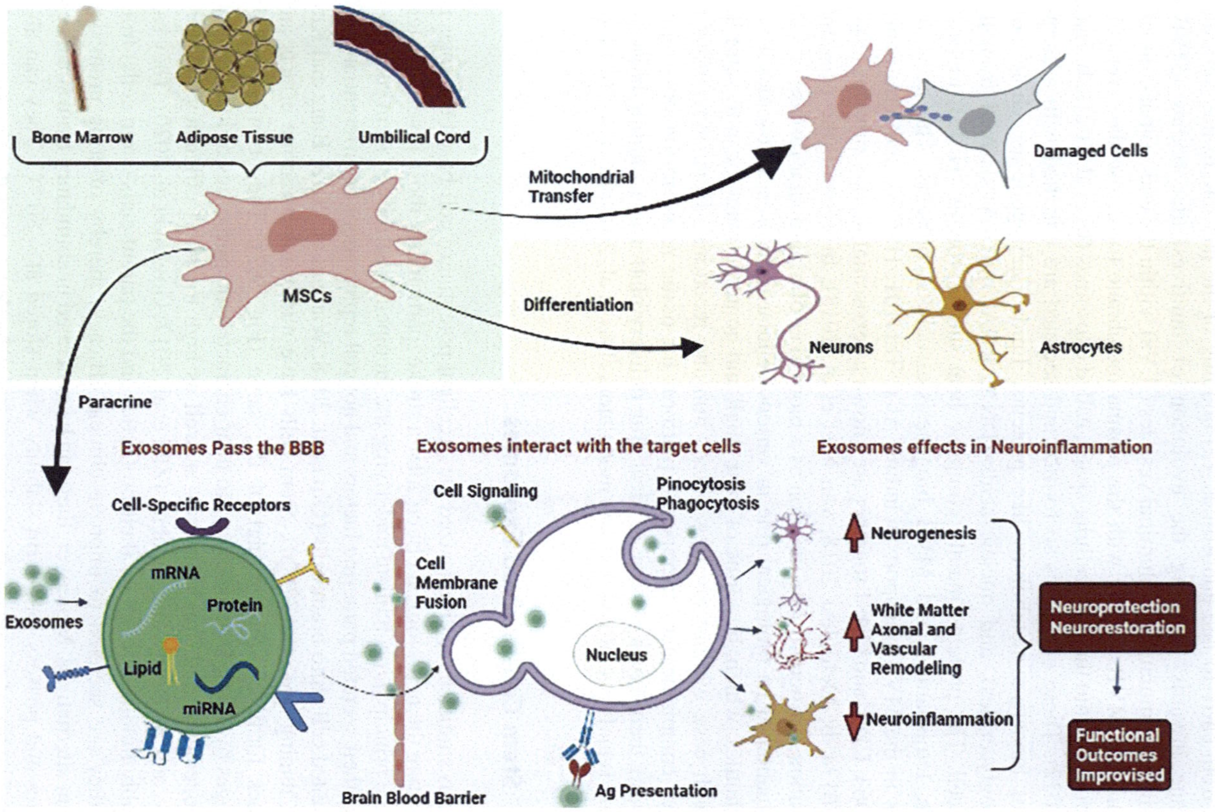

Fig. 9.2 Schematic figure summarizing the pathway of MSC differentiation and therapeutic response of derived exosomes for stroke (taken from Chen and Chopp 2018 and Li et al. 2021a, b; modified using biorender.com). It is noteworthy that the exosomes secreted as paracrine factors from MSCs, and not the MSC mitochondrial transfer to injured cells or direct differentiation of transplanted cells into brain cells, play a central role in the therapeutic activity of stem cells against stroke

producers of exosomes, MSCs being the best known, which primarily comprise BMMSCs, adipose-derived mesenchymal stem cells (ADMSCs), human umbilical cord-derived mesenchymal stem cells (HUCMSCs), and neural-derived mesenchymal stem cells (NMSCs) (Cosenza et al. 2018; Wang et al. 2018). Distinct cell types secrete exosomes containing distinct molecules; however, it is the marker proteins (most commonly CD9, CD63, and CD81) of these extracellular vesicles which give them their unique attributes (Hassanpour et al. 2020). MSCs are the preferred choice since they are the only human cell type capable of producing exosomes in large quantities for the delivery of therapeutic biomolecules (Vakhshiteh et al. 2019). They possess remarkable self-renewal and regeneration capacity, as well as can differentiate in impaired tissues, stimulate microglia, and modulate inflammation (Castro-Manrreza and Montesinos 2015; Li et al. 2015). Further on, they possess potent immunomodulatory characteristics and high differentiation capacity and can be easily manipulated and suitably cultured (Vakhshiteh et al. 2019). Also, their ability to overcome ethical problems (involving fetuses unlike neural and embryonic stem cells), the abundance with which they are found in the bone marrow, and reduced tumorigenicity in comparison to iPSCs pave way for effective new treatments for numerous human diseases (Stonesifer et al. 2017). Even though there is limited data on the restorative effects mediated by MSCs, the paracrine factors released by transplanted MSC-derived exosomes are primarily responsible for the perceived clinical outcomes (Chen et al. 2022a, b; Xiong et al. 2022).

9.4.2.1 Potential of Stem Cell-Derived Exosomes in Neurovascular Remodeling

Exosomes cater to numerous attributes of normal brain physiology owing to their essential role in the mediation of intercellular communication and transmission of biological signals between brain cells, the brain, and the periphery, thereby facilitating crucial interactions between them. Latest studies have demonstrated the ability of exosomes in influencing new blood vessel formation and glial participation by modulating proteins and nucleotide secretion, allowing them to take part in the regeneration of nervous tissue following cerebral ischemia (Luo et al. 2019; Kim et al. 2020). Neurovascular remodeling refers to the interaction between neurovascular unit components, i.e., neurons, endothelial cells, astrocytes, and glial cells, utilizing various trophic factors and signaling molecules in a manner that establishes a suitable biological microenvironment for facilitating neurological recovery and healing (Hermann et al. 2015; Gandhi and Tsoumpas 2019). Stem cell-derived exosomes, particularly those harbored from MSCs, have become the prime target of present-day research since they possess greater potency in tissue regeneration and repair compared to their parent stem cell counterparts (Kalani and Tyagi 2015). They replace the therapeutic advantages of administering parent MSCs in hastening recovery after stroke and neural injury (Zerna et al. 2018). They are internalized specifically by the target cells, thereby avoiding a variety of prospective problems linked with the administration of living cells, and offer curative benefits comparable to their cellular source (Gervois et al. 2016; Powers et al. 2019). The characteristic features of exosome-based therapy that give it an advantage over stem cell therapy include the following:

1. Low immunogenicity, oncogenicity, and toxicity: this property of exosomes makes them excellent candidates for drug delivery as they can easily escape the defense mechanism of the host immune system and release therapeutic particles to the target cells (Yang et al. 2017a, b).
2. They can overcome vascular obstructive effects thereby decreasing the likelihood of secondary microvascular thrombosis (Xin et al. 2014).
3. Their small size enables them to evade the phagocytosis of macrophages and sustain lysosomal degradation more naturally and stably compared to polymer nanoparticles and liposomes.
4. They can cross the BBB and enter cerebral parenchyma when systemically injected, delivering functional biomolecules to the recipient cells for controlling their gene expression. Moreover, since they can easily traverse the BBB and reach the CSF or peripheral blood, they can prospectively be used as biomarkers for stroke diagnosis and prognosis (Li et al. 2017, 2021a, b).
5. Potential for large-scale development of cellular factories harboring tailored therapeutic vesicles.
6. Ability to enhance the transfer of biomolecules from stem cells to target tissues as well as intensify ligand-gated signaling pathways.
7. They produce therapeutic effects by transferring their cargos, particularly miRNAs, to the target cells/organs, modulating a variety of physiological pathways to improve clinical efficacy. Inside the CNS, they can activate the regenerative and defensive pathways more efficiently since they can be readily engineered and enriched with modified cargos (miRNAs) (Chen and Chopp 2018; Ghoreishy et al. 2019).

Although the extent to which clinical evaluation of exosome therapeutics has been performed is extremely limited, promising efficacy of the same has been associated with and observed in animal replicas of ischemic stroke (Huang et al. 2020). MSCs produce more exosomes than other cell types. Otero-Ortega et al. discovered >2000 proteins through proteomic analysis in MSC-extracted exosomes, the majority of which are essential contributors to neurological remodeling (Otero-Ortega et al. 2021). Experimental data has confirmed a considerable improvement in ischemic recovery along with induction of anti-inflammation and inhibition of cell death via MSC-exosomes. Yan et al. found that the mir-421/circhipk3/foxo3a pathway is successful in ceasing pyrolysis and relieving ischemic muscle injury in exosomes produced from HUCMSCs (Yan et al. 2020). Similar experimental evidence demonstrating the potential of hypoxia-inducible factor-1 (HIF1) in enhancing fracture repair and angiogenesis was reported by Zhang et al. (Zhang et al. 2019a, b, c). Furthermore, while exosome-based therapy from ADMSCs reduced brain infarct size, ameliorated white matter repair and integrity of fiber tract, and showed enhancement of neurological performance restoration in stroke-prone rats (Chen et al. 2016; Otero-Ortega et al. 2017a, b), analysis by Doeppner et al.

illustrated the efficacy of exosomes isolated from BMMSCs in decreasing suppression of the peripheral immune system, boosting cerebrovascular reorganization, and enhancing motor activity 28 days post-ischemia (Doeppner et al. 2015). In addition to MSC-harbored exosomes, exosomes secreted by neuro-endothelial cells and astrocytes also aid in the defense mechanism of the neural system against stroke (Jiang et al. 2022).

As highlighted above, engineered MSC-derived exosomes with specific miRNA sequences exhibit more powerful therapeutic outcomes in stroke and TBI compared with naive MSC-derived exosomes (Table 9.2) (Zhang et al. 2019a, b, c). After a stroke, brain remodeling and functional recovery were improved by MSC-derived exosomes that overexpressed miR-133b, were rich in the miR-17-92 cluster, or contained miR-138-5p (Xin et al. 2017; Deng et al. 2019; Jiang et al. 2022). Thus, stem cell-derived exosomes may possess therapeutic advantages that are enhanced by modifications directed towards ≥2 major miRNAs. Moreover, the reduction of apoptosis and ischemic injury can be positively impacted by the overexpression of pigment epithelium-derived factor (PEDF) in exosomes (Huang et al. 2018). The promising ability of exosomes in therapeutic settings strongly showing VEGF, hepatocyte growth factor (HGF), and BDNF expression has also been implicated. Further on, 98% of small biomolecular agents fail to penetrate the BBB effectively to reach the brain (Lakhal and Wood 2011). Hence, exosomes will be useful in delivering such biopharmaceuticals (like curcumin and enkephalin) to the brain and simultaneously demonstrating significant therapeutic effects (Tian et al. 2018; Liu et al. 2019).

Exosome surface alterations may also improve the ability to target a particular cell. Rabies virus glycoprotein (RVG) was coupled and engineered with protein lysosome-associated membrane glycoprotein 2b (Lamp2b) to bind exosomes, enabling specific neuronal targeting (Alvarez-Erviti et al. 2011). RVG-exosomes were found by Yang et al. to effectively transport miR-124 to ischemic sites and improve the protection of neurons following an infarct injury (Zhang et al. 2017a, b). There is considerable evidence of the prospective significance of MSC-isolated exosomes in promoting the positive outcomes of cell-based treatments for stroke and TBI. Thus, exosome-based therapy is a powerful approach emerging as a remedy for neurological ailments; however, further studies and elaborate research on the modification of exosomes for stroke therapy/recovery are still required.

9.4.2.2 Therapeutic Effects of Stem Cell-Derived Exosomes in Neuroinflammation and Stroke

Extensive preclinical examination has revealed that systemic administration of stem cell-isolated exosomes mediates favorable outcomes by escalating several endogenous brain repair processes, thus enabling ischemic rehabilitation (Manuel et al. 2017; Bang and Kim 2019). Brain restoration following an ischemic stroke entails a sequence of these distinctly interactive events—formation of new blood vessels, neurons, and oligodendrocytes, white matter remodeling, anti-apoptosis, anti-inflammation, and immune responses, among others—that work in unison to facilitate the restoration of neurovascular units and promote neurological recovery

Table 9.2 Proposed effects of exosomal miRNAs in ischemic stroke remission

miRNAs	Origin	Advocated clinical responses	Pathway involved
miR-133b (Xin et al. 2017; Deng et al. 2019; Jiang et al. 2022)	Mesenchymal stem cells	Neural remodeling	Connective tissue growth factor (CTGF)
miR-17-92 (Xin et al. 2017, 2021)		Regeneration of neurons	PTEN/Akt/mTOR
miR-138-5p (Deng et al. 2019, 2022; Hou et al. 2020)		Reduction in inflammatory response and cell death	Lipocalin 2
miR-134 (Huang et al. 2015; Xiao et al. 2018)		Reduced cell death	Caspase-8
miR-30d-5p (Jiang et al. 2018; Zhang et al. 2021a, b)		Reduction in inflammatory response and cell death	Beclin-1/ autophagy-related proteins (Atg5)
miR-223-3p (Zhao et al. 2020a, b)		Reduction in inflammatory response	CysLT2R-ERK1/2
miR-1906 (Haupt et al. 2021)		Anti-inflammation	TLR4
miR-21-3p (Li et al. 2019)		Blood-brain barrier protection Reduction in inflammatory response and cell death	Methionine adenosyltransferase 2β (MAT2B)
miR-146a-5p (Wang et al. 2020a, b; Zhang et al. 2021a, b)		Anti-inflammation	IRAK1/TRAF6
miR-22-3p (Jiang et al. 2018; Zhang et al. 2021a, b)		Anti-apoptosis	KDM6B/BMP2/ BMF axis
miR-126 (Wang et al. 2008, 2020a, b)	Endothelial progenitor cells	Regeneration of neurons and new blood vessels generation	Caspase-3; VEGF receptor 2 (VEGFR 2)
miR-124 (Yang et al. 2017a, b; Vizoso et al. 2017; Song et al. 2019; Wei et al. 2022)	Mesenchymal stem cells M2 microglia	Regeneration of neurons Anti-apoptosis	GLI3 STAT3 USP14
miR-34c (Wu et al. 2020)	Astrocyte	Reduction in inflammatory response and cell death	TLR/MAPK; NF-κB/MAPK pathways

(Dabrowska et al. 2019). Present-day research illustrated that after ischemic events, the exosomal content as well as their synthesis and secretion is drastically altered, proposing them as novel targets of the disease. Through the abovementioned regenerative mechanisms, exosomes harbored from various cell types, such as MSCs, NSCs, neurons, bioengineered cells, microglia, astrocytes, and endothelial as well as their progenitor cells, can functionally enhance the reconstruction and recovery of the neurovascular unit as well as regulate poststroke inflammation. miRNAs are key to influencing exosome-based repair processes. Exosomes bioengineered with modified cargo or altered surfaces generally show stronger therapeutic effects (Yang et al. 2017a, b; Jiang et al. 2022).

Exosomes from stem cells mediate their therapeutic effects either directly or indirectly:

- *Direct effects:* Activation of healthy resident neural cells is the main goal of neuroregenerative therapy so as to promote inherent brain remodeling operations (Fisher and Saver 2015). MSC-exosomes or fibroblasts fulfill this restorative requirement by accelerating dendritic and axonal development directly, according to in vitro studies (Tassew et al. 2017; Yang et al. 2017a, b). Clinical evidence, for instance, demonstrates that exosomes from human and rodent cerebral endothelial cells as well as adult mouse NSCs initiate angiogenesis and neurogenesis seen during stroke rehabilitation (Haqqani et al. 2013; Pan et al. 2016; Yang et al. 2017a, b).
- *Indirect effects:* Studies have shown that exogenously delivered exosomes have indirect neurorestorative effects added to the direct impacts that they exhibit on cerebro-parenchymal cell activity. The secretion of inhabitant exosomes in a focal brain ischemia rat model, from astrocytes, is stimulated by the administration of exosomes originating from MSCs, which supplemented the neurite outgrowth of cultured cortical neurons, suggesting their contribution to the curative effects of MSC-derived exosomes on ischemic brain plasticity (Xin et al. 2017). Similarly, another study implicated the importance of the recipient cells' secretomes in the therapeutic responses of exosomes. The findings demonstrated a variation in the cargo profile of glial fibrillary acidic protein (GFAP)-positive astrocyte-derived exosomes in the ischemic neural tissue in both untreated and MSC-derived exosome-treated astrocytes, with the latter enhancing brain plasticity more superiorly than the former (Xin et al. 2017). This led to the conclusion that exogenously supplied exosomes promote neurorestoration by interacting with recipient brain cells, inducing them to discharge their exosomes for carrying out restorative functions through essential crosstalk of the recipient cells with other brain cells (Zhang et al. 2019a, b, c).

9.4.2.2.1 Promotion of Angiogenesis

A common strategy for overcoming inadequate blood supply following an ischemic stroke is to administer drugs that pharmacologically promote angiogenesis and reestablish blood flow. The administration of miRNAs, proteins, and lipid-rich exosomes could be an unconventional approach for the same. It was discovered in animal ischemic models that exosomes derived from BMMSCs improved the proliferation of cerebral endothelial cells (Xin et al. 2013a, b; Moon et al. 2019). Proteomic analysis has revealed that cerebral ischemic tissue confers protection by MSC-exosomes having angiogenic paracrine effects that augment the blood flow to the ischemic penumbra (Ghafouri-Fard et al. 2020; Chen et al. 2022a, b). These exosomes via their miRNAs encourage angiogenesis in ischemic tissue, while cerebrovascular endothelial cells can migrate to the ischemic region and control vascular integrity when CXCL12 released by exosomes binds to CXCR4 (Liu et al. 2022). Additionally, exosome miRNAs stimulate angiogenesis by activating the TRPM7-TIPM3/HIF-1/VEGF or PTEN-PIK3-Akt signaling cascades (Xiong et al. 2022). A known target of miR-126 for regulating endothelial cell activity and angiogenesis is vascular cell adhesion protein 1 (VCAM1) (Wang et al. 2008). Exosomes from endothelial progenitor cells (EPCs) were found to encourage angiogenesis and neurogenesis in diabetic ischemic stroke mice, according to research by Wang et al. The generated EPC-derived exosomes showed increased therapeutic activity after miR-126 enrichment (Wang et al. 2020a, b). Furthermore, exosomes rich in miRNA 181b-5p released by ADMSCs, through the suppression of transient receptor potential melastatin 7 (TRPM7) coupled with upregulation of HIF-1α and VEGF, demonstrated favorable effects in controlling angiogenesis after stroke (Yang et al. 2018). An essential factor governing BBB integrity and angiogenesis in endothelial cells and pericytes is the Dll4-Notch signaling pathway because exosomes containing Dll4 protein from human microvascular endothelial cells also influenced angiogenesis. However, it is necessary to conduct further studies on stroke models for the validation of these experiments (Sharghi-Namini et al. 2014).

9.4.2.2.2 Stimulation of Neurogenesis and Cell Structure Growth

Neurogenesis along with angiogenesis is a vital process essential for ischemic stroke rehabilitation. Numerous findings have reported that exosome-based therapy alters brain stem cells and encourages neurogenesis (Jiang et al. 2022). Following an ischemic injury, exosomes isolated from BMMSCs may encourage the growth of cerebral neurons (Xin et al. 2013a, b). MiR-124 plays a critical role in neurogenesis since it is widely expressed in neural tissues. The upregulation of miR-124 in ischemic regions post middle cerebral artery occlusion (MCAO) led to its overexpression, resulting in neuronal differentiation (Åkerblom et al. 2012; Sun et al. 2013). According to Yang et al., exosomes supplemented with miR-124 transform the neural progenitor cells into neuronal lineages, considerably alleviating ischemic injury (Yang et al. 2017a, b; Song et al. 2019). Moreover, while overexpression of miR-133b in BMMSC-exosomes triggered the astrocytes to secrete secondary exosomes as well as decreased glial scar thickness, enrichment of Ex-Zeb2/Axin2 in the same reduced Wnt/−catenin expression in MCAO rats, promoting neurological recovery and enhancing neural plasticity (Vizoso et al. 2017; Wei et al. 2022). Further on, heightened miR-17-92 levels in MSC-harvested exosomes led to the downregulation of PTEN and activation of the PI3K/Akt/mTOR signaling pathway, causing the emergence of primary cortical neurons with axonal growth, oligodendrogenesis, and neuroplasticity in the MCAO rats (Xin et al.

2017). Stimulation of oligodendrogenesis during neural development further increases stroke-induced neurogenesis (Xin et al. 2021), whereas inactivation of GSK-3β, a serine/threonine protein kinase crucial for axon regeneration, promotes functional neurological recovery in the CNS (Arciniegas Ruiz and Eldar-Finkelman 2022). To conclude, utilizing exosomes enriched with active components can significantly increase the chances of improving neurological function and rejuvenating the neural tissue after an ischemic stroke.

9.4.2.2.3 Anti-Apoptosis (Inhibition of Cell Death)

Apoptosis has emerged as a pivotal target for designing therapeutic interventions due to its noteworthy involvement in the pathogenesis of stroke as several neurons die as a result of brain ischemia and ensuing neuroinflammation (Uzdensky 2019). Substantial evidence suggests that exosomes can inhibit apoptosis and relieve cerebral injury in stroke models (Pei et al. 2019). Exosomes isolated from various cell types show neuroprotective effects in ischemia-induced neuronal death (Jiang et al. 2022). MiR-134 plays a vital role in regulating neuronal apoptosis after ischemia-reperfusion injury (Huang et al. 2015). Since caspase-8 and CREB protein are specific targets of miR-134, BMMSC-derived exosomes enriched with miR-134 inhibited oligodendrocyte apoptosis by caspase-8-dependent downregulation of the apoptotic pathway, making it an essential novel target for stroke therapy (Xiao et al. 2018). Furthermore, neutrophil gelatinase-associated lipocalin (LCN2), which is abundantly expressed after a stroke and aggravates neuronal death and cerebral injury, is strongly suppressed by miR-138-5p or miR-29b-3p. While overexpression of these miRNAs in BMMSC-exosomes reportedly inhibits apoptosis of oxygen-glucose-deprived injured astrocytes and enhances neuronal survival by downregulating caspase-3, LCN2, and Bax levels while upregulating Bcl-2 levels (Deng et al. 2019; Hou et al. 2020), miR-26a-5p-rich exosomes from MSCs show microglial cell death inhibition by CDK6 targeting (Cheng et al. 2021). Moreover, ischemic models also demonstrate the protective effects of miR-30d-5p- and miR-22-3p-abundant exosomes against brain infarction and neuronal apoptosis, the latter mediating them via the KDM6B/BMP2/BMF axis (Jiang et al. 2018; Zhang et al. 2021a, b). Additionally, PEDF overexpression in ADMSC-derived exosomes averted brain ischemia-reperfusion injury in stroke rat models by controlling apoptotic parameters and stimulating autophagy (Huang et al. 2018). Following a rat stroke, enkephalin delivery via exosomes from BMMSCs led to improved neuronal density and hastened neurological recovery by suppressing p53, caspase-3, and nitric oxide expression (Liu et al. 2019). Overall, the critical role played by exosomes in the inhibition of apoptosis following ischemic stroke makes it worthy of further research and analysis for developing therapeutic agents for stroke in the future.

9.4.2.2.4 Reduction of Neuroinflammation by Regulation of Inflammatory Mediators

Inflammation is a key pathogenic outcome of cerebral ischemia that exacerbates injury and damage to the brain. Post an ischemic attack, reperfusion triggers an inflammatory cascade, an event crucial for causing secondary neuronal damage and apoptosis (Xiong et al. 2022). Acute ischemic-reperfusion injury can be abrogated by MSC-derived exosomes by modulating IL-6, TNF-α, IL-1β (pro-inflammatory), and IL-4, IL-10 (anti-inflammatory) cytokine levels and suppressing inflammation in microglia (Jiang et al. 2018). As mentioned earlier, the transformation of microglia to M1 or M2 phenotypes is essentially influenced by alterations in the microenvironment of brain tissue (Spellicy and Stice 2021). Synchronizing the M1–M2 microglial transition could be a prospective treatment plan for stroke and TBI. MiR-30d-5p-supplemented exosomes, by suppressing Beclin-1, Atg5-expression and inhibiting activation of polarization in M1 microglia after ischemia, significantly decreased inflammatory cytokine levels and affected the microglial phenotypes (Jiang et al. 2018, 2022). Furthermore, experimental evidence shows that even miR-26b-5p-loaded HUCMSC-derived exosomes inhibit M1 polarization and promote M2 polarization by inactivating the TLR pathway, thereby alleviating autophagy-induced cerebral injury (Li et al. 2020). Zhao et al. illustrated that miR-223-3p-enhanced exosomes isolated from BMMSCs reversed the CysLT2R-ERK1/2-induced M1 microglial polarization, thereby exhibiting anti-inflammatory properties, made possible because miR-223-3p in the exosomes significantly suppressed the CysLT2R expression (cysteinyl leukotriene that acts as a pivotal inflammatory medium for aggravating ischemic tissue injury and necrosis) in microglia after stroke, leading to inhibition of pro-inflammatory cytokine production and stimulation of anti-inflammatory and neurotrophic factor secretion (Zhao et al. 2020a, b). Moreover, the most abundant miRNA in MSC-exosomes, miR-223, present in a highly encapsulated form, stimulates the anti-inflammatory cytokines by effectively mediating the NMLTC4/ltd4-induced transformation of harmful M1 phenotype to useful M2 phenotype of microglia (Morales et al. 2022; Xiong et al. 2022). Thus, all accumulating evidence indicates that a high cerebral M2/M1 ratio corresponds to decreased inflammation and increased neurorestorative outcomes (Gao et al. 2016; Kumar et al. 2016). However, several other anti-inflammatory mechanisms exist in addition to the M1–M2 microglial phenotype interconversion. For instance, Liu et al. demonstrated that exosomes released by BMMSCs play an essential role in abrogating ischemia-reperfusion injury by hindering inflammation and pyroptosis induced by the NLRP3 inflammasome (Liu et al. 2021). Overexpression of miR-138-5p in these exosomes leads to the downregulation of the iron transporter LCN2 (secreted in response to cerebral injury), thereby delaying inflammation (Deng et al. 2022). MiR-138-5p- or miR-1906-abundant exosomes, through the suppression of pro-inflammatory signaling cascades, were found to inhibit inflammatory responses and improve stroke rehabilitation (Deng et al. 2019; Haupt et al. 2021). Furthermore, ADMSC-derived exosomes attenuate neuroinflammation and inhibit apoptosis by upregulating MAT2B and downregulating miR-21-3p expression in hypoxia/reoxygenation-treated cells (Li et al. 2019). Studies show that while miR-34c-rich astrocyte-derived exosomes demonstrate neuroprotective properties by downscaling TLR7 and NF-κB/MAPK pathways (Wu et al. 2020), miR-542-3p-enhanced exosomes by moderating TLR4 in glial cells, show suppression of inflammatory responses, stroke-ensued apoptosis, and ROS (Cai et al. 2019). Additionally, exosomes engineered with miR-126 promote functional recovery after stroke by alleviating neuroinflammation and increasing neuronal development (Geng et al. 2019). Moreover, experimental data revealed that exosomes secreted by HUCMSCs high in miR-146a-5p attenuated microglia-induced neuroinflammation and neurological deficits via the IRAK1/TRAF6 pathway by downregulating the levels of IRAK1 and TRAF6, both of which are crucial for activating pro-inflammatory gene expression (Wang et al. 2020a, b; Zhang et al. 2021a, b). Hence, in light of the above data, it can be concluded that targeting specific exosomes for their anti-inflammatory properties may prove to be a key defense mechanism against ischemic injury and stroke.

9.5 Conclusion and Future Considerations

Ischemic stroke is the primary reason of sickness and death worldwide, and complications may make an early diagnosis and treatment challenging. Stroke, the third most prevalent cause of death, is evolving as a result of the introduction of new viewpoints on neurodegeneration. Historically, stroke was believed to be only a blood vessel issue, but research has expanded to take into account the interaction of glia, neurons, vascular cells, and matrix elements, conjointly termed as the neurovascular unit. As a result of acute strokes, which are primarily ischemic, secondary neuroinflammation occurs, which promotes further damage and death of cells, while also promoting healing. By stimulating local cells and inducing the influx of many types of inflammatory cells into the ischemic region (neutrophils, monocytes/macrophages, and several T-cell subtypes), immune mediators contribute to brain injury.

Transplantation of MSCs into ischemic stroke animal models has been shown to alter immune response; operate as a neuroprotector; induce neurogenesis, astrogenesis, and oligodendrogenesis; and activate angiogenesis, according to experimental research. Recanalization, ischemic penumbra preservation, and infarct size reduction are the three main focuses of ischemic stroke therapy. Mechanical recanalization and intravenous thrombolysis are two vascular reconstruction techniques for acute ischemic stroke. While both treatments are clinically successful for some patients, their clinical effectiveness is constrained by the rigorous selection criteria associated with their short therapeutic timeframes.

In addition to the treatment of neurodegenerative disorders, MSCs show great potential in treating ischemic injuries. They exhibit homing—translocation to areas of inflammation and injury—to reduce inflammation and demonstrate neuroprotective effects. Clinical trials on patients after the onset of ischemic stroke demonstrated no significant side effects of treatment with MSCs, showing the beneficial outcome of cell transplantation in some trials. Due to their properties of self-renewability and multipotency as well as easy accessibility (can be obtained without invasive procedures) in addition to their genomic stability that makes them culturally expandable in vitro with few ethical issues, MSCs are important tools in cell therapy, regenerative medicine, and tissue repair.

Exosomes are a promising therapeutic substitute for cell-based treatments, and MSCs are deemed the preferred cell choice for harvesting exosomes owing to their scalable capacity, tissue regeneration, antitumor, anti-inflammatory, immunomodulatory, and paracrine secretory properties (Vakhshiteh et al. 2019). While BMMSCs and ADMSCs are more frequently used for extracting exosomes compared to the HUCMSCs, the latter has demonstrated greater viability and lower susceptibility to graft rejection as well as easier acceptance among the patients (Norouzi-Barough et al. 2022). MSC-exosomes perform multifaceted roles in the regulation of pathophysiological processes; even though they retain most of the functions of their parent MSCs, they have additional advantages over cell-based therapies including small size; BBB permeability; low immunogenicity, oncogenicity, and toxicity; ability to transfer desired cargos to the target sites; and potential to be exploited as personalized targeted drug delivery vehicles, among others. These exosomes, upon extensive evaluation of their therapeutic potential in animal stroke models, were found to significantly reduce neuroinflammation, inhibit apoptosis, and boost angiogenesis, neurogenesis, and white matter remodeling (Chen and Chopp 2018). Of all the cargos, miRNAs are of prime importance in mediating the exosome-induced neurorestorative and neuroprotective effects following an ischemic stroke; hence, exosomes engineered with modified miRNA show better therapeutic efficacy compared to their naïve counterparts (Zhang et al. 2019a, b, c). In addition, a high cerebral M2/M1 ratio also corresponds to neuroregenerative and anti-inflammatory outcomes. Therefore, it can be clearly said that stem cell-derived exosomes are prospective treatment tools for stroke in clinical practice, but whose therapeutic efficacy needs to be warranted in humans through further research, by extensively conducting clinical trials aimed at investigating the safety, time window, and dose responses in patients. Issues like scaling the production of exosomes for human studies, enhancing their targeting ability, overcoming drug resistance, and extension of exosome half-lives also need to be addressed in the future. Further on, it is necessary to understand how transplanted cells can survive longer, how most exosomes derived from MSCs can lead to infarction, and the manner in which their distribution can be identified without causing damage to any tissue (Chen and Chopp 2018; Jiang et al. 2022). Additionally, future studies need to shift their attention towards the anti-inflammatory benefits of the M2 microglial phenotype along with the role of miRNAs in regulating the inflammatory network (Lian et al. 2021). Through this chapter, we briefed on stroke pathophysiology and discussed the significance of MSCs and the promising effects of MSC-derived exosomes as novel therapeutic agents in promoting functional neurological recovery, in the hope that it aids in providing future perspective and an insight into the potential benefits of MSC-exosomes in stroke therapy.

References

Ahmad M, Dar NJ, Bhat ZS, Hussain A, Shah A, Liu H, Graham SH (2014) Inflammation in ischemic stroke: mechanisms, consequences and possible drug targets. CNS Neurol Disord Drug Targets 13(8):1378–1396. https://doi.org/10.2174/1871527313666141023094720

Åkerblom M, Sachdeva R, Barde I, Verp S, Gentner B, Trono D, Jakobsson J (2012) MicroRNA-124 is a subventricular zone neuronal fate determinant. J Neurosci Off J Soc Neurosci 32(26): 8879–8889. https://doi.org/10.1523/JNEUROSCI.0558-12.2012

Alhadidi Q, Bin Sayeed MS, Shah ZA (2016) Cofilin as a promising therapeutic target for ischemic and Hemorrhagic stroke. Transl Stroke Res 7(1):33–41. https://doi.org/10.1007/s12975-015-0438-2

Alvarez-Erviti L, Seow Y, Yin H, Betts C, Lakhal S, Wood MJ (2011) Delivery of siRNA to the mouse brain by systemic injection of targeted exosomes. Nat Biotechnol 29(4):341–345. https://doi.org/10.1038/nbt.1807

An SJ, Kim TJ, Yoon BW (2017) Epidemiology, risk factors, and clinical features of intracerebral Hemorrhage: an update. J Stroke 19(1):3–10. https://doi.org/10.5853/jos.2016.00864

Andersen KK, Olsen TS, Dehlendorff C, Kammersgaard LP (2009) Hemorrhagic and ischemic strokes compared: stroke severity, mortality, and risk factors. Stroke 40(6):2068–2072. https://doi.org/10.1161/STROKEAHA.108.540112

Anthony S, Cabantan D, Monsour M, Borlongan CV (2022) Neuroinflammation, stem cells, and stroke. Stroke 53(5):1460–1472. https://doi.org/10.1161/STROKEAHA.121.036948

Arciniegas Ruiz SM, Eldar-Finkelman H (2022) Glycogen synthase Kinase-3 inhibitors: preclinical and clinical focus on CNS-A decade onward. Front Mol Neurosci 14:792364. https://doi.org/10.3389/fnmol.2021.792364

Baksh D, Yao R, Tuan RS (2007) Comparison of proliferative and multilineage differentiation potential of human mesenchymal stem cells derived from umbilical cord and bone marrow. Stem Cells (Dayton, Ohio) 25(6):1384–1392. https://doi.org/10.1634/stemcells.2006-0709

Bang OY, Kim EH (2019) Mesenchymal stem cell-derived extracellular vesicle therapy for stroke: challenges and Progress. Front Neurol 10:211. https://doi.org/10.3389/fneur.2019.00211

Bang OY, Lee JS, Lee PH, Lee G (2005) Autologous mesenchymal stem cell transplantation in stroke patients. Ann Neurol 57(6):874–882. https://doi.org/10.1002/ana.20501

Bhaskar S, Stanwell P, Cordato D, Attia J, Levi C (2018) Reperfusion therapy in acute ischemic stroke: dawn of a new era? BMC Neurol 18(1):8. https://doi.org/10.1186/s12883-017-1007-y

Boltze J, Wagner DC, Henninger N, Plesnila N, Ayata C (2016) Phase III preclinical trials in translational stroke research: community response on framework and guidelines. Transl Stroke Res 7(4):241–247. https://doi.org/10.1007/s12975-016-0474-6

Bonaguidi MA, Wheeler MA, Shapiro JS, Stadel RP, Sun GJ, Ming GL, Song H (2011) In vivo clonal analysis reveals self-renewing and multipotent adult neural stem cell characteristics. Cell 145(7):1142–1155. https://doi.org/10.1016/j.cell.2011.05.024

Bond AM, Ming GL, Song H (2015) Adult mammalian neural stem cells and neurogenesis: five decades later. Cell Stem Cell 17(4):385–395. https://doi.org/10.1016/j.stem.2015.09.003

Bronckaers A, Hilkens P, Martens W, Gervois P, Ratajczak J, Struys T, Lambrichts I (2014) Mesenchymal stem/stromal cells as a pharmacological and therapeutic approach to accelerate angiogenesis. Pharmacol Ther 143(2):181–196. https://doi.org/10.1016/j.pharmthera.2014.02.013

Bruyneel AA, Sehgal A, Malandraki-Miller S, Carr C (2016) Stem cell therapy for the heart: blind alley or magic bullet? J Cardiovasc Transl Res 9(5–6):405–418. https://doi.org/10.1007/s12265-016-9708-y

Cai W, Dai X, Chen J, Zhao J, Xu M, Zhang L, Yang B, Zhang W, Rocha M, Nakao T, Kofler J, Shi Y, Stetler RA, Hu X, Chen J (2019) STAT6/Arg1 promotes microglia/macrophage efferocytosis and inflammation resolution in stroke mice. JCI Insight 4(20):e131355. https://doi.org/10.1172/jci.insight.131355

Castro-Manrreza ME, Montesinos JJ (2015) Immunoregulation by mesenchymal stem cells: biological aspects and clinical applications. J Immunol Res 2015:394917. https://doi.org/10.1155/2015/394917

Catanese L, Tarsia J, Fisher M (2017) Acute ischemic stroke therapy overview. Circ Res 120(3): 541–558. https://doi.org/10.1161/CIRCRESAHA.116.309278

Chamorro Á, Dirnagl U, Urra X, Planas AM (2016) Neuroprotection in acute stroke: targeting excitotoxicity, oxidative and nitrosative stress, and inflammation. Lancet Neurol 15(8): 869–881. https://doi.org/10.1016/S1474-4422(16)00114-9

Chen H, Xue R, Huang P, Wu Y, Fan W, He X, Dong Y, Liu C (2022a) Modified exosomes: a good transporter for miRNAs within stem cells to treat ischemic heart disease. J Cardiovasc Transl Res 15(3):514–523. https://doi.org/10.1007/s12265-022-10216-1

Chen J, Chopp M (2018) Exosome therapy for stroke. Stroke 49(5):1083–1090. https://doi.org/10.1161/STROKEAHA.117.018292

Chen J, Li Y, Katakowski M, Chen X, Wang L, Lu D, Lu M, Gautam SC, Chopp M (2003) Intravenous bone marrow stromal cell therapy reduces apoptosis and promotes endogenous cell proliferation after stroke in female rat. J Neurosci Res 73(6):778–786. https://doi.org/10.1002/jnr.10691

Chen J, Li Y, Wang L, Zhang Z, Lu D, Lu M, Chopp M (2001) Therapeutic benefit of intravenous administration of bone marrow stromal cells after cerebral ischemia in rats. Stroke 32(4): 1005–1011. https://doi.org/10.1161/01.str.32.4.1005

Chen KH, Chen CH, Wallace CG, Yuen CM, Kao GS, Chen YL, Shao PL, Chen YL, Chai HT, Lin KC, Liu CF, Chang HW, Lee MS, Yip HK (2016) Intravenous administration of xenogenic adipose-derived mesenchymal stem cells (ADMSC) and ADMSC-derived exosomes markedly reduced brain infarct volume and preserved neurological function in rat after acute ischemic stroke. Oncotarget 7(46):74537–74556. https://doi.org/10.18632/oncotarget.12902

Chen W, Huang W, Yang Y, Li K (2022b) Methylglyoxal scavengers attenuate angiogenesis dysfunction induced by methylglyoxal and oxygen-glucose deprivation. Oxidative Med Cell Longev 2022:8854457. https://doi.org/10.1155/2022/8854457

Cheng C, Chen X, Wang Y, Cheng W, Zuo X, Tang W, Huang W (2021) MSCs-derived exosomes attenuate ischemia-reperfusion brain injury and inhibit microglia apoptosis might via exosomal miR-26a-5p mediated suppression of CDK6. Mol Med 27(1):67. https://doi.org/10.1186/s10020-021-00324-0

Christian KM, Song H, Ming GL (2014) Functions and dysfunctions of adult hippocampal neurogenesis. Annu Rev Neurosci 37:243–262. https://doi.org/10.1146/annurev-neuro-071013-014134

Cosenza S, Toupet K, Maumus M, Luz-Crawford P, Blanc-Brude O, Jorgensen C, Noël D (2018) Mesenchymal stem cells-derived exosomes are more immunosuppressive than microparticles in inflammatory arthritis. Theranostics 8(5):1399–1410. https://doi.org/10.7150/thno.21072

Crigler L, Robey RC, Asawachaicharn A, Gaupp D, Phinney DG (2006) Human mesenchymal stem cell subpopulations express a variety of neuro-regulatory molecules and promote neuronal cell survival and neuritogenesis. Exp Neurol 198(1):54–64. https://doi.org/10.1016/j.expneurol.2005.10.029

Cunningham CJ, Wong R, Barrington J, Tamburrano S, Pinteaux E, Allan SM (2020) Systemic conditioned medium treatment from interleukin-1 primed mesenchymal stem cells promotes recovery after stroke. Stem Cell Res Ther 11(1):32. https://doi.org/10.1186/s13287-020-1560-y

Dabrowska S, Andrzejewska A, Lukomska B, Janowski M (2019) Neuroinflammation as a target for treatment of stroke using mesenchymal stem cells and extracellular vesicles. J Neuroinflammation 16(1):178. https://doi.org/10.1186/s12974-019-1571-8

Das CK, Jena BC, Banerjee I, Das S, Parekh A, Bhutia SK, Mandal M (2019) Exosome as a novel shuttle for delivery of therapeutics across biological barriers. Mol Pharm 16(1):24–40. https://doi.org/10.1021/acs.molpharmaceut.8b00901

Davis RL, Weintraub H, Lassar AB (1987) Expression of a single transfected cDNA converts fibroblasts to myoblasts. Cell 51(6):987–1000. https://doi.org/10.1016/0092-8674(87)90585-x

Deng Y, Chen D, Gao F, Lv H, Zhang G, Sun X, Liu L, Mo D, Ma N, Song L, Huo X, Yan T, Zhang J, Miao Z (2019) Exosomes derived from microRNA-138-5p-overexpressing bone marrow-derived mesenchymal stem cells confer neuroprotection to astrocytes following ischemic stroke via inhibition of LCN2. J Biol Eng 13:71. https://doi.org/10.1186/s13036-019-0193-0

Deng Y, Chen D, Gao F, Lv H, Zhang G, Sun X, Liu L, Mo D, Ma N, Song L, Huo X, Yan T, Zhang J, Miao Z (2022) Correction to: exosomes derived from microRNA-138-5poverexpressing bone marrow-derived mesenchymal stem cells confer neuroprotection to astrocytes following ischemic stroke via inhibition of LCN2. J Biol Eng 16(1):8. https://doi.org/10.1186/s13036-022-00285-w

Doeppner TR, Herz J, Görgens A, Schlechter J, Ludwig AK, Radtke S, de Miroschedji K, Horn PA, Giebel B, Hermann DM (2015) Extracellular vesicles improve post-stroke Neuroregeneration and prevent postischemic immunosuppression. Stem Cells Transl Med 4(10):1131–1143. https://doi.org/10.5966/sctm.2015-0078

Dong BC, Li MX, Wang XY, Cheng X, Wang Y, Xiao T, Jolkkonen J, Zhao CS, Zhao SS (2020) Effects of CXCR7-neutralizing antibody on neurogenesis in the hippocampal dentate gyrus and cognitive function in the chronic phase of cerebral ischemia. Neural Regen Res 15(6): 1079–1085. https://doi.org/10.4103/1673-5374.270416

Dong Y, Benveniste EN (2001) Immune function of astrocytes. Glia 36(2):180–190. https://doi.org/10.1002/glia.1107

Dugue R, Nath M, Dugue A, Barone FC (2017) Roles of pro-and anti-inflammatory cytokines in traumatic brain injury and acute ischemic stroke. Mech Neuroinflamm 211:4901

Estrada R, Li N, Sarojini H, An J, Lee MJ, Wang E (2009) Secretome from mesenchymal stem cells induces angiogenesis via Cyr61. J Cell Physiol 219(3):563–571. https://doi.org/10.1002/jcp.21701

Eulalio A, Huntzinger E, Izaurralde E (2008) Getting to the root of miRNA-mediated gene silencing. Cell 132(1):9–14. https://doi.org/10.1016/j.cell.2007.12.024

Feigin VL, Stark BA, Johnson CO, Roth GA, Bisignano C, Abady GG, Hamidi S et al (2021) Global, regional, and national burden of stroke and its risk factors, 1990–2019: a systematic analysis for the global burden of disease study 2019. Lancet Neurol 20(10):795–820

Fisher M, Saver JL (2015) Future directions of acute ischaemic stroke therapy. Lancet Neurol 14(7): 758–767. https://doi.org/10.1016/S1474-4422(15)00054-X

Fugate JE, Rabinstein AA (2014) Update on intravenous recombinant tissue plasminogen activator for acute ischemic stroke. Mayo Clin Proc 89(7):960–972. https://doi.org/10.1016/j.mayocp.2014.03.001

Gandhi R, Tsoumpas C (2019) Preclinical imaging biomarkers for Postischaemic neurovascular remodelling. Contrast Media Mol Imaging 2019:3128529. https://doi.org/10.1155/2019/3128529

Gao B, Zhou S, Sun C, Cheng D, Zhang Y, Li X, Zhang L, Zhao J, Xu D, Bai Y (2020) Brain endothelial cell-derived exosomes induce neuroplasticity in rats with ischemia/reperfusion injury. ACS Chem Neurosci 11(15):2201–2213. https://doi.org/10.1021/acschemneuro.0c00089

Gao J, Grill RJ, Dunn TJ, Bedi S, Labastida JA, Hetz RA, Xue H, Thonhoff JR, DeWitt DS, Prough DS, Cox CS Jr, Wu P (2016) Human neural stem cell transplantation-mediated alteration of microglial/macrophage phenotypes after traumatic brain injury. Cell Transplant 25(10): 1863–1877. https://doi.org/10.3727/096368916X691150

GBD 2017 DALYs and HALE Collaborators (2018) Global, regional, and national disability-adjusted life-years (DALYs) for 359 diseases and injuries and healthy life expectancy (HALE) for 195 countries and territories, 1990-2017: a systematic analysis for the global burden of disease study 2017. Lancet (London, England) 392(10159):1859–1922. https://doi.org/10.1016/S0140-6736(18)32335-3

Geng W, Tang H, Luo S, Lv Y, Liang D, Kang X, Hong W (2019) Exosomes from miRNA-126-modified ADSCs promotes functional recovery after stroke in rats by improving neurogenesis and suppressing microglia activation. Am J Transl Res 11(2):780–792

Gervois P, Wolfs E, Ratajczak J, Dillen Y, Vangansewinkel T, Hilkens P, Bronckaers A, Lambrichts I, Struys T (2016) Stem cell-based therapies for ischemic stroke: preclinical results and the potential of imaging-assisted evaluation of donor cell fate and mechanisms of brain regeneration. Med Res Rev 36(6):1080–1126. https://doi.org/10.1002/med.21400

Ghafouri-Fard S, Shoorei H, Taheri M (2020) Non-coding RNAs participate in the ischemia-reperfusion injury. Biomed Pharmacother 129:110419. https://doi.org/10.1016/j.biopha.2020.110419

Ghoreishy A, Khosravi A, Ghaemmaghami A (2019) Exosomal microRNA and stroke: a review. J Cell Biochem 120(10):16352–16361. https://doi.org/10.1002/jcb.29130

Glennie S, Soeiro I, Dyson PJ, Lam EW, Dazzi F (2005) Bone marrow mesenchymal stem cells induce division arrest anergy of activated T cells. Blood 105(7):2821–2827. https://doi.org/10.1182/blood-2004-09-3696

Gould E (2007) How widespread is adult neurogenesis in mammals? Nature reviews. Neuroscience 8(6):481–488. https://doi.org/10.1038/nrn2147

Gowen A, Shahjin F, Chand S, Odegaard KE, Yelamanchili SV (2020) Mesenchymal stem cell-derived extracellular vesicles: challenges in clinical applications. Front Cell Dev Biol 8:149. https://doi.org/10.3389/fcell.2020.00149

Gurusamy N, Alsayari A, Rajasingh S, Rajasingh J (2018) Adult stem cells for regenerative therapy. Prog Mol Biol Transl Sci 160:1–22. https://doi.org/10.1016/bs.pmbts.2018.07.009

Han M, Cao Y, Guo X, Chu X, Li T, Xue H, Xin D, Yuan L, Ke H, Li G, Wang Z (2021) Mesenchymal stem cell-derived extracellular vesicles promote microglial M2 polarization after subarachnoid hemorrhage in rats and involve the AMPK/NF-κB signaling pathway. Biomed Pharmacother 133:111048. https://doi.org/10.1016/j.biopha.2020.111048

Haqqani AS, Delaney CE, Tremblay TL, Sodja C, Sandhu JK, Stanimirovic DB (2013) Method for isolation and molecular characterization of extracellular microvesicles released from brain endothelial cells. Fluids Barriers CNS 10(1):4. https://doi.org/10.1186/2045-8118-10-4

Hasegawa Y, Nakagawa T, Uekawa K, Ma M, Lin B, Kusaka H, Katayama T, Sueta D, Toyama K, Koibuchi N, Kim-Mitsuyama S (2016) Therapy with the combination of amlodipine and irbesartan has persistent preventative effects on stroke onset associated with BDNF preservation on cerebral vessels in hypertensive rats. Transl Stroke Res 7(1):79–87. https://doi.org/10.1007/s12975-014-0383-5

Hassanpour M, Rezabakhsh A, Rezaie J, Nouri M, Rahbarghazi R (2020) Exosomal cargos modulate autophagy in recipient cells via different signaling pathways. Cell Biosci 10:92. https://doi.org/10.1186/s13578-020-00455-7

Haupt M, Zheng X, Kuang Y, Lieschke S, Janssen L, Bosche B, Jin F, Hein K, Kilic E, Venkataramani V, Hermann DM, Bähr M, Doeppner TR (2021) Lithium modulates miR-1906 levels of mesenchymal stem cell-derived extracellular vesicles contributing to poststroke neuroprotection by toll-like receptor 4 regulation. Stem Cells Transl Med 10(3):357–373. https://doi.org/10.1002/sctm.20-0086

Hermann DM, Buga AM, Popa-Wagner A (2015) Neurovascular remodeling in the aged ischemic brain. J Neural Transm (Vienna, Austria: 1996) 122(Suppl 1):S25–S33. https://doi.org/10.1007/s00702-013-1148-0

Hira K, Ueno Y, Tanaka R, Miyamoto N, Yamashiro K, Inaba T, Urabe T, Okano H, Hattori N (2018) Astrocyte-derived exosomes treated with a Semaphorin 3A inhibitor enhance stroke recovery via prostaglandin D_2 synthase. Stroke 49(10):2483–2494. https://doi.org/10.1161/STROKEAHA.118.021272

Hong SB, Yang H, Manaenko A, Lu J, Mei Q, Hu Q (2019) Potential of exosomes for the treatment of stroke. Cell Transplant 28(6):662–670. https://doi.org/10.1177/0963689718816990

Hou K, Li G, Zhao J, Xu B, Zhang Y, Yu J, Xu K (2020) Bone mesenchymal stem cell-derived exosomal microRNA-29b-3p prevents hypoxic-ischemic injury in rat brain by activating the

PTEN-mediated Akt signaling pathway. J Neuroinflammation 17(1):46. https://doi.org/10.1186/s12974-020-1725-8. (Retraction published J Neuroinflammation. 2020 Nov 23;17(1):352)

Huang M, Hong Z, Xiao C, Li L, Chen L, Cheng S, Lei T, Zheng H (2020) Effects of exosomes on neurological function recovery for ischemic stroke in pre-clinical studies: a meta-analysis. Front Cell Neurosci 14:593130. https://doi.org/10.3389/fncel.2020.593130

Huang W, Liu X, Cao J, Meng F, Li M, Chen B, Zhang J (2015) miR-134 regulates ischemia/reperfusion injury-induced neuronal cell death by regulating CREB signaling. J Mol Neurosci 55(4):821–829. https://doi.org/10.1007/s12031-014-0434-0

Huang X, Ding J, Li Y, Liu W, Ji J, Wang H, Wang X (2018) Exosomes derived from PEDF modified adipose-derived mesenchymal stem cells ameliorate cerebral ischemia-reperfusion injury by regulation of autophagy and apoptosis. Exp Cell Res 371(1):269–277. https://doi.org/10.1016/j.yexcr.2018.08.021

Jahan S, Kumar D, Kumar A, Rajpurohit CS, Singh S, Srivastava A, Pandey A, Pant AB (2017) Neurotrophic factor mediated neuronal differentiation of human cord blood mesenchymal stem cells and their applicability to assess the developmental neurotoxicity. Biochem Biophys Res Commun 482(4):961–967. https://doi.org/10.1016/j.bbrc.2016.11.140

Jayaraj RL, Azimullah S, Beiram R, Jalal FY, Rosenberg GA (2019) Neuroinflammation: friend and foe for ischemic stroke. J Neuroinflammation 16(1):142. https://doi.org/10.1186/s12974-019-1516-2

Jian Z, Liu R, Zhu X, Smerin D, Zhong Y, Gu L, Fang W, Xiong X (2019) The involvement and therapy target of immune cells after ischemic stroke. Front Immunol 10:2167. https://doi.org/10.3389/fimmu.2019.02167

Jiang L, Chen W, Ye J, Wang Y (2022) Potential role of exosomes in ischemic stroke treatment. Biomol Ther 12(1):115. https://doi.org/10.3390/biom12010115

Jiang M, Wang H, Jin M, Yang X, Ji H, Jiang Y, Zhang H, Wu F, Wu G, Lai X, Cai L, Hu R, Xu L, Li L (2018) Exosomes from MiR-30d-5p-ADSCs reverse acute ischemic stroke-induced, autophagy-mediated brain injury by promoting M2 microglial/macrophage polarization. Cell Physiol Biochem 47(2):864–878. https://doi.org/10.1159/000490078

Kalani A, Tyagi N (2015) Exosomes in neurological disease, neuroprotection, repair and therapeutics: problems and perspectives. Neural Regen Res 10(10):1565–1567. https://doi.org/10.4103/1673-5374.165305

Keyser KA, Beagles KE, Kiem HP (2007) Comparison of mesenchymal stem cells from different tissues to suppress T-cell activation. Cell Transplant 16(5):555–562. https://doi.org/10.3727/000000007783464939

Kim HY, Kim TJ, Kang L, Kim YJ, Kang MK, Kim J, Ryu JH, Hyeon T, Yoon BW, Ko SB, Kim BS (2020) Mesenchymal stem cell-derived magnetic extracellular nanovesicles for targeting and treatment of ischemic stroke. Biomaterials 243:119942. https://doi.org/10.1016/j.biomaterials.2020.119942

Kowal J, Tkach M, Théry C (2014) Biogenesis and secretion of exosomes. Curr Opin Cell Biol 29:116–125. https://doi.org/10.1016/j.ceb.2014.05.004

Krishnamurthi RV, Ikeda T, Feigin VL (2020) Global, regional and country-specific burden of ischaemic stroke, intracerebral haemorrhage and subarachnoid haemorrhage: a systematic analysis of the global burden of disease study 2017. Neuroepidemiology 54(2):171–179. https://doi.org/10.1159/000506396

Kumar A, Barrett JP, Alvarez-Croda DM, Stoica BA, Faden AI, Loane DJ (2016) NOX2 drives M1-like microglial/macrophage activation and neurodegeneration following experimental traumatic brain injury. Brain Behav Immun 58:291–309. https://doi.org/10.1016/j.bbi.2016.07.158

Lakhal S, Wood MJ (2011) Exosome nanotechnology: an emerging paradigm shift in drug delivery: exploitation of exosome nanovesicles for systemic in vivo delivery of RNAi heralds new horizons for drug delivery across biological barriers. Bioessays 33(10):737–741. https://doi.org/10.1002/bies.201100076

Larpthaveesarp A, Pathipati P, Ostrin S, Rajah A, Ferriero D, Gonzalez FF (2021) Enhanced mesenchymal stromal cells or erythropoietin provide long-term functional benefit after neonatal stroke. Stroke 52(1):284–293. https://doi.org/10.1161/STROKEAHA.120.031191

Lee JS, Hong JM, Moon GJ, Lee PH, Ahn YH, Bang OY, STARTING collaborators (2010) A long-term follow-up study of intravenous autologous mesenchymal stem cell transplantation in patients with ischemic stroke. Stem Cells (Dayton, Ohio) 28(6):1099–1106. https://doi.org/10.1002/stem.430

Lee TY, Murphy BD, Aviv RI, Fox AJ, Black SE, Sahlas DJ, Symons S, Lee DH, Pelz D, Gulka IB, Chan R, Beletsky V, Hachinski V, Hogan MJ, Goyal M, Demchuk AM, Coutts SB (2006) Cerebral blood flow threshold of ischemic penumbra and infarct core in acute ischemic stroke: a systematic review. Stroke 37(9):2201–2203. https://doi.org/10.1161/01.STR.0000237068.25105.aa

Leong WK, Henshall TL, Arthur A, Kremer KL, Lewis MD, Helps SC, Field J, Hamilton-Bruce MA, Warming S, Manavis J, Vink R, Gronthos S, Koblar SA (2012) Human adult dental pulp stem cells enhance poststroke functional recovery through non-neural replacement mechanisms. Stem Cells Transl Med 1(3):177–187. https://doi.org/10.5966/sctm.2011-0039

Li CY, Wu XY, Tong JB, Yang XX, Zhao JL, Zheng QF, Zhao GB, Ma ZJ (2015) Comparative analysis of human mesenchymal stem cells from bone marrow and adipose tissue under xeno-free conditions for cell therapy. Stem Cell Res Ther 6(1):55. https://doi.org/10.1186/s13287-015-0066-5

Li C, Fei K, Tian F, Gao C, Yang S (2019) Adipose-derived mesenchymal stem cells attenuate ischemic brain injuries in rats by modulating miR-21-3p/MAT2B signaling transduction. Croat Med J 60(5):439–448. https://doi.org/10.3325/cmj.2019.60.439

Li DB, Liu JL, Wang W, Li RY, Yu DJ, Lan XY, Li JP (2017) Plasma Exosomal miR-422a and miR-125b-2-3p serve as biomarkers for ischemic stroke. Curr Neurovasc Res 14(4):330–337. https://doi.org/10.2174/1567202614666171005153434

Li G, Xiao L, Qin H, Zhuang Q, Zhang W, Liu L, Di C, Zhang Y (2020) Exosomes-carried microRNA-26b-5p regulates microglia M1 polarization after cerebral ischemia/reperfusion. Cell Cycle (Georgetown, Tex) 19(9):1022–1035. https://doi.org/10.1080/15384101.2020.1743912

Li W, Shi L, Hu B, Hong Y, Zhang H, Li X, Zhang Y (2021a) Mesenchymal stem cell-based therapy for stroke: current understanding and challenges. Front Cell Neurosci 15:628940. https://doi.org/10.3389/fncel.2021.628940

Li Y, Chen J, Wang L, Lu M, Chopp M (2001) Treatment of stroke in rat with intracarotid administration of marrow stromal cells. Neurology 56(12):1666–1672. https://doi.org/10.1212/wnl.56.12.1666

Li Y, Cheng Q, Hu G, Deng T, Wang Q, Zhou J, Su X (2018) Extracellular vesicles in mesenchymal stromal cells: a novel therapeutic strategy for stroke. Exp Ther Med 15(5):4067–4079. https://doi.org/10.3892/etm.2018.5993

Li Y, Chopp M, Chen J, Wang L, Gautam SC, Xu YX, Zhang Z (2000) Intrastriatal transplantation of bone marrow nonhematopoietic cells improves functional recovery after stroke in adult mice. J Cereb Blood Flow Metab 20(9):1311–1319. https://doi.org/10.1097/00004647-200009000-00006

Li Y, Tang Y, Yang GY (2021b) Therapeutic application of exosomes in ischaemic stroke. Stroke Vasc Neurol 6(3):483–495. https://doi.org/10.1136/svn-2020-000419

Lian L, Zhang Y, Liu L, Yang L, Cai Y, Zhang J, Xu S (2021) Neuroinflammation in ischemic stroke: focus on MicroRNA-mediated polarization of microglia. Front Mol Neurosci 13:612439. https://doi.org/10.3389/fnmol.2020.612439

Liu QW, He Y, Xu WW (2022) Molecular functions and therapeutic applications of exosomal noncoding RNAs in cancer. Exp Mol Med 54(3):216–225. https://doi.org/10.1038/s12276-022-00744-w

Liu X, Zhang M, Liu H, Zhu R, He H, Zhou Y, Zhang Y, Li C, Liang D, Zeng Q, Huang G (2021) Bone marrow mesenchymal stem cell-derived exosomes attenuate cerebral ischemia-

reperfusion injury-induced neuroinflammation and pyroptosis by modulating microglia M1/M2 phenotypes. Exp Neurol 341:113700. https://doi.org/10.1016/j.expneurol.2021.113700
Liu Y, Fu N, Su J, Wang X, Li X (2019) Rapid enkephalin delivery using exosomes to promote neurons recovery in ischemic stroke by inhibiting neuronal p53/Caspase-3. Biomed Res Int 2019:4273290. https://doi.org/10.1155/2019/4273290
Luo X, Wang W, Li D, Xu C, Liao B, Li F, Zhou X, Qin W, Liu J (2019) Plasma Exosomal miR-450b-5p as a possible biomarker and therapeutic target for transient ischaemic attacks in rats. J Mol Neurosci 69(4):516–526. https://doi.org/10.1007/s12031-019-01341-9
Ma Y, Liu Y, Zhang Z, Yang GY (2019) Significance of complement system in ischemic stroke: a comprehensive review. Aging Dis 10(2):429–462. https://doi.org/10.14336/AD.2019.0119
Majumdar MK, Keane-Moore M, Buyaner D, Hardy WB, Moorman MA, McIntosh KR, Mosca JD (2003) Characterization and functionality of cell surface molecules on human mesenchymal stem cells. J Biomed Sci 10(2):228–241. https://doi.org/10.1007/BF02256058
Majumdar MK, Thiede MA, Haynesworth SE, Bruder SP, Gerson SL (2000) Human marrow-derived mesenchymal stem cells (MSCs) express hematopoietic cytokines and support long-term hematopoiesis when differentiated toward stromal and osteogenic lineages. J Hematother Stem Cell Res 9(6):841–848. https://doi.org/10.1089/152581600750062264
Manuel GE, Johnson T, Liu D (2017) Therapeutic angiogenesis of exosomes for ischemic stroke. Int J Physiol Pathophysiol Pharmacol 9(6):188–191
Mead B, Logan A, Berry M, Leadbeater W, Scheven BA (2014) Paracrine-mediated neuroprotection and neuritogenesis of axotomised retinal ganglion cells by human dental pulp stem cells: comparison with human bone marrow and adipose-derived mesenchymal stem cells. PLoS One 9(10):e109305. https://doi.org/10.1371/journal.pone.0109305
Meldolesi J (2018) Exosomes and ectosomes in intercellular communication. Curr Biol 28(8): R435–R444. https://doi.org/10.1016/j.cub.2018.01.059
Ming GL, Song H (2005) Adult neurogenesis in the mammalian central nervous system. Annu Rev Neurosci 28:223–250. https://doi.org/10.1146/annurev.neuro.28.051804.101459
Ming GL, Song H (2011) Adult neurogenesis in the mammalian brain: significant answers and significant questions. Neuron 70(4):687–702. https://doi.org/10.1016/j.neuron.2011.05.001
Moon GJ, Sung JH, Kim DH, Kim EH, Cho YH, Son JP, Bang OY et al (2019) Application of mesenchymal stem cell-derived extracellular vesicles for stroke: biodistribution and microRNA study. Transl Stroke Res 10(5):509–521
Morales L, Oliveros JC, Enjuanes L, Sola I (2022) Contribution of host miRNA-223-3p to SARS-CoV-induced lung inflammatory pathology. MBio 13(2):e0313521. https://doi.org/10.1128/mbio.03135-21
Nakamura K, Shichita T (2019) Cellular and molecular mechanisms of sterile inflammation in ischaemic stroke. J Biochem 165(6):459–464. https://doi.org/10.1093/jb/mvz017
Nogueira RG, Jadhav AP, Haussen DC, Bonafe A, Budzik RF, Bhuva P, Yavagal DR, Ribo M, Cognard C, Hanel RA, Sila CA, Hassan AE, Millan M, Levy EI, Mitchell P, Chen M, English JD, Shah QA, Silver FL, Pereira VM, Trial Investigators DAWN et al (2018) Thrombectomy 6 to 24 hours after stroke with a mismatch between deficit and infarct. N Engl J Med 378(1): 11–21. https://doi.org/10.1056/NEJMoa1706442
Norouzi-Barough L, Asgari Khosroshahi A, Gorji A, Zafari F, Shahverdi Shahraki M, Shirian S (2022) COVID-19-induced stroke and the potential of using mesenchymal stem cells-derived extracellular vesicles in the regulation of neuroinflammation. Cell Mol Neurobiol 43:1–10. Advance online publication. https://doi.org/10.1007/s10571-021-01169-1
Ojaghihaghighi S, Vahdati SS, Mikaeilpour A, Ramouz A (2017) Comparison of neurological clinical manifestation in patients with hemorrhagic and ischemic stroke. World J Emerg Med 8(1):34–38. https://doi.org/10.5847/wjem.j.1920-8642.2017.01.006
Otero-Ortega L, Alonso-López E, Pérez-Mato M, Laso-García F, Gómez-de Frutos MC, Diekhorst L, García-Bermejo ML, Conde-Moreno E, Fuentes B, de Leciñana MA, Bravo SB, Díez-Tejedor E, Gutiérrez-Fernández M (2021) Circulating extracellular vesicle proteins and

MicroRNA profiles in subcortical and cortical-subcortical ischaemic stroke. Biomedicine 9(7): 786. https://doi.org/10.3390/biomedicines9070786

Otero-Ortega L, Gómez-de Frutos MC, Laso-García F, Sánchez-Gonzalo A, Martínez-Arroyo A, Díez-Tejedor E, Gutiérrez-Fernández M (2017a) NogoA neutralization promotes axonal restoration after white matter injury in subcortical stroke. Sci Rep 7(1):9431. https://doi.org/10.1038/s41598-017-09705-0

Otero-Ortega L, Laso-García F, Gómez-de Frutos MD, Rodríguez-Frutos B, Pascual-Guerra J, Fuentes B, Díez-Tejedor E, Gutiérrez-Fernández M (2017b) White matter repair after extracellular vesicles Administration in an Experimental Animal Model of subcortical stroke. Sci Rep 7: 44433. https://doi.org/10.1038/srep44433

Palmer TD, Markakis EA, Willhoite AR, Safar F, Gage FH (1999) Fibroblast growth factor-2 activates a latent neurogenic program in neural stem cells from diverse regions of the adult CNS. J Neurosci 19(19):8487–8497. https://doi.org/10.1523/JNEUROSCI.19-19-08487.1999

Pan W, Zhang X, Liu X, Li C, Hu J, Chopp M (2016) Exosomes derived from ischemic cerebral endothelial cells and neural progenitor cells enhance neurogenesis and angiogenesis. Stroke 47(47):AWMP39

Pei X, Li Y, Zhu L, Zhou Z (2019) Astrocyte-derived exosomes suppress autophagy and ameliorate neuronal damage in experimental ischemic stroke. Exp Cell Res 382(2):111474. https://doi.org/10.1016/j.yexcr.2019.06.019

Powers WJ, Rabinstein AA, Ackerson T, Adeoye OM, Bambakidis NC, Becker K, Biller J, Brown M, Demaerschalk BM, Hoh B, Jauch EC, Kidwell CS, Leslie-Mazwi TM, Ovbiagele B, Scott PA, Sheth KN, Southerland AM, Summers DV, Tirschwell DL (2019) Guidelines for the early Management of Patients with Acute Ischemic Stroke: 2019 update to the 2018 guidelines for the early Management of Acute Ischemic Stroke: a guideline for healthcare professionals from the American Heart Association/American Stroke Association. Stroke 50(12):e344–e418. https://doi.org/10.1161/STR.0000000000000211

Prabhakaran S, Naidech AM (2012) Ischemic brain injury after intracerebral hemorrhage: a critical review. Stroke 43(8):2258–2263. https://doi.org/10.1161/STROKEAHA.112.655910

Qiu G, Zheng G, Ge M, Wang J, Huang R, Shu Q, Xu J (2018) Mesenchymal stem cell-derived extracellular vesicles affect disease outcomes via transfer of microRNAs. Stem Cell Res Ther 9(1):320. https://doi.org/10.1186/s13287-018-1069-9

Qureshi AI, Mendelow AD, Hanley DF (2009) Intracerebral haemorrhage. Lancet (London, England) 373(9675):1632–1644. https://doi.org/10.1016/S0140-6736(09)60371-8

Reynolds BA, Weiss S (1992) Generation of neurons and astrocytes from isolated cells of the adult mammalian central nervous system. Science (New York, NY) 255(5052):1707–1710. https://doi.org/10.1126/science.1553558

Rodin S, Antonsson L, Niaudet C, Simonson OE, Salmela E, Hansson EM, Domogatskaya A, Xiao Z, Damdimopoulou P, Sheikhi M, Inzunza J, Nilsson AS, Baker D, Kuiper R, Sun Y, Blennow E, Nordenskjöld M, Grinnemo KH, Kere J, Betsholtz C, Tryggvason K et al (2014) Clonal culturing of human embryonic stem cells on laminin-521/E-cadherin matrix in defined and xeno-free environment. Nat Commun 5:3195. https://doi.org/10.1038/ncomms4195

Roy-O'Reilly M, McCullough LD (2014) Sex differences in stroke: the contribution of coagulation. Exp Neurol 259:16–27. https://doi.org/10.1016/j.expneurol.2014.02.011

Russell AL, Lefavor R, Durand N, Glover L, Zubair AC (2018) Modifiers of mesenchymal stem cell quantity and quality. Transfusion 58(6):1434–1440. https://doi.org/10.1111/trf.14597

Rymer MM (2011) Hemorrhagic stroke: intracerebral hemorrhage. Mo Med 108(1):50–54

Samai AA, Martin-Schild S (2015) Sex differences in predictors of ischemic stroke: current perspectives. Vasc Health Risk Manag 11:427–436. https://doi.org/10.2147/VHRM.S65886

Sharghi-Namini S, Tan E, Ong LL, Ge R, Asada HH (2014) Dll4-containing exosomes induce capillary sprout retraction in a 3D microenvironment. Sci Rep 4:4031. https://doi.org/10.1038/srep04031

Shen LH, Li Y, Chen J, Zacharek A, Gao Q, Kapke A, Lu M, Raginski K, Vanguri P, Smith A, Chopp M (2007) Therapeutic benefit of bone marrow stromal cells administered 1 month after stroke. J Cereb Blood Flow Metab 27(1):6–13. https://doi.org/10.1038/sj.jcbfm.9600311

Siegel G, Schäfer R, Dazzi F (2009) The immunosuppressive properties of mesenchymal stem cells. Transplantation 87(9 Suppl):S45–S49. https://doi.org/10.1097/TP.0b013e3181a285b0

Song Y, Li Z, He T, Qu M, Jiang L, Li W, Shi X, Pan J, Zhang L, Wang Y, Zhang Z, Tang Y, Yang GY (2019) M2 microglia-derived exosomes protect the mouse brain from ischemia-reperfusion injury via exosomal miR-124. Theranostics 9(10):2910–2923. https://doi.org/10.7150/thno.30879

Spellicy SE, Stice SL (2021) Tissue and stem cell sourced extracellular vesicle communications with microglia. Stem Cell Rev Rep 17(2):357–368. https://doi.org/10.1007/s12015-020-10011-y

Stonesifer C, Corey S, Ghanekar S, Diamandis Z, Acosta SA, Borlongan CV (2017) Stem cell therapy for abrogating stroke-induced neuroinflammation and relevant secondary cell death mechanisms. Prog Neurobiol 158:94–131. https://doi.org/10.1016/j.pneurobio.2017.07.004

Sun Y, Gui H, Li Q, Luo ZM, Zheng MJ, Duan JL, Liu X (2013) MicroRNA-124 protects neurons against apoptosis in cerebral ischemic stroke. CNS Neurosci Ther 19(10):813–819. https://doi.org/10.1111/cns.12142

Takahashi K, Yamanaka S (2006) Induction of pluripotent stem cells from mouse embryonic and adult fibroblast cultures by defined factors. Cell 126(4):663–676. https://doi.org/10.1016/j.cell.2006.07.024

Takahashi K, Tanabe K, Ohnuki M, Narita M, Ichisaka T, Tomoda K, Yamanaka S (2007) Induction of pluripotent stem cells from adult human fibroblasts by defined factors. Cell 131(5):861–872. https://doi.org/10.1016/j.cell.2007.11.019

Tassew NG, Charish J, Shabanzadeh AP, Luga V, Harada H, Farhani N, D'Onofrio P, Choi B, Ellabban A, Nickerson P, Wallace VA, Koeberle PD, Wrana JL, Monnier PP (2017) Exosomes mediate mobilization of autocrine Wnt10b to promote axonal regeneration in the injured CNS. Cell Rep 20(1):99–111. https://doi.org/10.1016/j.celrep.2017.06.009

Tian T, Zhang HX, He CP, Fan S, Zhu YL, Qi C, Huang NP, Xiao ZD, Lu ZH, Tannous BA, Gao J (2018) Surface functionalized exosomes as targeted drug delivery vehicles for cerebral ischemia therapy. Biomaterials 150:137–149. https://doi.org/10.1016/j.biomaterials.2017.10.012

Tondreau T, Lagneaux L, Dejeneffe M, Massy M, Mortier C, Delforge A, Bron D (2004) Bone marrow-derived mesenchymal stem cells already express specific neural proteins before any differentiation. Differentiation 72(7):319–326. https://doi.org/10.1111/j.1432-0436.2004.07207003.x

Tsuyama J, Nakamura A, Ooboshi H, Yoshimura A, Shichita T (2018) Pivotal role of innate myeloid cells in cerebral post-ischemic sterile inflammation. Semin Immunopathol 40(6):523–538. https://doi.org/10.1007/s00281-018-0707-8

Uccelli A, Moretta L, Pistoia V (2008) Mesenchymal stem cells in health and disease. Nat Rev Immunol 8(9):726–736. https://doi.org/10.1038/nri2395

Uzdensky AB (2019) Apoptosis regulation in the penumbra after ischemic stroke: expression of pro- and antiapoptotic proteins. Apoptosis 24(9–10):687–702. https://doi.org/10.1007/s10495-019-01556-6

Vakhshiteh F, Atyabi F, Ostad SN (2019) Mesenchymal stem cell exosomes: a two-edged sword in cancer therapy. Int J Nanomedicine 14:2847–2859. https://doi.org/10.2147/IJN.S200036

van Niel G, D'Angelo G, Raposo G (2018) Shedding light on the cell biology of extracellular vesicles. Nat Rev Mol Cell Biol 19(4):213–228. https://doi.org/10.1038/nrm.2017.125

Vizoso FJ, Eiro N, Cid S, Schneider J, Perez-Fernandez R (2017) Mesenchymal stem cell Secretome: toward cell-free therapeutic strategies in regenerative medicine. Int J Mol Sci 18(9):1852. https://doi.org/10.3390/ijms18091852

Wang F, Tang H, Zhu J, Zhang JH (2018) Transplanting mesenchymal stem cells for treatment of ischemic stroke. Cell Transplant 27(12):1825–1834. https://doi.org/10.1177/0963689718795424

Wang J, Chen S, Zhang W, Chen Y, Bihl JC (2020a) Exosomes from miRNA-126-modified endothelial progenitor cells alleviate brain injury and promote functional recovery after stroke. CNS Neurosci Ther 26(12):1255–1265. https://doi.org/10.1111/cns.13455

Wang S, Aurora AB, Johnson BA, Qi X, McAnally J, Hill JA, Richardson JA, Bassel-Duby R, Olson EN (2008) The endothelial-specific microRNA miR-126 governs vascular integrity and angiogenesis. Dev Cell 15(2):261–271. https://doi.org/10.1016/j.devcel.2008.07.002

Wang Y, Zhang S, Li H, Wang H, Zhang T, Hutchinson MR, Yin H, Wang X (2020b) Small-molecule modulators of toll-like receptors. Acc Chem Res 53(5):1046–1055. https://doi.org/10.1021/acs.accounts.9b00631

Wei R, Zhang L, Hu W, Shang X, He Y, Zhang W (2022) Zeb2/Axin2-enriched BMSC-derived exosomes promote post-stroke functional recovery by enhancing neurogenesis and neural plasticity. J Mol Neurosci 72(1):69–81. https://doi.org/10.1007/s12031-021-01887-7

Williams GR, Jiang JG, Matchar DB, Samsa GP (1999) Incidence and occurrence of total (first-ever and recurrent) stroke. Stroke 30(12):2523–2528. https://doi.org/10.1161/01.str.30.12.2523

Wu W, Liu J, Yang C, Xu Z, Huang J, Lin J (2020) Astrocyte-derived exosome-transported microRNA-34c is neuroprotective against cerebral ischemia/reperfusion injury via TLR7 and the NF-κB/MAPK pathways. Brain Res Bull 163:84–94. https://doi.org/10.1016/j.brainresbull.2020.07.013

Xiao Y, Geng F, Wang G, Li X, Zhu J, Zhu W (2018) Bone marrow-derived mesenchymal stem cells-derived exosomes prevent oligodendrocyte apoptosis through exosomal miR-134 by targeting caspase-8. J Cell Biochem 120(2):2109–2118. https://doi.org/10.1002/jcb.27519

Xin H, Li Y, Chopp M (2014) Exosomes/miRNAs as mediating cell-based therapy of stroke. Front Cell Neurosci 8:377. https://doi.org/10.3389/fncel.2014.00377

Xin H, Li Y, Cui Y, Yang JJ, Zhang ZG, Chopp M (2013a) Systemic administration of exosomes released from mesenchymal stromal cells promote functional recovery and neurovascular plasticity after stroke in rats. J Cereb Blood Flow Metab 33(11):1711–1715. https://doi.org/10.1038/jcbfm.2013.152

Xin H, Li Y, Liu Z, Wang X, Shang X, Cui Y, Zhang ZG, Chopp M (2013b) MiR-133b promotes neural plasticity and functional recovery after treatment of stroke with multipotent mesenchymal stromal cells in rats via transfer of exosome-enriched extracellular particles. Stem Cells (Dayton, Ohio) 31(12):2737–2746. https://doi.org/10.1002/stem.1409

Xin H, Liu Z, Buller B, Li Y, Golembieski W, Gan X, Wang F, Lu M, Ali MM, Zhang ZG, Chopp M (2021) MiR-17-92 enriched exosomes derived from multipotent mesenchymal stromal cells enhance axon-myelin remodeling and motor electrophysiological recovery after stroke. J Cereb Blood Flow Metab 41(5):1131–1144. https://doi.org/10.1177/0271678X20950489

Xin H, Wang F, Li Y, Lu QE, Cheung WL, Zhang Y, Zhang ZG, Chopp M (2017) Secondary release of exosomes from astrocytes contributes to the increase in neural plasticity and improvement of functional recovery after stroke in rats treated with exosomes harvested from MicroRNA 133b-overexpressing multipotent mesenchymal stromal cells. Cell Transplant 26(2):243–257. https://doi.org/10.3727/096368916X693031

Xiong Y, Song J, Huang X, Pan Z, Goldbrunner R, Stavrinou L, Lin S, Hu W, Zheng F, Stavrinou P (2022) Exosomes derived from mesenchymal stem cells: novel effects in the treatment of ischemic stroke. Front Neurosci 16:899887. https://doi.org/10.3389/fnins.2022.899887

Xu J, Chen Z, Yu F, Liu H, Ma C, Xie D, Hu X, Leak RK, Chou S, Stetler RA, Shi Y, Chen J, Bennett M, Chen G (2020) IL-4/STAT6 signaling facilitates innate hematoma resolution and neurological recovery after hemorrhagic stroke in mice. Proc Natl Acad Sci U S A 117(51): 32679–32690. https://doi.org/10.1073/pnas.2018497117

Yan B, Zhang Y, Liang C, Liu B, Ding F, Wang Y, Zhu B, Zhao R, Yu XY, Li Y (2020) Stem cell-derived exosomes prevent pyroptosis and repair ischemic muscle injury through a novel exosome/circHIPK3/ FOXO3a pathway. Theranostics 10(15):6728–6742. https://doi.org/10.7150/thno.42259

Yang J, Zhang X, Chen X, Wang L, Yang G (2017a) Exosome mediated delivery of miR-124 promotes neurogenesis after ischemia. Mol Ther Nucleic Acids 7:278–287. https://doi.org/10.1016/j.omtn.2017.04.010

Yang Y, Cai Y, Zhang Y, Liu J, Xu Z (2018) Exosomes secreted by adipose-derived stem cells contribute to angiogenesis of brain microvascular endothelial cells following oxygen-glucose deprivation in vitro through MicroRNA-181b/TRPM7 Axis. J Mol Neurosci 65(1):74–83. https://doi.org/10.1007/s12031-018-1071-9

Yang Y, Ye Y, Su X, He J, Bai W, He X (2017b) MSCs-derived exosomes and neuroinflammation, neurogenesis and therapy of traumatic brain injury. Front Cell Neurosci 11:55. https://doi.org/10.3389/fncel.2017.00055

Zerna C, Thomalla G, Campbell BCV, Rha JH, Hill MD (2018) Current practice and future directions in the diagnosis and acute treatment of ischaemic stroke. The Lancet 392(10154):1247–1256. https://doi.org/10.1016/S0140-6736(18)31874-9

Zhang J, Li S, Li L, Li M, Guo C, Yao J, Mi S (2015a) Exosome and exosomal microRNA: trafficking, sorting, and function. Genomics Proteomics Bioinformatics 13(1):17–24. https://doi.org/10.1016/j.gpb.2015.02.001

Zhang Y, Chopp M, Liu XS, Katakowski M, Wang X, Tian X, Wu D, Zhang ZG (2017a) Exosomes derived from mesenchymal stromal cells promote axonal growth of cortical neurons. Mol Neurobiol 54(4):2659–2673. https://doi.org/10.1007/s12035-016-9851-0

Zhang Y, Chopp M, Meng Y, Katakowski M, Xin H, Mahmood A, Xiong Y (2015b) Effect of exosomes derived from multipluripotent mesenchymal stromal cells on functional recovery and neurovascular plasticity in rats after traumatic brain injury. J Neurosurg 122(4):856–867. https://doi.org/10.3171/2014.11.JNS14770

Zhang Y, Dong N, Hong H, Qi J, Zhang S, Wang J (2022) Mesenchymal stem cells: therapeutic mechanisms for stroke. Int J Mol Sci 23(5):2550. https://doi.org/10.3390/ijms23052550

Zhang Y, Hao Z, Wang P, Xia Y, Wu J, Xia D, Fang S, Xu S (2019a) Exosomes from human umbilical cord mesenchymal stem cells enhance fracture healing through HIF-1α-mediated promotion of angiogenesis in a rat model of stabilized fracture. Cell Prolif 52(2):e12570. https://doi.org/10.1111/cpr.12570

Zhang Y, Kim MS, Jia B, Yan J, Zuniga-Hertz JP, Han C, Cai D (2017b) Hypothalamic stem cells control ageing speed partly through exosomal miRNAs. Nature 548(7665):52–57. https://doi.org/10.1038/nature23282

Zhang Y, Liu J, Su M, Wang X, Xie C (2021a) Exosomal microRNA-22-3p alleviates cerebral ischemic injury by modulating KDM6B/BMP2/BMF axis. Stem Cell Res Ther 12(1):111. https://doi.org/10.1186/s13287-020-02091-x

Zhang ZG, Buller B, Chopp M (2019b) Exosomes–beyond stem cells for restorative therapy in stroke and neurological injury. Nat Rev Neurol 15(4):193–203. https://doi.org/10.1038/s41582-018-0126-4

Zhang Z, Huang S, Wu S, Qi J, Li W, Liu S, Cong Y, Chen H, Lu L, Shi S, Wang D, Chen W, Sun L (2019c) Clearance of apoptotic cells by mesenchymal stem cells contributes to immunosuppression via PGE2. EBioMedicine 45:341–350. https://doi.org/10.1016/j.ebiom.2019.06.016

Zhang Z, Zou X, Zhang R, Xie Y, Feng Z, Li F, Han J, Sun H, Ouyang Q, Hua S, Lv B, Hua T, Liu Z, Cai Y, Zou Y, Tang Y, Jiang X (2021b) Human umbilical cord mesenchymal stem cell-derived exosomal miR-146a-5p reduces microglial-mediated neuroinflammation via suppression of the IRAK1/TRAF6 signaling pathway after ischemic stroke. Aging 13(2):3060–3079. https://doi.org/10.18632/aging.202466

Zhao LR, Duan WM, Reyes M, Keene CD, Verfaillie CM, Low WC (2002) Human bone marrow stem cells exhibit neural phenotypes and ameliorate neurological deficits after grafting into the ischemic brain of rats. Exp Neurol 174(1):11–20. https://doi.org/10.1006/exnr.2001.7853

Zhao Y, Gan Y, Xu G, Hua K, Liu D (2020a) Exosomes from MSCs overexpressing microRNA-223-3p attenuate cerebral ischemia through inhibiting microglial M1 polarization mediated inflammation. Life Sci 260:118403. https://doi.org/10.1016/j.lfs.2020.118403

Zhao Y, Gan Y, Xu G, Yin G, Liu D (2020b) MSCs-derived exosomes attenuate acute brain injury and inhibit microglial inflammation by reversing CysLT2R-ERK1/2 mediated microglia M1 polarization. Neurochem Res 45(5):1180–1190. https://doi.org/10.1007/s11064-020-02998-0

Zhou Z, Lu J, Liu WW, Manaenko A, Hou X, Mei Q, Huang JL, Tang J, Zhang JH, Yao H, Hu Q (2018) Advances in stroke pharmacology. Pharmacol Ther 191:23–42. https://doi.org/10.1016/j.pharmthera.2018.05.012

Zhu X, Badawi M, Pomeroy S, Sutaria DS, Xie Z, Baek A, Jiang J, Elgamal OA, Mo X, Perle K, Chalmers J, Schmittgen TD, Phelps MA (2017) Comprehensive toxicity and immunogenicity studies reveal minimal effects in mice following sustained dosing of extracellular vesicles derived from HEK293T cells. J Extracell Vesicles 6(1):1324730. https://doi.org/10.1080/20013078.2017.1324730

Role of Stem Cells and Derived Exosomes as a Novel Therapeutic Agent against Alzheimer's and Parkinson's Disease

10

Shaheen Ali, Shouvik Mukherjee, Divya Goel, Anindita Ghosh, and Mohammed Faruq

Abstract

The significant rise in the global incidence of neurological diseases, such as Alzheimer's disease (AD) and Parkinson's disease (PD), has been attributed to the aging population. There are no efficient therapies for these disorders, regardless of medical improvements. Therefore, there is an urgent need for innovative treatments for these diseases. Over the years, stem cell-based therapy has transformed regenerative medicine by providing crucial and compelling possibilities to treat a variety of diseases (e.g., cancer), including neurological

S. Ali
Department of Molecular Medicine, SIST, Jamia Hamdard, New Delhi, India

Medicinal Chemistry and Molecular Modeling Lab, Department of Pharmaceutical Chemistry, SPER, Jamia Hamdard, New Delhi, India

S. Mukherjee
Department of Biotechnology, School of Chemical and Life Sciences, Jamia Hamdard, New Delhi, India

D. Goel
Genomics and Molecular Medicine Division, CSIR-Institute of Genomics and Integrated Biology, New Delhi, India

Department of Pharmacology, SPER, Jamia Hamdard, New Delhi, India

A. Ghosh
Genomics and Molecular Medicine Division, CSIR-Institute of Genomics and Integrated Biology, New Delhi, India

M. Faruq (✉)
Genomics and Molecular Medicine Division, CSIR-Institute of Genomics and Integrated Biology, New Delhi, India

Academy of Scientific and Innovative Research (AcSIR), Ghaziabad, India
e-mail: faruq.mohd@igib.in

S. Jahan, A. J. Siddiqui (eds.), *Applications of Stem Cells and Derived Exosomes in Neurodegenerative Disorders*, https://doi.org/10.1007/978-981-99-3848-3_10

diseases. Stem cell therapy, referred to as regenerative therapy, uses stem cells or their derivatives to enhance the curative response of dysfunctional and injured tissue. Hence, mesenchymal stem cells (MSCs) and embryonic stem cells (ESCs) are two stem cell types that are used. Recent evidence supports stem cell transplantation as promising therapeutic potential, which could be regarded as stem cells' mechanistic actions. In addition to stem cells, exosomes are a type of nanovesicles with a wide range of functionalities and possibilities for diagnosing and treating. It is shown that exosomes are implicated in cell-cell communication and have been explored as candidates for possible biomarkers, which are particularly relevant in AD and PD. Exosomes are also employed as a drug delivery vehicle at a target site; thus, their inherent ability to cross the blood-brain barrier and selectively adorn it with the ligand depends on the treatments of the targeted brain regions. This chapter aims to shed light on the different roles of stem cells and derived exosomes as therapeutic agents in the treatment of neurodegenerative disorders.

Keywords

Stem cells · Exosomes · Neurodegenerative diseases · Alzheimer's disease · Parkinson's disease

10.1 Introduction

Two of the most prevalent neurodegenerative diseases are Alzheimer's disease (AD) and Parkinson's disease (PD), with PD affecting 1% of adults over 60 years of age, while AD has affected approximately 50 million people in the United States (Han et al. 2018). AD or PD currently has no known treatments for complete cure; however, there are treatments to reduce the disease's progression and relieve some symptoms (e.g., rivastigmine, a cholinesterase inhibitor for PD) (Sun and Armstrong 2021). The common clinical symptoms of AD and PD include memory loss, motor dysfunction, and cognitive decline. Aggregates of proteins in cells cause the death of neurons in both PD and AD, although distinct brain regions are affected in each condition (Kim et al. 2003). For the creation of functional human neural cells for cell-based therapy and in vitro modeling, stem cells represent an unlimited cell source. Exosomes alone have a neuroprotective significance, as seen when oligodendrocyte-derived exosomes are added to cultured neurons, increasing cell viability even under stressful circumstances (Soares Martins et al. 2021).

This chapter highlights the pathophysiology of AD and PD, stem cell-derived cells, exosome technologies, and, the most recent, cutting-edge approaches to treating the conditions. Particularly, it focuses on the promising role or functions of stem cells in the differentiation of the neurons and implantation in vivo for potential treatment and drug testing of neurodegenerative diseases (Fig. 10.1). Additionally, derived exosomes serve as targeted cargoes and biomarkers in both AD and PD (Lakshmi et al. 2020).

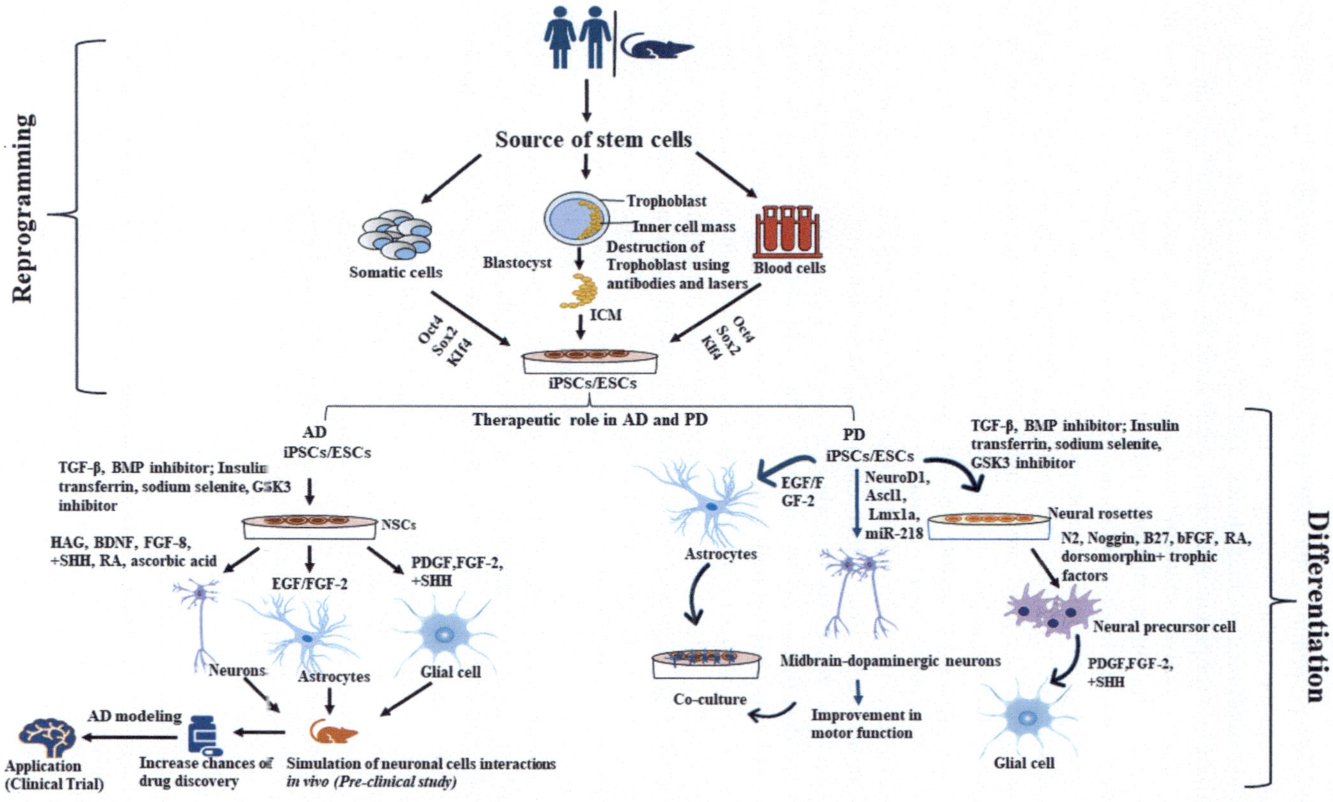

Fig. 10.1 The general method for reprogramming somatic/blood samples to differentiate ESCs from blastocysts is by destroying the trophoblast layer with antibodies or lasers and iPSCs by different transcription factors. The most often used differentiation factors to form neural rosettes are sodium selenite GSK3

10.1.1 What Are Stem Cells?

Stem cells are progenitor cells with the capability to regenerate or self-renew unlimitedly and the ability to develop further into differentiated cell types under appropriate conditions (Chagastelles and Nardi 2011). Stem cells can be characterized into totipotent stem cells which have the potential to form into every cell type of an organism (extraembryonic and embryonic structures) and have the highest capacity for differentiation. Pluripotent stem cells usually give rise to the cells of the three germ layers (endoderm, mesoderm, and ectoderm), and multipotent stem cells develop into multiple specialized cell types which exist in a particular tissue or organ (e.g., hematopoietic stem cells) (Zakrzewski et al. 2019). Unipotent stem cells, on the other hand, are distinct by having the most limited capability for differentiation. Having a special tendency to divide repeatedly, these cells can form only one cell type, e.g., satellite cells of skeleton muscles, making them a prime requisite for therapeutic agents (Dulak et al. 2015). After fertilization, the embryo divides and reaches the stage of the blastocyst, at which point it loses totipotency and acquires pluripotency, from which embryonic stem cells (ESCs) can be isolated (Zakrzewski et al. 2019). By acquiring the inner cell mass (ICM), which results in the potential destruction of the embryos, ESCs are obtained. The cells keep on dividing until it reaches the multipotent stage, thus becoming adult stem cells with the constrained potential to differentiate into cell types within a specific lineage (Vatsa et al. 2022). Adult cells such as the fibroblast or peripheral blood mononuclear cells which are isolated from the blood, can be reprogrammed into pluripotent stem cells termed as induced pluripotent stem cells (iPSCs). Reprogramming is established by the expression of certain reprogramming transcription factors, also termed as "Yamanaka factors" or "OKSM factors," Oct 3/4 (octamer-binding transcription factor 3/4), Sox2 (Sex-determining region Y) box 2, and Klf4 (Kruppel-like factor 4) (Takahashi et al. 2007; Takahashi and Yamanaka 2006). These transcription factors modulate the stemness and differentiation potential of stem cells (Yamanaka et al. 2007).

Adult or somatic stem cells are undifferentiated and are present in the differentiated cells of different tissues or organs. These cells serve to enhance the growth, repair, and replacement of the cells that are shed daily. These cells are multipotent or unipotent stem cells because they have limited differentiation potential into different cell kinds of their genesis tissue (Prochazkova et al. 2015; Romito and Cobellis 2016). Somatic stem cells exist in a range of distinct types, such as

Fig. 10.1 (continued) inhibitor, RA, insulin, transferrin, TGF-β, and BMP inhibitors. Additionally, N2, bFGF, Noggin, B27, RA, trophic factors, and dorsomorphin promote neuronal development. FGF8, SHH, RA, ascorbic acid, BDNF, and HAG influence the distinct differentiation by the lineage of iPSC-NSCs into the motor neurons, glial cells, and astrocytes (SHH, FGF2, and PDGF, respectively). These models and reprogramming methods can advance the research of neurodegenerative diseases, drug discovery, and therapeutic applications (Amoroso et al. 2013; Hayashi et al. 2011; Kim et al. 2012; Nutt et al. 2013)

MSCs, neural stem cells (NSCs), hematopoietic stem cells, and skin stem cells. The bone marrow is where these cells largely mature into fat, bone, or even cartilage cells. The NSCs develop into oligodendrocytes and astrocytes, which are cells that support nerve cells. Compared to ESCs, somatic stem cells multiply longer although adult stem cells can also be reprogrammed to regain their pluripotency. The first cellular reprogramming was achieved by Sir John Gurdon in 1962 by transplanting the nucleus from the intestinal epithelial somatic cells of tadpoles into enucleated unfertilized frog egg cells and reported generation of tadpoles (Gurdon 1962). This unique technique is termed somatic nuclear transfer (SCNT) for reprogramming of the somatic stem cells to the pluripotent embryonic state with a similar or the same genetic makeup; as a result, it led to the invention of cloning. The first mammal to be generated through somatic cloning was Dolly the sheep, which was cloned in the year 1997 by Sir Ian Wilmut and his team using the same method (Wilmut et al. 1997).

10.1.2 Exosomes

About 50 years ago, Peter Wolf first referred to extracellular vesicles (EVs) in plasma as "platelet dust" (Wolf 1967). Exosomes, one of the three forms of EVs, are distinguished by their varied sizes as well as by their biogenesis and release mechanisms. All cell types, including blood cells, neurons, epithelial cells, immunological or cancer cells, etc., release exosomes. Several biological fluids, such as cell culture supernatants, cerebrospinal fluid (CSF), serum, plasma, saliva, semen, urine, breast milk, and amniotic fluid, can be used to extract exosomes (Théry et al. 2006). Exosomal vesicles are formed by the inward growth of early endosomes that restrict membranes, which grow into multivesicular bodies (MVBs) (Raposo and Stoorvogel 2013). Their density ranges from around 1.13 g/mL to 1.19 g/mL (Bobrie et al. 2011; Zakharova et al. 2007). The sorting of protein as well as recycling, storage, transportation, and release is facilitated by exosomes in their intercellular signaling. MVBs are either destroyed with all their elements in the lysosome or may get fused with the plasma membrane of the cell in order to liberate their materials, including the exosomes, into the extracellular space (Simons and Raposo 2009). Exosomes were discovered to be released in a similar manner by the B-lymphocytes as well as dendritic cells. Eventually, it was revealed that exosomes were also released in an analogous manner by platelets, cytotoxic T cells, neurons, Schwann cells, mast cells, oligodendrocytes, intestinal epithelial cells, and endothelial cells, as well as stem cells; however, some cells, like the immune cells and MSCs, release more than others (Budnik et al. 2016; Jahan et al. 2022; Muller 2020).

Exosomes are particularly promising in regenerative medicine because of the range of components they contain, and their lipid bilayer membranes ensure stability and durability as well as include distinct marker proteins that link them to specific cells. The endosomal sorting complex required for transport (ESCRT) process initiated through the ubiquitin-binding subunits of ESCRT-0 to identify and sequester complex will merge with ESCRT-III, which is a protein complex involved in encouraging the process of budding, after contact with ESCRT-I and II complexes. After cleaving the buds to create intraluminal vesicles (ILVs), the ESCRT-III complex subsequently separates from the MVB membrane with the aid of energy given by the sorting protein Vps4 (Henne et al. 2011). The microdomains based on the raft seem to be required for the ESCRT-independent lateral segregation of the cargo present inside the endosomal membrane. These microdomains have a large number of sphingomyelinases, which can hydrolyze the phosphocholine moiety to produce ceramides (Airola and Hannun 2013). Furthermore, both ESCRT-dependent and ESCRT-independent pathways engage in exosome formation as well as the packing of biological cargo into exosomes, therefore including the utilization of cellular signals like phosphatidic acid, diglycerides, and ceramides as well as exosomal membrane lipid messengers (Stahl and Barbieri 2002). As lipid rafts, which are highly concentrated regions of sphingolipids and cholesterol in the membrane that are necessary for cell communication and endocytosis, are a component of exosome packaging, their presence on exosomal membranes makes them easily distinguishable as endosomes and can be used to detect exosomes rather than other vesicular products (Théry 2011). Exosomes can transport many molecules, such as particular proteins, RNA, and miRNA. The exosomal system also supports the horizontal transfer of the mRNA as well as proteins, with the genetic information successfully translated into suitable proteins, according to various studies (Bruno et al. 2009; Ratajczak et al. 2006).

10.2 Role of Stem Cells and Derived Exosomes in Neurodegenerative Disease

The progressive degradation of neurons in the central and peripheral nervous systems is the characteristic feature of the heterogeneous group of disorders known as neurodegenerative diseases (Przedborski et al. 2003). They manifest as a consequence of loss of function, structure, and/or several neurons, including the death of neurons in the spinal cord or brain, which results in loss of cognition and different degrees of motor disability (Poddar et al. 2021). Neurodegeneration is linked with the disruption of a neural network, the synapse, and the accumulation of physiochemically altered proteins (Lamptey et al. 2022).

The most common neurodegenerative conditions are Huntington's disease, spinal muscular atrophy, Alzheimer's disease (AD), Parkinson's disease (PD), amyotrophic lateral sclerosis (ALS), motor neuron disease, prion disease, and spinocerebellar ataxia (SCA) (Lamptey et al. 2022). According to reports, AD is the sixth most prevalent cause of death in the United States (Kumar et al. 2021). Up to 5.8 million Americans were expected to have AD in 2020 ((Matthews et al. 2019). Up to 24 million people worldwide are estimated to have dementia; by 2050, that number suffered from PD, most of which are idiopathic (Zafar and Yaddanapudi 2021). Around 10% of them have a genetic cause and affect the young age group. ALS, also known by the name "Lou Gehrig," exists in two forms, familial, comprising 90–95% of cases, and sporadic. It has been estimated that ALS has a prevalence of 5.2 per 100,000 within the United States and 1.6 cases per 100,000 persons annually worldwide (Brotman et al. 2022). SCA is a subset of hereditary cerebellar ataxia, is a comparatively rare disease, and has a global prevalence of 1 to 5 in 10,000. It has many subtypes, with SCA3 the most common worldwide and SCA2 the most common subtype in South Korea and India (Bhandari et al. 2022).

According to a report by Feigin et al., neurological diseases were the second-most common incidence of deaths globally in 2016 and the leading cause of DALYs (disability-adjusted life years; the sum of the years of life lost and the years spent with disability) (Feigin and Vos 2019). Migraine, stroke, Alzheimer's, and other dementias were the four largest contributors to neurological DALY. In general, over the past 27 years, it was shown that the prevalence of neurological disorders has grown and it is anticipated to continue to increase due to the exponential growth of the population as well as aging, placing even higher pressure on heavily overburdened resources and services for people with neurological disorders (Feigin and Vos 2019).

The absence of effective treatments for neurodegenerative diseases places a heavy burden on society, as well as the cost of care, and has a significant influence on the quality of life. Current clinical approaches focus on addressing symptom alleviation and disease management but fail to halt the progress of neurodegeneration (Poddar et al. 2021). Neither pharmaceutical nor neurosurgical treatments are effective at arresting the course of the neurodegenerative processes (Sakthiswary and Raymond 2012). Furthermore, due to variations in mammalian genomes and embryonic development, numerous therapies developed in animal models have not yet been successfully translated into clinical trials (Dawson et al. 2018). On this basis, human-based studies involving stem cells present a clear model for studying the pathophysiological process, signaling pathways, growth control, and disease mechanism in previously inaccessible human brain tissue.

10.2.1 Stem Cells Therapy in Neurodegenerative Diseases

Regenerative cell therapy, otherwise called stem cell therapy, over the past two decades has provided a significant opportunity to explore potentially powerful innovative strategies to treat diseases associated with neurodegeneration. This is because stem cells have the ability to restore damaged neural tissue by mostly replacing lost or damaged cells with differentiated ones, creating an environment that promotes regeneration, stabilizes neuronal networks, or protects already healthy neurons and glial cells from more harm. Different stem cell types have a differential capacity to descend into a specific cell lineage/type, as discussed above. This seemingly unlimited potential of stem cells has opened unprecedented opportunities for developing novel medical therapies for degenerative diseases and injuries, and diseases like AD, PD, HD, spinal cord injury, diabetes, and a few heart diseases have few or no treatment options; hence, stem cell-based therapy is a beneficial option, as stem cell uses their derivatives to enhance the repair reaction of damaged and dysfunctional tissue (Institute of Medicine 2002). Since every stem cell type comes with its own unique traits and benefits, the rationale for using a specific type depends on the applications as well as its results that are desired. Subsequently, human embryonic stem cell (hESC) research involves destruction of human embryos, thus raising ethical as well as political concern (Lo and Parham 2009a). Because of these limitations and ethical concerns surrounding ESCs, scientists have developed techniques for inducing pluripotency in non-pluripotent cells or somatic cells. The generated cells, thus named iPSCs, open the possibility of offering customized models to study and treat neurodegenerative diseases using patients' own somatic cells through reprogramming (Dantuma Elise et al. 2010).

The primary types of stem cells utilized for neurodegenerative therapies are embryonic, progenitors, mesenchymal, and iPSCs (Singh et al. 2016). ESCs being pluripotent hold remarkable potential to restore the damage that occurred due to injury or neurodegeneration (Singh et al. 2016). An additional restriction is an immunological incompatibility between donor and recipient cells, which can lead to the of transplanted cells. The spectrum of their clinical application is however constrained by their capacity for unrestricted self-renewal, which carries a significant danger for the development of tumors after engraftment (Sivandzade and Cucullo 2021).

MSCs are immunomodulating, that is, derived from the same source (and hence do not elicit host immune response) and multipotent and find high applicability in neurodegenerative diseases, as it promotes neural growth, reduce free radicals, decrease apoptosis, and repress inflammation (Sakthiswary and Raymond 2012). However, their use gets restricted in genetic diseases because the autologous source holds the same genetic predisposition to the disease. In the next part, the prevalence of stem cells is discussed in different neurodegenerative diseases.

10.2.1.1 Huntington's Disease

Transplantation of median spiny neurons (MSNs) derived from both iPSCs and ESCs has shown successful integration and neural circuit development (Aubry et al. 2008), but most studies are restricted to animal models. Reliable data indicates that three out of five participants had the advantage of transplantation in the Créteil pilot experiment, nonrandomized, open-label, monocentric cell transplant trial employing fetal donor cells of humans obtained from the fetal ganglionic eminence (GE). Therefore, the two participants exhibited the typical HD deterioration; however, at the end it was hypothesized that among one of them, the disease was too advanced in order to allow for effective graft vascularization (Bachoud-Lévi et al. 2021).

10.2.1.2 Amyotrophic Lateral Sclerosis

Stem cell transplantation has demonstrated great potential in recent clinical trials in ALS patients. Six studies have shown proof that stem cell therapy largely has a positive effect in slowing the progression of disease (L. Xu et al. 2006). Numerous studies have revealed the efficacy of NSC therapy on ALS rats, as a result of the transplanted NSCs' ability to differentiate into neurons and form synaptic connections in addition to their potential to prevent the disease's onset and progress, thereby increasing the survival of animal models (Xu et al. 2006). Considering the preclinical evidence for NSC-based treatments, the Food and Drug Administration (FDA) approved a clinical trial in 2009 regarding the safety as well as tolerability of the surgical introduction of stem cells and any therapy following cell toxicity. In this first-in-human phase I clinical trial, researchers injected fetal-derived NSCs into the lumber spinal cord of 12 ALS patients (Glass et al. 2012). Clinical evaluations performed around 6–18 months after the transplantation showed no indications of further progression of the disease. By exploring the intraspinal injections in the cervical spinal cord, the researchers aim to advance this clinical trial and potentially prolong the lives of ALS patients by protecting the motor neuron groups that affect respiratory function. The therapeutic potential of MSCs has been evaluated in several research employing ALS animal models by either injecting the cells peripherally or directly into the spinal cord (Mao et al. 2015). In 2003, Mazzini et al. examined the safety as well as tolerability of MSC transplantation directly intraparenchymal for the treatment of ALS (Mazzini et al. 2003). Although there was no functional improvement, follow-up studies also did not show detrimental effects.

10.2.1.3 Spinocerebellar Ataxia

MSCs serve as good candidates for SCA treatment because of their ability to differentiate between lines and immunomodulatory properties. In a study done by Chang et al., the human MSCs were administered intravenously and intracranially to transgenic mice having a poly-glutamine mutation in the ataxin-2 gene before and after the onset of loss of motor function. Intravenous transplantation successfully enhanced the rotarod function of the SCA2 mice as well as postponed the onset of neuronal deterioration, whereas intracranial transplantation was unable to procure any kind of neuroprotective effect (Chang et al. 2011). In phase I/IIa of the clinical trials conducted in Taiwan, six patients having type 3 SCA and one with multiple-system atrophy-cerebellar were intravenously administered MSC derived from allogenic adipose tissue from healthy donors (Tsai et al. 2017). Upon a year of follow-up, the intravenously injected MSCs seemed to be well tolerated and no adverse effects were observed. The study concludes the safety and tolerability of allogenic MSCs through intravenous injection.

10.2.2 Exosomes in Neurodegenerative Diseases

Recent preclinical research has indicated that EVs produced from stem cells can be utilized as a possible alternative to stem cell therapy to treat brain disorders. EV-based therapy outperforms cell therapy by means of biodistribution, scalability, and safety profiles; it can be used to treat neurological disorders as a potential substitute to stem cell-based therapy. Furthermore, EVs produced from stem cells have superior biocompatibility, immunogenicity, and safety characteristics compared to minute chemicals and macromolecules (Bang and Kim 2022).

Stem cell-derived exosomes are considered an inherent drug delivery system and a natural therapeutic agent for the potential treatments of brain diseases. EVs consist of molecules with heterogeneous functions such as cellular proteins, DNA and RNA.

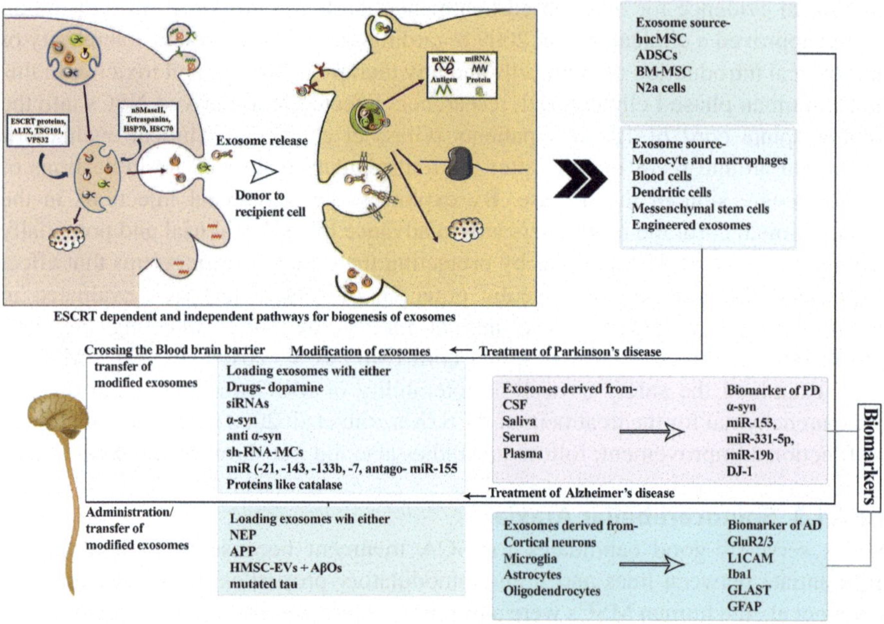

Fig. 10.2 Biogenesis of exosomes from different cell sources and acting as drug delivery vehicle and biomarkers in AD and PD

EVs have been revealed to be able to cross the BBB as well as can target specific cell types effectively (Fig. 10.2). Contrary to cell-based therapeutics, EVs usually are not affected by the first-pass effect or by cell-mediated side effects like coagulopathy, tumor growth, or arterial blockage. Additionally, the several components of the EVs' copious nucleic acid are protected from RNase in the blood because lipids are required for the development of EVs' membranes (Mirzaaghasi et al. 2021; Wen et al. 2019).

In a study incorporating MSC transplant, it was shown that the number or amount of circulating EVS increased rapidly in number after the transplantation of MSCs (Bang et al. 2022). Also, in patients receiving the same dosage of MSCs, the quantity of the circulating EVS differed among patients, which was further associated with the improvement of motor function. Given the quantity of EVs that influence the results of MSC-based therapy, these studies suggested the possibility of using MSC-EVs rather than MSCs in and of themselves (Allan et al. 2020).

Henceforth, EVs have a role in maintaining CNS cellular function, also contributing to the pathophysiology underlying neurodegenerative disease. EVS that are derived from native stem cells can cure neurological disorders, but their shot half-lives, restricted targeting, quick clearance after the application, and insufficient payload make them difficult to use successfully in clinical settings (Khan et al. 2021). The heterogeneity of the donors is a significant problem that hinders the clinical applicability of EVs. Due to the usage of different donors, the cell variability is huge which might affect the therapeutic potential of EVs due to differences in donor age, comorbidity, artificial niches of MSCs, and culture conditions. In addition to creating a production/developmental process that reduces the donor-to-donor as well as batch-to-batch variances, each EV production lot needs to have a strong quality control system (Bang and Kim 2022).

10.3 Therapeutic Role of Stem Cells and Derived Exosomes in AD

Alzheimer's disease (AD) is a fatal neurological disorder characterized by behavioral problems, progressive cognitive decline, and loss of daily functioning. Globally, between 50% and 70% of cases with dementia are caused by AD. Moreover, dementia affects 50 million individuals worldwide and costs around $818 billion. By 2050, this number is expected to increase to 132 million because of age being the primary risk factor and the fact that national populations are gradually aging (Association 2016; Ferri et al. 2005). Clinical symptoms of AD develop rapidly. Early neuroinflammation, learning, and memory issues are prominent features, followed by visuospatial function, executive function, complex attention, praxis (learned motor activity), gnosis (recognized previously learned information), language, behavior, and/or social abnormalities (Si and Wang 2021).

Additionally, the two primary neuropathological markers for the diagnosis of AD are beta-amyloid extracellular deposits (senile plaques) and hyperphosphorylated tau intercellular deposits (neurofibrillary tangles) (Kent et al. 2020). Many cases of AD have a late onset and are sporadic. Moreover, there are established risk factors for the illness besides age, such as the apolipoprotein-E4 (ApoE4) gene, cardiovascular disease, depression, and low education levels (Piers et al. 2021). In addition, under 5% of AD cases are familial and are caused by highly penetrant autosomal mutations in the PSEN1, PSEN2, and, less frequently, the APP genes (Sabayan and Sorond 2017). Moreover, genome-wide association studies (GWAS) revealed that mutations in various genes can promote the development of AD (Kang et al. 2016). Positron emission tomography (PET), CSF biomarkers, and some comparatively recent clinical standards can all be used to diagnose living patients, even though only a postmortem autopsy can provide a diagnosis with certainty. These criteria underline that etiological diagnosis still relies heavily on neuropathological evaluation (Kang et al. 2016). Numerous pharmaceutical strategies, including vaccination and secretase inhibition, have been investigated to improve amyloid clearance and reduce production (Duncan and Valenzuela 2017).

New therapies and medications are suggested every year to decrease the cognitive decline and neuronal death linked to AD. Nonetheless, only five drugs, including the glutamate receptor antagonist memantine and cholinesterase inhibitors galantamine, rivastigmine, tacrine, and donepezil, have received FDA approval for the clinical treatment of AD (Si and Wang 2021). Sadly, these drugs can only treat symptoms and do nothing to alter the primary pathologic aspects of AD (Si and Wang 2021).

To ameliorate the pathogenic state of the disease, enhance neural precursors, prevent nerve death, and promote structural plasticity, effective, novel treatments must be developed. These therapies may include eliminating toxic deposits and replacing damaged neurons (Liu et al. 2020). Hope for the treatment of refractory neurodegenerative disorders like Alzheimer's has been ignited by recent developments in stem cell preclinical research and clinical trials. For providing the uniform and cell replacement therapy that call required, stem cells are the greatest alternative (Liu et al. 2020).

10.3.1 Stem Cells as Therapeutics for AD

In AD research today, iPSCs, MSCs, brain-derived NSCs, and ESCs are most frequently exploited. Since the main reason for neurodegenerative diseases, for example, AD, is aging, it appears counterintuitive to research AD using stem cells. iPSC-derived neurons can develop electrophysiologically active synaptic networks and are structurally and functionally mature. It is also feasible to control the differentiation of iPSCs into various neuronal subgroups, including dopaminergic neurons, by using additional transcription factors during the induction phase (Duncan and Valenzuela 2017). Among multipotent stem cells, NSCs are a subset which can differentiate into oligodendrocytes, astrocytes, neurons, and microglia (Si and Wang 2021). In a study, neurons generated using iPSCs with FAD mutations or from AD patients displayed similar AD characteristics at the earliest possible developmental stages. The amyloid precursor protein (APP) V717I gene mutation increased Aβ and tau phosphorylation in neurons, whereas the APP A673T mutation suppressed β-secretase cleavage of APP and Aβ synthesis (Maloney et al. 2014; Muratore et al. 2014). Astrocytes with the PSEN1 ΔE9 mutation also produced more reactive oxygen species (ROS), had defective fatty acid oxidation, and elevated the production of Aβ42. When compared to astrocytes with APOE3 mutations, those with APOE4 mutations show substantial changes in gene expression and a decreased capacity to absorb amyloid beta 42. Lipopolysaccharide treatment increased the release of certain cytokines and altered phagocytosis in iPSC-derived microglia from SAD patients (Xu et al. 2019; Si and Wang 2021). When compared to isogenic APOE3 controls, microglia with mutated APOE4 had a decreased impaired capacity to internalize Aβ and morphologic complexity. The phenotypes of SAD-derived iPSCs and FAD mutation carriers are frequently similar. Another recent research discovered that human iPSC-derived cholinergic neuronal progenitors recovered from intra-hippocampal transplantation into transgenic AD mice model matured into phenotypically adult cholinergic neurons and repaired the spatial memory loss (Liu et al. 2020).

Additionally, human iPSC-derived NSCs improved neurological function and decreased pro-inflammatory markers in a mouse model of ischemic stroke through a neurotrophin-associated bystander effect (Duncan and Valenzuela 2017). The secreted neurotrophic factors by transplanted NSCs improved the memory function and overexpression of a β-degrading enzyme by NSCs reduced Aβ aggregation.

Alleviated Aβ synthesis and acetylcholinesterase activity were seen following NSC transplantation into these Tg2576 mice who carry human Swedish APP mutation (isoform 695; KM670/671NL). Additionally, early-stage NSC transplantation enhanced anti-inflammatory cytokine levels in microglial cells and may reduce the generation of Aβ, increasing the Aβ clearance rate. It also increased vascular endothelial growth factor (VEGF), synaptic density, and neurogenesis. However, prompt action is required because later NSC transplantation into Tg2576 mice did not provide comparable outcomes (Liu et al. 2020). Also, in NSC administration in APP/PS1 mice, tropomyosin receptor kinase B and BDN levels increased. NSC-derived cholinergic neurons were also introduced into APP/PS1 mice showing an increase in the concentration of cholinergic acetyltransferase and its activity, and there were enhanced operational dendrites (Gu et al. 2015). The genetic modification of NSCs to optimally release the Aβ-degrading enzyme neprilysin may promote synaptic plasticity and potential Aβ-pathogenic characteristics (Liu et al. 2020).

MSCs are not predicted to replace injured neurons and integrate into neural network, in contrast to iPSCs and NSCs, because it is not known if they can differentiate into endodermal or ectodermal cells. The neuroprotective properties of MSCs are mediated by a variety of pathways. To increase brain cell survival, MSCs can secrete neurotrophic growth factors, including brain-derived neurotrophic factor (BDNF) and glial cell-derived neurotrophic factors. Ucb-MSCs have been demonstrated in prior research utilizing AD murine models to enhance spatial learning and prevent memory loss. Numerous additional pathways have also been proposed, such as the decrease in Aβ plaques, the hyperphosphorylation of BACE and tau, the restoration of microglial inflammation, and the activation of anti-inflammatory cytokines (Lee et al. 2012; Liu et al. 2020). Two Aβ-degrading factors including neprilysin and insulin-degrading enzyme rose following treatment with neuron-like cells generated by MSCs of the umbilical cord (human). Bone marrow-derived MSCs were implanted in the APP/PS1 mouse model of AD by tail vein injection. These mice had fewer microglia but the amounts of amyloid plaques remained unchanged (Naaldijk et al. 2017). Contrarily, Cartel et al. discovered that intracerebral injection of bone marrow-derived MSCs resulted in a substantial decrease over the course of 2 months compared to PBS-treated controls (Bae et al. 2013). In a study that evaluated Aβ therapy alone, brain progenitor cells co-cultured with MSCs showed considerably greater expression of GFAP, nestin, Ki-67, HuD, and SOX2. Furthermore, treatment with Aβ showed increased expression of β-catenin and Ngn1 in neural progenitor cells co-cultured with MSCs (Oh et al. 2015).

An in vivo study showed that ESCs can be differentiated into astrocytes and cells that mimic neurons, which can be used to treat neurodegenerative diseases. Researchers have demonstrated that ESCs, when transformed into gamma-aminobutyric acid and basal forebrain cholinergic neurons, can enhance spatial learning and memory in AD-affected mice (Liu et al. 2020). Nonetheless, direct ESC transplantation resulted in the development of teratomas in vivo rather than neurons, limiting their clinical utility (Liu et al. 2020). With and without pretreatment, NPCs derived from mESCs were transplanted to the unilateral Meynert basal nucleus, enhancing learning and memory in the AD rat model. The transplanted NPC cells show a cholinergic phenotype to a degree of about 40%, whereas the remaining cells continue to maintain a neuronal phenotype (Moghadam et al. 2009). Contrarily, the control group's ESCs produced teratomas, which are tumors incapable of generating neurons, resulting in a drastic decline in working memory (Wang et al. 2006). Despite the limitation of evidence regarding hESCs' ability to treat AD, hESCs can be considered a new therapeutic alternative for various neurological conditions and degenerative diseases. However, ethical issues should be addressed prior to the administration in clinical trials that have acquired approval from the FDA (Liu et al. 2020).

10.3.2 Exosomes as AD Biomarkers

Exosomes derived from the brain in peripheral blood have shown remarkable promise as the perfect "liquid biopsy" for Alzheimer's. It is interesting to note that brain-derived exosomes can cross the BBB and enter the peripheral blood circulation; however, their quantity is lower than in CSF. Researchers use immunoprecipitation techniques to enrich plasma brain-derived exosomes to overcome these restrictions. According to earlier studies, the lysosomal and synaptic protein concentrations of neuron-derived exosomes (NDEs) can be used to predict dementia before it manifests as mild cognitive impairment (MCI) (Winston et al. 2016). Reduced levels of synaptic proteins with specific functions in NDE may signify the progression degree of AD. Additionally, complement protein levels in astrocyte-derived exosomes (ADEs) seem to be correlated with the disease's stage (Goetzl et al. 2018). Because plasma ADEs contain far more cargo proteins than NDEs, BACE-1 inhibitors may be able to specifically target ADEs. A systematic method for the preparation and qualification of biomarkers remains extremely difficult, and to evaluate the diagnostic relevance of exosomes, large cohort studies are needed (Guo et al. 2020).

Small compounds and macromolecules are only partially effective against brain diseases because of the single method of action and intricate pathophysiology of these conditions. For instance, despite promising results in preclinical investigations, around 1000 neuroprotective drugs for acute stroke failed in human trials. Consequently, pleiotropic multi-target therapies may be more effective if they employ a variety of strategies to obstruct various stages. Additionally, almost no macromolecules and nearly all small compounds cross the BBB. Adeno-associated viral capsids and polymer- or lipid-based nanoparticles were employed to get around this restriction. Contrarily, using a drug delivery method raises the risk of infection, immunogenicity, and toxicity. For the treatment of brain diseases, stem cell-derived EVs are regarded as inherent drug delivery mechanisms and naturally therapeutic molecules (Guo et al. 2020; Kang et al. 2016; Liu et al. 2020).

The first theory linking exosomes to AD postulates that exosomes can produce Aβ peptides from cultured cells, and exosomal proteins including Alix and flotillin-1 have been observed to aggregate within amyloid plaques in the brains of AD patients. The full-length amyloid precursor protein (flAPP), Aβ peptide, amyloid precursor protein C-terminal fragments (APP-CTFs), and amyloid intracellular domain (AICD), as well as enzymes that cleave flAPP and APP CTFs, have been discovered to be present in exosomes from humans with the autosomal dominant Swedish mutation (BACE1, PS1, PS2, and ADAM10) (Laulagnier et al. 2017). According to these results, APP and its catabolites may be transported across cells via neuronal exosomes. More importantly, several in vivo experiments showed linkage of exosomes to AD. Exosomes from AD patients' bodily fluids, such as blood and CSF, show an increase in soluble Aβ1-42, p-T181-tau, and p-S396, among other things. Exosomes can one day be employed as a diagnostic tool for AD as this rise could be observed many years before a diagnosis (Guo et al. 2020; Lee et al. 2019).

Exosomes derived from plasma neuronal membranes of AD patients cause AD-like neuropathology in normal mouse brains by increasing tau aggregation. Exosomes from AD patients with BIN1-related genetic variations in their CSF encourage tau spreading in mouse, but microglia depletion or exosome synthesis suppression greatly reduces tau propagation in vivo and in vitro. Furthermore, these toxic species can be transferred to recipient neurons in culture by brain-derived exosomes of AD patients with high amounts of Aβ oligomer, leading to neurotoxicity. It was revealed that ESCRT proteins TSG101 and VPS4A knockdown decreased toxicity and oligomer spread by blocking exosome synthesis, secretion, or uptake (Sardar Sinha et al. 2018).

Exosomes have a role in the etiology of AD by spreading amyloid beta and tau, causing neuroinflammation, affecting neuronal functioning, and ultimately resulting in cell death. Neuron-derived exosomes from AD patients substantially contained less heat shock factor-1 (HFS1), repressor element 1-silencing transcription factor (REST), and low-density lipoprotein receptor-related protein 6 (LRP6) compared to controls. As compared to controls, AD patients had considerably greater amounts of the pro-inflammatory substances IL-1, TNF-α, and IL-1 in ADEs as well as the complement components C1q, C4b, and C3d. The subsequent research showed that patients at the CE 2 stage had higher levels of complement proteins and lower regulatory proteins than those in the CE 1 preclinical stage.

The exosomes released by astrocytes that have been exposed to Aβ contain the protein known as the proapoptotic prostate apoptosis 4 (PAR-4), which causes astrocytes in the culture to undergo apoptosis (Liu et al. 2020). Additionally, serum from AD patients as well as brain tissue and serum from the 5XFAD mice model includes ceramide-enriched and astrocyte-derived exosomes, both of which are linked to the Aβ pathology. These exosomes migrated to the mitochondria both in vivo and in vitro, caused mitochondrial clustering, and elevated the amount of the fission protein DRP1 in neurons. Further, Aβ-associated exosomes encouraged Aβ-binding to voltage-dependent anion channel (VDAC1), which led to neurite rupturing, caspase activation, and neuronal death (Elsherbini et al. 2020a, b; Zhang et al. 2021).

Even though exosomes first seemed to be damaging to AD, a growing corpus of current research indicates that exosomes may be beneficial against the condition.

Exosomes may include a range of molecules that primarily work by restoring neuronal function or Aβ clearance to mediate protective effects against AD. For instance, a protein with significant neuroprotective effects like cystatin-C is present in exosomes (Pérez-González et al. 2019; Zhang et al. 2021). Statins were acknowledged in lowering the possibility of AD and might change the content as well as secretion of exosomes. Exosomes produced by BV-2 microglial cells and neuroblastoma cells treated with statins have been shown to encourage extracellular Aβ degradation through exosome IDE (Zhang et al. 2021). Exosomes can release metalloproteases to the extracellular environment to help in the breakdown of Aβ, such as endothelin-converting enzymes (ECE) 1 and 2. The inhibition of metalloprotease disrupts Aβ catabolism, which raises Aβ levels, and promotes the production of Aβ oligomers intracellularly (Pacheco-Quinto et al. 2019). Aβ oligomerization has been demonstrated to be inhibited in vitro by exosome secretion from neuronal cells via enhancing microglia-mediated Aβ clearance.

10.4 Therapeutic Role of Stem Cells and Derived Exosomes in PD

In 1817, James Parkinson described "shaking palsy," which is known as Parkinson's disease (PD) and is associated with dyskinesias and dystonias, and several other motor and nonmotor symptoms (Jankovic and Tan 2020). In addition to multiple-system atrophy, progressive supranuclear palsy, chorea, and ataxia, PD is the most prevalent mobility disorder. PD has affected more than ten million people globally, and it was reported that approximately 6.2 million individuals have PD in 2015 (Dorsey et al. 2018). Although its prevalence increases with age, sex being a contributing factor as men are predominantly afflicted by PD, research also indicated that the load of PD would improve significantly in the coming decades (Van Den Eeden et al. 2003; Wanneveich et al. 2018). Consequently, this chronic neurodegenerative disease PD has severely challenged healthcare systems. It is linked with several risk factors, such as oxidative stress, the production of free radicals, and several environmental pollutants, and genetic abnormalities (Chen and Ritz 2018; Zhou et al. 2008). The clinical hallmarks of PD are a premature selective loss of midbrain dopamine neurons (DA) and a buildup of Lewy bodies, which are composed of misfolded α-synuclein and accumulate in several systems in patients with PD (Rizek et al. 2016). Before the 1990s, PD was doubted significantly to be heritable due to sporadic genesis; nonetheless, it was discovered that 5–10% of PD patients have a conventional Mendelian inheritance pattern and that 15% of patients have a family history of the disease (Duvoisin 1984; Lesage and Brice 2009). Mutations in SNCA, LRRK2, and VPS35 are the root causes of inherited monogenic and idiopathic cases of PD. Premature PD cases, before 40 years of age, are linked to autosomal recessive variations such as PARKIN, PINKI, and DJ1, although autosomal dominant mutation variations such as LRRK2 and GBA are associated with delayed PD, after 50 years of age (Cook et al. 2021). The commencement age, diagnosis age, ethnicity, and family history are significant considerations in the genetic diagnosis of PD. Therefore, single-gene screening is only useful in certain situations, such as those with a family history of Gaucher disease and African-Berber ancestry (as they have an increased risk of carrying LRRK2 mutations); in other cases, it is not reliable.

The phenotype of PD cannot often be articulated, even though a multigene panel can determine the cause (Cook et al. 2021). The list keeps growing as more genes are found to be correlated to the onset of PD. As a result, screening of only these genes is available in commercial laboratories. Additionally, whether the patient has the known mutation or not, there is currently no difference in the therapy or administration of PD. It could be achieved by the advent of clinical trials targeting specific mutations. Deaths attributed to PD rise with age. Deaths frequently occur before advanced disease stages, and causes (such as aspiration pneumonia) of death of PD patients are like non-PD patients' cohorts, as written on the death certificates (Armstrong and Okun 2020). The L-DOPA treatment for PD was developed by Hornykiewicz et al. based on Carlsson's discoveries (Lees et al. 2015). This strategy remunerates for lower dopaminergic levels via encouraging dopaminergic synthesis in dopaminergic neurons in the midbrain; however, effects of L-DOPA were frequently erratic, even among the same patients, and frequently brought on deep and unbearable side effects like emotional disturbances, motor fluctuations, dyskinesia, and psychiatric conditions (Iarkov et al. 2020). Although several drugs can reduce the disease's motor symptoms, therapies for PD do not cure the condition. To look over this, many therapeutics are emerging for the prognosis of PD; however, they always result in side effects or are less effective (Clarke 2008; Jagadeesan et al. 2017). So the need for a novel targeted therapy is required for efficient outcomes; in this part, we will talk about the novelty of stem cells and derived exosomes in treating PD.

10.4.1 Stem Cells as Therapeutic for PD

There is a potential pool of cells that can be employed for neural grafting since it is possible to influence the outcome of these cells to become dopaminergic neurons. ESCs and iPSCs are the most promising stem cell types when it comes to treating PD (Stoker 2018). By using in vitro fertilization techniques, many human ESC cell lines have been generated that are derived using the early blastocyst's ICM. Techniques to direct the development of cells into dopaminergic neurons emerged over the following decade (Kriks et al. 2011). Since ESCs are believed to be the most pluripotent, they can, under specific circumstances, develop cells from all three primary germ layers (Murry and Keller 2008). Although the yield of TH-positive cells was quite variable, studies have indicated that expression of tyrosine hydroxylase (TH), the rate-limiting enzyme for the synthesis of dopamine, could be increased. However, it was shown that these cells can be transplanted into rodents and generate some degree of motor recovery (Brederlau et al. 2006; Grealish et al. 2014; Roy et al. 2006). PD treatment was recently reported to use the first dopamine neuron cell product produced from ESCs. Scientists have long been a driving force in the creation of methods to create dopamine neurons from hPSCs (human pluripotent stem cells) (Piao et al. 2021). These characteristics result from telomerase, which extends telomeres and slows the aging in hESCs (Hiyama and Hiyama 2007). When it comes to developmental potential, murine ESCs and human ESCs differ from each other. hESCs have a limited capacity for development compared to murine ESCs, which can differentiate into tissues from the three germ layers (Pera et al. 2000). Additionally, ESCs cannot be implanted directly into the substantia nigra due to their strong tumorigenic potential in undifferentiated conditions. It is generally believed to use a combination of fibroblast growth-8, sonic hedgehog, and brain-derived neurotrophic factor to develop ECS into DA neurons. Furthermore, MS5 cells and ESCs must be co-cultured (Kriks et al. 2011; Guo et al. 2021). However, due to involvement in the destruction of human embryos, ethical and political controversies are in hESC research (Lo and Parham 2009b).

iPSCs provide a platform to investigate how genetic defects affect the likelihood of developing the disease as they have the complete genomic sequence of a patient. By adopting methods like those used with ESCs, the iPSCs produced in a manner that can be differentiated into DA, which could serve as the foundation for efficient cell-based therapy for PD (Soldner et al. 2009). iPSC-derived grafts have an advantage over ESC-derived grafts in that a patient's fibroblasts can be used to generate a neural graft product, negating the need for immunosuppression in ESC-derived grafts (Stoker 2018). Several investigations have demonstrated the potential of autologous iPSC-derived dopaminergic neurons or precursors to survive in vivo through transplantation to primates or murine PD models (Song et al. 2020; Hallett et al. 2015). Improvements in motor function were seen after iPSC-derived dopaminergic neurons that had undergone cultured differentiation were implanted into the putamen of Parkinsonian Cygnus monkeys (Hallett et al. 2015). From an iPSC cell line, Song et al. produced clinical-grade dopaminergic neural progenitors of the midbrain. Following the implantation of 100,000–300,000 of these cells into the striatum, the motor function of mice with immunodeficiency-induced PD significantly improved after 14 weeks; further improvement persisted for at least 52 weeks. These cells were created under good manufacturing practices (GMPs) with special procedure following the crucial components: an episomal vector and specific miRNA improved the efficiency of converting fibroblasts to iPSCs (Song et al. 2020).

The first iPSCs harboring SNCA triple replication and iPSCs that have been differentiated into dopaminergic neurons were obtained by Devine et al. These iPSC-derived dopaminergic neurons successfully reproduced the α-syn accumulation characteristic of PD that was not seen in PD patient skin fibroblasts (Devine et al. 2011). PD is characterized by the accumulation of α-syn, the intrinsic upregulation of oxidative stress markers, and peroxide-induced oxidation in SNCA triplication is linked to increased expression and accumulation of α-syn (Byers et al. 2011). SNCA triplication iPSC-derived neural stem cells are more vulnerable and sensitive to oxidative stress under environment toxins or oxidative stress conditions. Importantly, knocking down endogenous α-syn allows for the reversal of this phenotype (Flierl et al. 2014).

10.4.2 Derived Exosomes in PD

Scientists are using new targeted strategies, such as derived exosomes, to overcome the problem of neurodegenerative diseases, as cell-cell communication is the main feature of it. Also, exosomes can be used as biomarkers for metabolic diseases to diagnose disease risk factors for disease and perhaps treat or prevent disease, and they are also important in the treatment of other diseases such as cancer (Dai et al. 2020).

The key diagnostic criteria for PD are the apparent clinical motor symptoms. However, several non-motor signs are present prior to the onset of motor symptoms (Chaudhuri et al. 2006). Many drugs are used to treat PD, but provide only temporary relief and have adverse side effects as the disease progresses (Müller 2012). Most drugs tested for CNS conditions during clinical trials failed because they would not be able to cross the BBB (Banks et al. 2020; Pardridge 2012). In both in vivo and in vitro experiments, the BBB was crossed by human blood exosomes and delivered dopamine to the brain due to a connection between transferrin and its receptor. When administered intravenously or systemically, dopamine-containing exosomes were less toxic than free dopamine and more therapeutically effective in the PD mouse model (Qu et al. 2018).

MSC-derived exosomes have demonstrated efficacy in treating many clinical conditions such as PD, multiple sclerosis, and osteoarthritis (Li et al. 2019; Vilaça-Faria et al. 2019). In 6-OHDA mouse models of PD, MSC-derived exosomes were found to repair dopaminergic neurons, offering a potential therapy for PD. MSC-derived exosomes associate with neuron cells in the PD animal model to reduce neuroinflammation and enhance neurogenesis along with the expression of advantageous miRNAs. In MSC-derived exosomes, miR-21 and miR-143 significantly influenced immune regulation and neural death (Mianehsaz et al. 2019). As one of the miRNAs downregulated in PD, miR-133b can be supplied to neural cells via MSC-derived exosomes to promote neurite outgrowth. Furthermore, mimic-miR-7 can enhance the neuroinflammation response in PD by modifying MSC-derived exosomes to reduce α-syn aggregation and repress NLRP3 inflammasome activation in SNpc and striatum. The modification in antago-miR-155 can also lessen neuroinflammation and microglia cell activation, therefore perchance beneficial for PD (Mianehsaz et al. 2019).

10.5 Conclusion

Recent research revealing therapeutic advantages for several neurodegenerative diseases in stem cells have promising translational significance. Numerous studies have described the fundamental processes, which range from cell replacement and paracrine actions at the region of neurodegeneration to proliferation, differentiation, and immunomodulation. Depending on the nature and origins of stem cells, preclinical investigations have revealed a variety of impacts. But there are some restrictions that prevent employing of regenerative cell therapy derived from stem cells. Studies indicated that grafts failed to show an advantage of transplantation (J. Y. Li and Li 2021). However, the advantages are noted in countless animal research and human pilot trials; therefore, an in-depth analysis of the sources, types, stages, dosages, and methods is needed to establish the optimum treatment outcome for stem cell transplantation in AD and PD models (Reuter et al. 2008). The different stages of AD and PD progression, along with other related diseases, may significantly affect how cell transplantation functions. Additionally, exosomes are subcellular components that contribute to the start, progression, and spread of AD and PD by releasing harmful substances including misfolded α-syn and inflammatory mediators into the environment. Through the isolation and identification of EV cargo, new biomarkers for AD and PD diagnosis have been discovered. Two main factors embody exosomes as the ideal drug delivery platforms in treating AD and PD and their ability to cross the BBB and low immunogenic activities. Nonetheless, further investigations are needed into the molecular mechanism in the pathology of AD and PD and directly categorize α-syn in PD.

Acknowledgments The authors thank Mr. Manish Kumar for taking the necessary time and effort to review the chapter. We sincerely acknowledge your insightful comments and recommendations, which enabled us to improve the quality of the chapter.

References

Airola MV, Hannun YA (2013) Sphingolipid metabolism and neutral sphingomyelinases. Handb Exp Pharmacol 215(215):57–76. https://doi.org/10.1007/978-3-7091-1368-4_3

Allan D, Tieu A, Lalu M, Burger D (2020) Mesenchymal stromal cell-derived extracellular vesicles for regenerative therapy and immune modulation: progress and challenges toward clinical application. Stem Cells Transl Med 9(1):39. https://doi.org/10.1002/SCTM.19-0114

Amoroso MW, Croft GF, Williams DJ, O'Keeffe S, Carrasco MA, Davis AR, Roybon L, Oakley DH, Maniatis T, Henderson CE, Wichterle H (2013) Accelerated high-yield generation of limb-innervating motor neurons from human stem cells. J Neurosci 33(2):574–586. https://doi.org/10.1523/JNEUROSCI.0906-12.2013

Armstrong MJ, Okun MS (2020) Diagnosis and treatment of Parkinson disease: a review. JAMA–J Am Med Assoc 323(6):548–560. https://doi.org/10.1001/JAMA.2019.22360

Association A (2016) 2016 Alzheimer's disease facts and figures. Alzheimers Dement 12(4): 459–509

Aubry L, Bugi A, Lefort N, Rousseau F, Peschanski M, Perrier AL (2008) Striatal progenitors derived from human ES cells mature into DARPP32 neurons in vitro and in quinolinic acid-lesioned rats. Proc Natl Acad Sci U S A 105(43):16707–16712. https://doi.org/10.1073/PNAS.0808488105/SUPPL_FILE/0808488105SI.PDF

Bachoud-Lévi A-C, Massart R, Rosser A, Anne-Catherine Bachoud-Lévi C, de Neurologie S, Henri Mondor H (2021) Cell therapy in Huntington's disease: taking stock of past studies to move the field forward. Stem Cells 39(2):144–155. https://doi.org/10.1002/STEM.3300

Bae J, Jin HK, Lee JK, Richardson JC, Carter JE (2013) Bone marrow-derived mesenchymal stem cells contribute to the reduction of amyloid-β deposits and the improvement of synaptic transmission in a mouse model of pre-dementia Alzheimer's disease. Curr Alzheimer Res 10(5):524–531

Bang OY, Kim EH, Cho YH, Oh MJ, Chung J-W, Chang WH, Kim Y-H, Yang SW, Chopp M (2022) Circulating extracellular vesicles in stroke patients treated with mesenchymal stem cells:

a biomarker analysis of a randomized trial. Stroke 53(7). https://doi.org/10.1161/STROKEAHA.121.036545

Bang OY, Kim JE (2022) Stem cell-derived extracellular vesicle therapy for acute brain insults and neurodegenerative diseases. BMB Rep 55(1):20–29. https://doi.org/10.5483/BMBREP.2022.55.1.162

Banks WA, Sharma P, Bullock KM, Hansen KM, Ludwig N, Whiteside TL (2020) Transport of extracellular vesicles across the blood-brain barrier: brain pharmacokinetics and effects of inflammation. Int J Mol Sci 21(12):1–21. https://doi.org/10.3390/IJMS21124407

Bhandari J, Thada PK, Samanta D (2022) Spinocerebellar ataxia. StatPearls. https://www.ncbi.nlm.nih.gov/books/NBK557816/

Bobrie A, Colombo M, Raposo G, Théry C (2011) Exosome secretion: molecular mechanisms and roles in immune responses. Traffic 12(12):1659–1668. https://doi.org/10.1111/J.1600-0854.2011.01225.X

Brederlau A, Correia AS, Anisimov SV, Elmi M, Paul G, Roybon L, Morizane A, Bergquist F, Riebe I, Nannmark U, Carta M, Hanse E, Takahashi J, Sasai Y, Funa K, Brundin P, Eriksson PS, Li J-Y (2006) Transplantation of human embryonic stem cell-derived cells to a rat model of Parkinson's disease: effect of in vitro differentiation on graft survival and Teratoma formation. Stem Cells 24(6):1433–1440. https://doi.org/10.1634/STEMCELLS.2005-0393

Brotman RG, Moreno-Escobar MC, Joseph J, Pawar G (2022) Amyotrophic lateral sclerosis. StatPearls. https://www.ncbi.nlm.nih.gov/books/NBK556151/

Bruno S, Grange C, Deregibus MC, Calogero RA, Saviozzi S, Collino F, Morando L, Busca A, Falda M, Bussolati B, Tetta C, Camussi G (2009) Mesenchymal stem cell-derived microvesicles protect against acute tubular injury. J Am Soc Nephrol: JASN 20(5):1053–1067. https://doi.org/10.1681/ASN.2008070798

Budnik V, Ruiz-Cañada C, Wendler F (2016) Extracellular vesicles round off communication in the nervous system. Nat Rev Neurosci 17(3):160–172. https://doi.org/10.1038/NRN.2015.29

Byers B, Cord B, Nguyen HN, Schüle B, Fenno L, Lee PC, Deisseroth K, Langston JW, Pera RR, Palmer TD (2011) SNCA triplication Parkinson's Patient's iPSC-derived DA neurons accumulate α-Synuclein and are susceptible to oxidative stress. PLoS One 6(11):e26159. https://doi.org/10.1371/JOURNAL.PONE.0026159

Chagastelles PC, Nardi NB (2011) Biology of stem cells: an overview. Kidney Int Suppl 1(3):63–67. https://doi.org/10.1038/KISUP.2011.15

Chang YK, Chen MH, Chiang YH, Chen YF, Ma WH, Tseng CY, Soong BW, Ho JH, Lee OK (2011) Mesenchymal stem cell transplantation ameliorates motor function deterioration of spinocerebellar ataxia by rescuing cerebellar Purkinje cells. J Biomed Sci 18(1). https://doi.org/10.1186/1423-0127-18-54

Chaudhuri KR, Healy DG, Schapira AHV (2006) Non-motor symptoms of Parkinson's disease: diagnosis and management. Lancet Neurol 5(3):235–245. https://doi.org/10.1016/S1474-4422(06)70373-8

Chen H, Ritz B (2018) The search for environmental causes of Parkinson's disease: moving forward. J Parkinsons Dis 8(Suppl 1):S9. https://doi.org/10.3233/JPD-181493

Clarke CE (2008) Medical Management of Parkinson's disease. Pharm Ther 33(10):590. https://doi.org/10.1136/jnnp.72.suppl_1.i22

Cook L, Schulze J, Kopil C, Hastings T, Naito A, Wojcieszek J, Payne K, Alcalay RN, Klein C, Saunders-Pullman R, Simuni T, Foroud T (2021) Genetic testing for Parkinson disease. Neurol: Clin Pract 11(1):69–77. https://doi.org/10.1212/CPJ.0000000000000831

Dai J, Su Y, Zhong S, Cong L, Liu B, Yang J, Tao Y, He Z, Chen C, Jiang Y (2020) Exosomes: key players in cancer and potential therapeutic strategy. Signal Transduct Target Ther 5(1):1–10. https://doi.org/10.1038/s41392-020-00261-0

Elise D, Stephanie M, Kiminobu S (2010) Stem cells for the treatment of neurodegenerative diseases. Stem Cell Res Ther. https://stemcellres.biomedcentral.com/track/pdf/10.1186/scrt37.pdf

Dawson TM, Golde TE, Lagier-Tourenne C (2018) Animal models of neurodegenerative diseases. Nat Neurosci 21(10):1370. https://doi.org/10.1038/S41593-018-0236-8

Devine MJ, Ryten M, Vodicka P, Thomson AJ, Burdon T, Houlden H, Cavaleri F, Nagano M, Drummond NJ, Taanman JW, Schapira AH, Gwinn K, Hardy J, Lewis PA, Kunath T (2011) Parkinson's disease induced pluripotent stem cells with triplication of the α-synuclein locus. Nat Commun 2(1):1–10. https://doi.org/10.1038/ncomms1453

Dorsey ER, Sherer T, Okun MS, Bloemd BR (2018) The emerging evidence of the Parkinson pandemic. J Parkinsons Dis 8(s1):S3–S8. https://doi.org/10.3233/JPD-181474

Dulak J, Szade K, Szade A, Nowak W, Józkowicz A (2015) Adult stem cells: hopes and hypes of regenerative medicine. Acta Biochim Pol 62(3):329–337. https://doi.org/10.18388/abp.2015_1023

Duncan T, Valenzuela M (2017) Alzheimer's disease, dementia, and stem cell therapy. Stem Cell Res Ther 8(1):111. https://doi.org/10.1186/s13287-017-0567-5

Duvoisin RC (1984) Is Parkinson's disease acquired or inherited? Can J Neurol Sci 11:151–155. https://doi.org/10.1017/S031716710004631X

Elsherbini A, Kirov AS, Dinkins MB, Wang G, Qin H, Zhu Z, Tripathi P, Crivelli SM, Bieberich E (2020a) Association of Aβ with ceramide-enriched astrosomes mediates Aβ neurotoxicity. Acta Neuropathol Commun 8(1):1–19

Elsherbini A, Qin H, Zhu Z, Tripathi P, Crivelli SM, Bieberich E (2020b) In vivo evidence of exosome-mediated Aβ neurotoxicity. Acta Neuropathol Commun 8(1):1–3

Feigin VL, Vos T (2019) Global burden of neurological disorders: from global burden of disease estimates to actions. Neuroepidemiology 52:1–2. https://doi.org/10.1159/000495197

Ferri CP, Prince M, Brayne C, Brodaty H, Fratiglioni L, Ganguli M, Hall K, Hasegawa K, Hendrie H, Huang Y (2005) Global prevalence of dementia: a Delphi consensus study. Lancet 366(9503):2112–2117

Flierl A, Oliveira LMA, Falomir-Lockhart LJ, Mak SK, Hesley J, Soldner F, Arndt-Jovin DJ, Jaenisch R, Langston JW, Jovin TM, Le BS (2014) Higher vulnerability and stress sensitivity of neuronal precursor cells carrying an alpha-Synuclein gene triplication. PLoS One 9(11): e112413. https://doi.org/10.1371/JOURNAL.PONE.0112413

Glass JD, Boulis NM, Johe K, Rutkove SB, Federici T, Polak M, Kelly C, Feldman EL (2012) Lumbar intraspinal injection of neural stem cells in patients with amyotrophic lateral sclerosis: results of a phase I trial in 12 patients. Stem Cells (Dayton, Ohio) 30(6):1144–1151. https://doi.org/10.1002/STEM.1079

Goetzl EJ, Schwartz JB, Abner EL, Jicha GA, Kapogiannis D (2018) High complement levels in astrocyte-derived exosomes of Alzheimer's disease. Ann Neurol 83(3):544. https://doi.org/10.1002/ANA.25172

Grealish S, Diguet E, Kirkeby A, Mattsson B, Heuer A, Bramoulle Y, Van Camp N, Perrier AL, Hantraye P, Björklund A, Parmar M (2014) Human ESC-derived dopamine neurons show similar preclinical efficacy and potency to fetal neurons when grafted in a rat model of Parkinson's disease. Cell Stem Cell 15(5):653–665. https://doi.org/10.1016/J.STEM.2014.09.017/ATTACHMENT/47BFF55A-6C68-46BB-B83D-B2C73155AAFD/MMC1.PDF

Gu G, Zhang W, Li M, Ni J, Wang P (2015) Transplantation of NSC-derived cholinergic neuron-like cells improves cognitive function in APP/PS1 transgenic mice. Neuroscience 291:81–92

Guo M, Yin Z, Chen F, Lei P (2020) Mesenchymal stem cell-derived exosome: a promising alternative in the therapy of Alzheimer's disease. Alzheimers Res Ther 12(1):109. https://doi.org/10.1186/s13195-020-00670-x

Guo X, Tang L, Tang X (2021) Current developments in cell replacement therapy for Parkinson's disease. Neuroscience 463:370–382. https://doi.org/10.1016/J.NEUROSCIENCE.2021.03.022

Gurdon JB (1962) The developmental capacity of nuclei taken from intestinal epithelium cells of feeding tadpoles. Development 10(4):622–640. https://doi.org/10.1242/DEV.10.4.622

Hallett PJ, Deleidi M, Astradsson A, Smith GA, Cooper O, Osborn TM, Sundberg M, Moore MA, Perez-Torres E, Brownell AL, Schumacher JM, Spealman RD, Isacson O (2015) Successful function of autologous iPSC-derived dopamine neurons following transplantation in a non-human primate model of Parkinson's disease. Cell Stem Cell 16(3):269–274. https://doi.org/10.1016/J.STEM.2015.01.018/ATTACHMENT/A5BD7CDE-4426-4170-89E7-7F91F73CE3F6/MMC1.PDF

Han Z, Tian R, Ren P, Zhou W, Wang P, Luo M, Jin S, Jiang Q (2018) Parkinson's disease and Alzheimer's disease: a Mendelian randomization study. BMC Med Genet 19(Suppl 1). https://doi.org/10.1186/S12881-018-0721-7

Hayashi K, Hashimoto M, Koda M, Naito AT, Murata A, Okawa A, Takahashi K, Yamazaki M (2011) Increase of sensitivity to mechanical stimulus after transplantation of murine induced pluripotent stem cell–derived astrocytes in a rat spinal cord injury model: laboratory investigation. J Neurosurg Spine 15(6):582–593. https://doi.org/10.3171/2011.7.SPINE10775

Henne WM, Buchkovich NJ, Emr SD (2011) The ESCRT pathway. Dev Cell 21(1):77–91. https://doi.org/10.1016/J.DEVCEL.2011.05.015

Hiyama E, Hiyama K (2007) Telomere and telomerase in stem cells. Br J Cancer 96(7):1020–1024. https://doi.org/10.1038/sj.bjc.6603671

Iarkov A, Barreto GE, Grizzell JA, Echeverria V (2020) Strategies for the treatment of Parkinson's disease: beyond dopamine. Front Aging Neurosci 12:4. https://doi.org/10.3389/FNAGI.2020.00004/BIBTEX

Institute of Medicine (2002) 2002. The National Academies Press, Stem Cells and the Future of Regenerative Medicine. https://doi.org/10.17226/10195

Jagadeesan AJ, Murugesan R, Vimala Devi S, Meera M, Madhumala G, Vishwanathan Padmaja M, Ramesh A, Banerjee A, Sushmitha S, Khokhlov AN, Marotta F, Pathak S (2017) Current trends in etiology, prognosis and therapeutic aspects of Parkinson's disease: a review. Acta Bio Medica : Atenei Parmensis 88(3):249. https://doi.org/10.23750/abm.v%vi%i.6063

Jahan S, Mukherjee S, Ali S, Bhardwaj U, Choudhary RK, Balakrishnan S, Naseem A, Mir SA, Banawas S, Alaidarous M, Alyenbaawi H, Iqbal D, Siddiqui AJ (2022) Pioneer role of extracellular vesicles as modulators of cancer initiation in progression, drug therapy, and vaccine prospects. Cell 11(3). https://doi.org/10.3390/CELLS11030490

Jankovic J, Tan EK (2020) Parkinson's disease: etiopathogenesis and treatment. J Neurol Neurosurg Psychiatry 91(8):795–808. https://doi.org/10.1136/JNNP-2019-322338

Kang JM, Yeon BK, Cho S-J, Suh Y-H (2016) Stem cell therapy for Alzheimer's disease: a review of recent clinical trials. J Alzheimers Dis 54(3):879–889

Kent SA, Spires-Jones TL, Durrant CS (2020) The physiological roles of tau and Aβ: implications for Alzheimer's disease pathology and therapeutics. Acta Neuropathol 140(4):417–447

Khan H, Pan JJ, Li Y, Zhang Z, Yang GY (2021) Native and bioengineered exosomes for ischemic stroke therapy. Front Cell Dev Biol 9:619565. https://doi.org/10.3389/FCELL.2021.619565

Kim DS, Lee DR, Kim HS, Yoo JE, Jung SJ, Lim BY, Jang J, Kang HC, You S, Hwang DY, Leem JW, Nam TS, Cho SR, Kim DW (2012) Highly pure and expandable PSA-NCAM-positive neural precursors from human ESC and iPSC-derived neural rosettes. PLoS One 7(7):e39715. https://doi.org/10.1371/JOURNAL.PONE.0039715

Kim SW, Jang YJ, Chang JW, Hwang O (2003) Degeneration of the nigrostriatal pathway and induction of motor deficit by tetrahydrobiopterin: an in vivo model relevant to Parkinson's disease. Neurobiol Dis 13(2):167–176. https://doi.org/10.1016/S0969-9961(03)00037-8

Kriks S, Shim JW, Piao J, Ganat YM, Wakeman DR, Xie Z, Carrillo-Reid L, Auyeung G, Antonacci C, Buch A, Yang L, Beal MF, Surmeier DJ, Kordower JH, Tabar V, Studer L (2011) Dopamine neurons derived from human ES cells efficiently engraft in animal models of Parkinson's disease. Nature 480(7378):547–551. https://doi.org/10.1038/nature10648

Kumar A, Sidhu J, Goyal A, Tsao JW (2021) Alzheimer disease. StatPearls:1–27. https://www.ncbi.nlm.nih.gov/books/NBK499922/

Lakshmi S, Essa MM, Hartman RE, Guillemin GJ, Sivan S, Elumalai P (2020) Exosomes in Alzheimer's disease: potential role as pathological mediators, biomarkers and therapeutic targets. Neurochem Res 45(11):2553–2559. https://doi.org/10.1007/S11064-020-03111-1

Lamptey RNL, Chaulagain B, Trivedi R, Gothwal A, Layek B, Singh J (2022) A review of the common neurodegenerative disorders: current therapeutic approaches and the potential role of nanotherapeutics. Int J Mol Sci 23(3):1851. https://doi.org/10.3390/ijms23031851

Laulagnier K, Javalet C, Hemming FJ, Chivet M, Lachenal G, Blot B, Chatellard C, Sadoul R (2017) Amyloid precursor protein products concentrate in a subset of exosomes specifically endocytosed by neurons. Cell Mol Life Sci 75(4):757–773. https://doi.org/10.1007/S00018-017-2664-0

Lee HJ, Lee JK, Lee H, Carter JE, Chang JW, Oh W, Yang YS, Suh J-G, Lee B-H, Jin HK (2012) Human umbilical cord blood-derived mesenchymal stem cells improve neuropathology and cognitive impairment in an Alzheimer's disease mouse model through modulation of neuroinflammation. Neurobiol Aging 33(3):588–602

Lee S, Mankhong S, Kang J-H (2019) Extracellular vesicle as a source of Alzheimer's biomarkers: opportunities and challenges. Int J Mol Sci 20(7):1728

Lees AJ, Tolosa E, Olanow CW (2015) Four pioneers of L-dopa treatment: Arvid Carlsson, Oleh Hornykiewicz, George Cotzias, and Melvin Yahr. Mov Disord 30(1):19–36. https://doi.org/10.1002/MDS.26120

Lesage S, Brice A (2009) Parkinson's disease: from monogenic forms to genetic susceptibility factors. Hum Mol Genet 18(R1):R48–R59. https://doi.org/10.1093/HMG/DDP012

Li JY, Li W (2021) Postmortem studies of fetal grafts in Parkinson's disease: what lessons have we learned? Front Cell Dev Biol 9. https://doi.org/10.3389/FCELL.2021.666675/FULL

Li Z, Liu F, He X, Yang X, Shan F, Feng J (2019) Exosomes derived from mesenchymal stem cells attenuate inflammation and demyelination of the central nervous system in EAE rats by regulating the polarization of microglia. Int Immunopharmacol 67:268–280. https://doi.org/10.1016/J.INTIMP.2018.12.001

Liu X-Y, Yang L-P, Zhao L (2020) Stem cell therapy for Alzheimer's disease. World J Stem Cells 12(8):787

Lo B, Parham L (2009a) Ethical issues in stem. Cell Res. https://doi.org/10.1210/er.2008-0031

Lo B, Parham L (2009b) Ethical issues in stem cell research. Endocr Rev 30(3):204. https://doi.org/10.1210/ER.2008-0031

Maloney JA, Bainbridge T, Gustafson A, Zhang S, Kyauk R, Steiner P, van der Brug M, Liu Y, Ernst JA, Watts RJ (2014) Molecular mechanisms of Alzheimer disease protection by the A673T allele of amyloid precursor protein. J Biol Chem 289(45):30990–31000

Mao Z, Zhang S, Chen H (2015) Stem cell therapy for amyotrophic lateral sclerosis. Cell Regen *4*(1):11. https://doi.org/10.1186/S13619-015-0026-7

Matthews KA, Xu W, Gaglioti AH, Holt JB, Croft JB, Mack D, McGuire LC (2019) Racial and ethnic estimates of Alzheimer's disease and related dementias in the United States (2015–2060) in adults aged ≥65 years. Alzheimers Dement 15(1):17–24. https://doi.org/10.1016/J.JALZ.2018.06.3063

Mazzini L, Fagioli F, Boccaletti R, Mareschi K, Oliveri G, Olivieri C, Pastore I, Marasso R, Madon E (2003) Stem cell therapy in amyotrophic lateral sclerosis: a methodological approach in humans. Amyotroph Lateral Scler Other Motor Neuron Disord 4(3):158–161. https://doi.org/10.1080/14660820310014653

Mianehsaz E, Mirzaei HR, Mahjoubin-Tehran M, Rezaee A, Sahebnasagh R, Pourhanifeh MH, Mirzaei H, Hamblin MR (2019) Mesenchymal stem cell-derived exosomes: a new therapeutic approach to osteoarthritis? Stem Cell Res Therapy 10(1):1–13. https://doi.org/10.1186/S13287-019-1445-0

Mirzaaghasi A, Han Y, Ahn SH, Choi C, Park JH (2021) Biodistribution and Pharmacokinectics of liposomes and exosomes in a mouse model of sepsis. Pharmaceutics 13(3). https://doi.org/10.3390/PHARMACEUTICS13030427

Moghadam FH, Alaie H, Karbalaie K, Tanhaei S, Esfahani MHN, Baharvand H (2009) Transplantation of primed or unprimed mouse embryonic stem cell-derived neural precursor cells improves cognitive function in Alzheimerian rats. Differentiation 78(2–3):59–68

Muller L (2020) Exosomes: nanodust? HNO 68(1):56–59. https://doi.org/10.1007/S00106-019-00786-Z

Müller T (2012) Drug therapy in patients with Parkinson's disease. Transl Neurodegen 1(1):1–13. https://doi.org/10.1186/2047-9158-1-10/TABLES/5

Muratore CR, Rice HC, Srikanth P, Callahan DG, Shin T, Benjamin LNP, Walsh DM, Selkoe DJ, Young-Pearse TL (2014) The familial Alzheimer's disease APPV717I mutation alters APP processing and tau expression in iPSC-derived neurons. Hum Mol Genet 23(13):3523–3536

Murry CE, Keller G (2008) Differentiation of embryonic stem cells to clinically relevant populations: lessons from embryonic development. Cell 132(4):661–680. https://doi.org/10.1016/J.CELL.2008.02.008

Naaldijk Y, Jaeger C, Fabian C, Leovsky C, Blüher A, Rudolph L, Hinze A, Stolzing A (2017) Effect of systemic transplantation of bone marrow-derived mesenchymal stem cells on neuropathology markers in APP/PS 1 Alzheimer mice. Neuropathol Appl Neurobiol 43(4):299–314

Nutt SE, Chang EA, Suhr ST, Schlosser LO, Mondello SE, Moritz CT, Cibelli JB, Horner PJ (2013) Caudalized human iPSC-derived neural progenitor cells produce neurons and glia but fail to restore function in an early chronic spinal cord injury model. Exp Neurol 248:491–503. https://doi.org/10.1016/J.EXPNEUROL.2013.07.010

Oh SH, Kim HN, Park H-J, Shin JY, Lee PH (2015) Mesenchymal stem cells increase hippocampal neurogenesis and neuronal differentiation by enhancing the Wnt signaling pathway in an Alzheimer's disease model. Cell Transplant 24(6):1097–1109

Pacheco-Quinto J, Clausen D, Pérez-González R, Peng H, Meszaros A, Eckman CB, Levy E, Eckman EA (2019) Intracellular metalloprotease activity controls intraneuronal Aβ aggregation and limits secretion of Aβ via exosomes. FASEB J 33(3):3758–3771

Pardridge WM (2012) Drug transport across the blood-brain barrier. J Cereb Blood Flow Metab 32(11):1959–1972. https://doi.org/10.1038/jcbfm.2012.126

Pera MF, Reubinoff B, Trounson A (2000) Human embryonic stem cells. J Cell Sci 113(1):5–10. https://doi.org/10.1242/JCS.113.1.5

Pérez-González R, Sahoo S, Gauthier SA, Kim Y, Li M, Kumar A, Pawlik M, Benussi L, Ghidoni R, Levy E (2019) Neuroprotection mediated by cystatin C-loaded extracellular vesicles. Sci Rep 9(1):1–10

Piao J, Zabierowski S, Dubose BN, Hill EJ, Navare M, Claros N, Rosen S, Ramnarine K, Horn C, Fredrickson C, Wong K, Safford B, Kriks S, El Maarouf A, Rutishauser U, Henchcliffe C, Wang Y, Riviere I, Mann S et al (2021) Preclinical efficacy and safety of a human embryonic stem cell-derived midbrain dopamine progenitor product, MSK-DA01. Cell Stem Cell 28(2): 217–229.e7. https://doi.org/10.1016/J.STEM.2021.01.004/ATTACHMENT/6B7AA27F-658E-4E60-99B0-F5AB59AD1824/MMC1.PDF

Piers RJ, Liu Y, Ang TFA, Tao Q, Au R, Qiu WQ (2021) Association between elevated depressive symptoms and cognitive function moderated by APOE4 status: framingham offspring study. J Alzheimers Dis: JAD 80(3):1269. https://doi.org/10.3233/JAD-200998

Poddar MK, Chakraborty A, Banerjee S (2021) Neurodegeneration: diagnosis, prevention, and therapy. Oxidoreductase. https://doi.org/10.5772/INTECHOPEN.94950

Prochazkova M, Chavez MG, Prochazka J, Felfy H, Mushegyan V, Klein OD (2015) Embryonic versus adult stem cells. Stem Cell Biol Tissue Engg Dental Sci 249–262. https://doi.org/10.1016/B978-0-12-397157-9.00020-5

Przedborski S, Vila M, Jackson-Lewis V (2003) Series introduction: neurodegeneration: what is it and where are we? J Clin Investig 111(1):3. https://doi.org/10.1172/JCI17522

Qu M, Lin Q, Huang L, Fu Y, Wang L, He S, Fu Y, Yang S, Zhang Z, Zhang L, Sun X (2018) Dopamine-loaded blood exosomes targeted to brain for better treatment of Parkinson's disease. J Control Release 287:156–166. https://doi.org/10.1016/J.JCONREL.2018.08.035

Raposo G, Stoorvogel W (2013) Extracellular vesicles: exosomes, microvesicles, and friends. J Cell Biol 200(4):373–383. https://doi.org/10.1083/JCB.201211138

Ratajczak J, Miekus K, Kucia M, Zhang J, Reca R, Dvorak P, Ratajczak MZ (2006) Embryonic stem cell-derived microvesicles reprogram hematopoietic progenitors: evidence for horizontal transfer of mRNA and protein delivery. Leukemia 20(5):847–856. https://doi.org/10.1038/sj.leu.2404132

Reuter I, Tai YF, Pavese N, Chaudhuri KR, Mason S, Polkey CE, Clough C, Brooks DJ, Barker RA, Piccini P (2008) Long-term clinical and positron emission tomography outcome of fetal striatal transplantation in Huntington's disease. J Neurol Neurosurg Psychiatry 79(8):948–951. https://doi.org/10.1136/JNNP.2007.142380

Rizek P, Kumar N, Jog MS (2016) An update on the diagnosis and treatment of Parkinson disease. CMAJ 188(16):1157–1165. https://doi.org/10.1503/CMAJ.151179/-/DC1

Romito A, Cobellis G (2016) Pluripotent stem cells: current understanding and future directions. Stem Cells Int 2016. https://doi.org/10.1155/2016/9451492

Roy NS, Cleren C, Singh SK, Yang L, Beal MF, Goldman SA (2006) Functional engraftment of human ES cell–derived dopaminergic neurons enriched by coculture with telomerase-immortalized midbrain astrocytes. Nat Med 12(11):1259–1268. https://doi.org/10.1038/nm1495

Sabayan B, Sorond F (2017) Reducing risk of dementia in older age. JAMA 317(19):2028

Sakthiswary R, Raymond AA (2012) Stem cell therapy in neurodegenerative diseases: from principles to practice. Neural Regen Res 7(23):1822. https://doi.org/10.3969/J.ISSN.1673-5374.2012.23.009

Sardar Sinha M, Ansell-Schultz A, Civitelli L, Hildesjö C, Larsson M, Lannfelt L, Ingelsson M, Hallbeck M (2018) Alzheimer's disease pathology propagation by exosomes containing toxic amyloid-beta oligomers. Acta Neuropathol 136(1):41–56. https://doi.org/10.1007/s00401-018-1868-1

Si Z, Wang X (2021) Stem cell therapies in Alzheimer's disease: applications for disease modeling. J Pharmacol Exp Ther 377(2):207–217

Simons M, Raposo G (2009) Exosomes–vesicular carriers for intercellular communication. Curr Opin Cell Biol 21(4):575–581. https://doi.org/10.1016/J.CEB.2009.03.007

Singh S, Srivastava A, Srivastava P, Dhuriya YK, Pandey A, Kumar D, Rajpurohit CS (2016) Advances in stem cell research- A ray of hope in better diagnosis and prognosis in neurodegenerative diseases. Front Mol Biosci 3(NOV):72. https://doi.org/10.3389/FMOLB.2016.00072/BIBTEX

Sivandzade F, Cucullo L (2021) Regenerative stem cell therapy for neurodegenerative diseases: an overview. Int J Mol Sci 22(4):1–21. https://doi.org/10.3390/IJMS22042153

Soares Martins T, Trindade D, Vaz M, Campelo I, Almeida M, Trigo G, da Cruz e Silva OAB, Henriques AG (2021) Diagnostic and therapeutic potential of exosomes in Alzheimer's disease. J Neurochem 156(2):162–181. https://doi.org/10.1111/JNC.15112

Soldner F, Hockemeyer D, Beard C, Gao Q, Bell GW, Cook EG, Hargus G, Blak A, Cooper O, Mitalipova M, Isacson O, Jaenisch R (2009) Parkinson's disease patient-derived induced pluripotent stem cells free of viral reprogramming factors. Cell 136(5):964–977. https://doi.org/10.1016/J.CELL.2009.02.013/ATTACHMENT/DCDE63C5-0DCA-4577-8025-E336FDFAE083/MMC2.XLS

Song B, Cha Y, Ko S, Jeon J, Lee N, Seo H, Park KJ, Lee IH, Lopes C, Feitosa M, Luna MJ, Jung JH, Kim J, Hwang D, Cohen BM, Teicher MH, Leblanc P, Carter BS, Kordower JH, Kim KS et al (2020) Human autologous iPSC–derived dopaminergic progenitors restore motor function in Parkinson's disease models. J Clin Invest 130(2):904–920. https://doi.org/10.1172/JCI130767

Stahl PD, Barbieri MA (2002) Multivesicular bodies and multivesicular endosomes: the "ins and outs" of endosomal traffic. Sci STKE 2002(141). https://doi.org/10.1126/STKE.2002.141.PE32

Stoker TB (2018) Stem cell treatments for Parkinson's disease. Parkinson's Disease: Pathogenesis and Clinical Aspects 161–175. https://doi.org/10.15586/CODONPUBLICATIONS.PARKINSONSDISEASE.2018.CH9

Sun C, Armstrong MJ (2021) Treatment of Parkinson's disease with cognitive impairment: current approaches and future directions. Behav Sci 11(4). https://doi.org/10.3390/BS11040054

Takahashi K, Tanabe K, Ohnuki M, Narita M, Ichisaka T, Tomoda K, Yamanaka S (2007) Induction of pluripotent stem cells from adult human fibroblasts by defined factors. Cell 131(5):861–872. https://doi.org/10.1016/J.CELL.2007.11.019

Takahashi K, Yamanaka S (2006) Induction of pluripotent stem cells from mouse embryonic and adult fibroblast cultures by defined factors. Cell 126(4):663–676. https://doi.org/10.1016/J.CELL.2006.07.024

Théry C (2011) Exosomes: secreted vesicles and intercellular communications. F1000 Biol Rep 3(1):15. https://doi.org/10.3410/B3-15

Théry C, Amigorena S, Raposo G, Clayton A (2006) Isolation and characterization of exosomes from cell culture supernatants and biological fluids. Curr Protoc Cell Biol 30(1):3.22.1-3.22.29. https://doi.org/10.1002/0471143030.CB0322S30

Tsai Y-A, Liu R-S, Lirng J-F, Yang B-H, Chang C-H, Wang Y-C, Wu Y-S, Hui-Chun Ho J, Lee OK, Soong B-W (2017) Treatment of spinocerebellar ataxia with mesenchymal stem cells: a phase I/IIa clinical study. Cell Transplant 26:503–512. https://doi.org/10.3727/096368916X694373

Van Den Eeden SK, Tanner CM, Bernstein AL, Fross RD, Leimpeter A, Bloch DA, Nelson LM (2003) Incidence of Parkinson's disease: variation by age, gender, and race/ethnicity. Am J Epidemiol 157(11):1015–1022. https://doi.org/10.1093/AJE/KWG068

Vatsa P, Jahan S, Ansari UA, Khan A, Mir SA, Alshehri B, Choudhary RK, Siddiqui AJ (2022) Stem cell safety and sterility testing: a promising approach in regenerative medicine. Stem Cell Prod 205–232. https://doi.org/10.1007/978-981-16-7589-8_9

Vilaça-Faria H, Salgado AJ, Teixeira FG (2019) Mesenchymal stem cells-derived exosomes: a new possible therapeutic strategy for Parkinson's disease? Cells 8:118. https://doi.org/10.3390/CELLS8020118

Wang Q, Matsumoto Y, Shindo T, Miyake K, Shindo A, Kawanishi M, Kawai N, Tamiya T, Nagao S (2006) Neural stem cells transplantation in cortex in a mouse model of Alzheimer's disease. J Med Investig 53(1, 2):61–69

Wanneveich M, Moisan F, Jacqmin-Gadda H, Elbaz A, Joly P (2018) Projections of prevalence, lifetime risk, and life expectancy of Parkinson's disease (2010-2030) in France. Movement Dis 33(9):1449–1455. https://doi.org/10.1002/MDS.27447

Wen S, Dooner M, Papa E, del Tatto M, Pereira M, Borgovan T, Cheng Y, Goldberg L, Liang O, Camussi G, Quesenberry P (2019) Biodistribution of mesenchymal stem cell-derived extracellular vesicles in a radiation injury bone marrow murine model. Int J Mol Sci 20(21). https://doi.org/10.3390/IJMS20215468

Wilmut I, Schnieke AE, McWhir J, Kind AJ, Campbell KHS (1997) Viable offspring derived from fetal and adult mammalian cells. Nature 385(6619):810–813. https://doi.org/10.1038/385810a0

Winston CN, Goetzl EJ, Akers JC, Carter BS, Rockenstein EM, Galasko D, Masliah E, Rissman RA (2016) Prediction of conversion from mild cognitive impairment to dementia with neuronally derived blood exosome protein profile. Alzheimers Dement 3:63. https://doi.org/10.1016/J.DADM.2016.04.001

Wolf P (1967) The nature and significance of platelet products in human plasma. Br J Haematol 13(3):269–288. https://doi.org/10.1111/J.1365-2141.1967.TB08741.X

Xu L, Yan J, Chen D, Welsh AM, Hazel T, Johe K, Hatfield G, Koliatsos VE (2006) Human neural stem cell grafts ameliorate motor neuron disease in SOD-1 transgenic rats. Transplantation 82(7):865–875. https://doi.org/10.1097/01.TP.0000235532.00920.7A

Xu M, Zhang L, Liu G, Jiang N, Zhou W, Zhang Y (2019) Pathological changes in Alzheimer's disease analyzed using induced pluripotent stem cell-derived human microglia-like cells. J Alzheimers Dis 67(1):357–368

Yamanaka S, Takahashi K, Okita K, Nakagawa M (2007) Induction of pluripotent stem cells from fibroblast cultures. Nat Protoc 2(12):3081–3089. https://doi.org/10.1038/NPROT.2007.418

Zafar S, Yaddanapudi SS (2021) Parkinson Disease. StatPearls 8:1–13. https://www.ncbi.nlm.nih.gov/books/NBK470193/

Zakharova L, Svetlova M, Fomina AF (2007) T cell exosomes induce cholesterol accumulation in human monocytes via phosphatidylserine receptor. J Cell Physiol 212(1):174–181. https://doi.org/10.1002/JCP.21013

Zakrzewski W, Dobrzyński M, Szymonowicz M, Rybak Z (2019) Stem cells: past, present, and future. Stem Cell Res Therapy 10(1):1–22. https://doi.org/10.1186/S13287-019-1165-5/FIGURES/8

Zhang T, Ma S, Lv J, Wang X, Afewerky HK, Li H, Lu Y (2021) The emerging role of exosomes in Alzheimer's disease. Ageing Res Rev 68:101321

Zhou C, Huang Y, Przedborski S (2008) Oxidative stress in Parkinson's disease: a mechanism of pathogenic and therapeutic significance. Ann N Y Acad Sci 1147:93–104. https://doi.org/10.1196/ANNALS.1427.023

Stem Cells, Derived Exosomes, and Associated Signaling Molecules in Neuroprotection

11

Sadaf Jahan, Dipak Kumar, Shouvik Mukherjee, Kalpeeta Roy, Arif Jamal Siddiqui, Andleeb Khan, Chetan Singh Rajpurohit, and Sahar Aldosari

Abstract

Neurodegenerative disorders (NDDs) are well known and major causes of disability, complications, and neuronal death worldwide. There are treatments available but less promising, expensive, and least effective. Therefore, researchers are focusing towards the reliable treatment with promising effect. There are numerous substances that have neuroprotective properties like free radical scavengers, anti-stimulants, anti-inflammatory agents, neurotrophic factors, iron chelators, stimulants, and gene therapy. They do have some drawbacks, however, such as the blood-brain barrier, which keeps infections and viruses out, but also drugs.

S. Jahan · S. Aldosari (✉)
Department of Medical Laboratory Sciences, College of Applied Medical Sciences, Majmaah University, Al-Majmaah, Saudi Arabia
e-mail: s.aldosari@mu.edu.sa

D. Kumar
Department of Zoology, KKM College, Munger University, Munger, India

S. Mukherjee
Department of Biotechnology, School of Chemical and Life Sciences, Jamia Hamdard, New Delhi, India

K. Roy
CliniRx Research Pvt. Ltd, New Delhi, India

A. J. Siddiqui
Department of Biology, College of Science, University of Hail, Hail City, Saudi Arabia

A. Khan
Department of Pharmacology and Toxicology, College of Pharmacy, Jazan University, Jazan, Saudi Arabia

C. S. Rajpurohit
Huffington Center on Aging, Baylor College of Medicine, One Baylor Plaza (Lab), Houston, USA

S. Jahan, A. J. Siddiqui (eds.), *Applications of Stem Cells and Derived Exosomes in Neurodegenerative Disorders*, https://doi.org/10.1007/978-981-99-3848-3_11

This makes it difficult to deliver treatments directly to the brain. This is where stem cell therapy and exosomes come into play. Exosomes are small membrane-bound nano vesicles which transport molecular cargo such as cellular proteins, lipids, micro-ribonucleic acid (miRNA) and messenger RNA (mRNA). Furthermore, signaling molecules also impacted the recovery of NDDs. This chapter mainly focused on the therapeutic approach of exosomes, signaling molecules against neuronal complications.

Keywords

Brain · Blood-brain barrier · Neurodegenerative disorders · Exosomes · Signaling Molecules · Neuroprotection

11.1 Introduction

Neuroprotection includes preservation of neuronal structures and/or functions and inhibition of pathophysiological processes that may lead to neuronal dysfunction and/or death (Duman 2022). Neuroprotection aims to limit neuronal death following central nervous system (CNS) injury and protect the CNS from premature degeneration and other causes of neuronal cell death. Neuroprotectants counteract the effects of neurodegeneration. Neuroprotection is needed during various events, such as traumatic brain injury (TBI), stroke, or neurological injury and support for people with neurological diseases such as Alzheimer's disease (AD), Parkinson's disease (PD) and multiple sclerosis (MS) (Jain and Jain 2019).

Current neuroprotective agents cannot reverse existing nerve damage, but they can prevent further nerve damage and slow CNS degeneration (Kumar et al. 2015; Sinha et al. 2020). There are numerous substances that have neuroprotective properties like free radical scavengers, anti-stimulants, anti-inflammatory agents, neurotrophic factors, iron chelators, stimulants, and gene therapy (Dergunova et al. 2021). They do have some drawbacks, however, such as the blood-brain barrier, which keeps infections and viruses out, but also drugs. This makes it difficult to deliver treatments directly to the brain. This is where stem cell therapy and exosomes come into play. Stem cells are undifferentiated cells, present in embryonic, fetal, and adult life stages that give rise to differentiated cells as building blocks of tissues and organs. Tissue-specific stem cells are present in differentiated organs after birth and in adulthood and play an important role in organ repair after injury (Ronaghi et al. 2010). Exosomes are small membrane-bound nanovesicles which transport molecular cargo such as cellular proteins, lipids, micro-ribonucleic acid (miRNA), and messenger RNA (mRNA). Exosomes are small membrane-bound nanovesicles that transport molecular cargo such as cellular proteins, lipids, micro-ribonucleic acid (miRNA), and messenger RNA (mRNA) (Jahan et al. 2022). The therapeutic effects of exosomes are mainly attributed to their surface markers, molecular content, and low immunogenicity, ability to cross the blood-brain barrier (BBB), and mediate neurogenesis (Kang et al. 2019).

Signaling molecules are molecules that allow cells to communicate with one another. Signaling molecules are required for coordinating bodily reactions, processing external inputs, and maintaining homeostasis in multicellular organisms (Kumar and Khanum 2012). Tiny signaling molecules, ranging from a few atoms grouped in a chemical structure to small peptide molecules, are common. There are various signaling molecules that have been found to have neuroprotective action like erythropoietin signaling pathways, glial Nrf2 signaling mediates neuroprotection, and ligand-gated ion channel interacting proteins also have been found for their role in neuroprotection (Broughton et al. 2009; Juybari et al. 2019). This chapter provides an overview of the literature on the roles of stem cells, exosomes, and associated signaling molecules in neuroprotection.

11.2 Stem Cells and Derived Factors

Stem cells are cells that can self-renew and differentiate. The ability of cells to proliferate without losing their differentiation potential or aging is called self-renewal. Stem cells that are known for their neuroprotective effects include embryonic stem cells (ESCs), mesenchymal stem cells (MSCs), brain-derived neural stem cells (NSCs), and induced pluripotent stem cells (iPSCs) (Rikhtegar et al. 2019). ESCs are pluripotent cells derived from the inner cell mass of the developing blastocyst (embryonic day 5–6) capable of giving rise to cell types from the ectoderm, mesoderm, and endoderm germ layers. MSCs play a role in the development of mesenchymal tissue types and can be obtained from umbilical cord blood (UCB-MSCs) or Wharton's jelly. They are also present in several adult stem cell niches, including bone marrow and adipose tissue. Stem cells are divided into three categories according to their ability to differentiate. The first are totipotent stem cells, which can be introduced into the uterus of a living animal and grow into a complete organism. Pluripotent stem cells such as embryonic stem cells (ESCs) and induced pluripotent stem cells (iPSCs) are the second category (Mummery et al. 2012). Multipotent stem cells are the third type. They are adult stem cells that divide asymmetrically to form specific lineages of neurons in the nervous system. These cells are endowed with the capacity to regenerate and can divide asymmetrically to produce all of the cell types of the nervous system (neurons, astrocytes, and oligodendrocytes) (Hu et al. 2015).

11.2.1 Exosomes

11.2.1.1 Biogenesis and Contents of Exosomes

Exosomes can contain a variety of different substances including proteins, fats, and nucleic acids. In order to understand how cells utilize exosomes to communicate with each other and alter their environments, it may be helpful to investigate the production and trafficking of exosomes. Exosome production is started when clathrin-coated vesicles are transported into the membrane, and they invade inwards

(Gurung et al. 2021). Invaginated vacuoles can evolve into early endosomes (EEs), which take in ubiquitinated cargo following invagination. This is facilitated by the endosomal sorting complex required for transport (ESCRT) (Bonifacino and Traub 2003). After that, intraluminal vesicles (ILVs) coalesce and become mature inside endosomes, which are now called giant multivesicular bodies. This is followed by a secondary inward invagination of the EEs. The multivesicular bodies can be digested into lysosomes for degradation (degradative MVBs) or they can fuse with the plasma membrane (exocytic MVBs), releasing ILVs into the extracellular space, which are known as exosomes (Maxfield et al. 2016). In oligodendrocytes (OLGs), Trajkovic's team demonstrated that the release of ILVs is dependent on the ESCRT machinery and that the distribution of the sphingolipid ceramide in the MVBs controls the extracellular release of ILVs as exosomes. The release of exosomes is documented in some studies to be dependent on Rab27 and Rab35 and can be halted with a neutral sphingomyelinase inhibitor. The release of exosomes is shown to be induced by Ca^{2+} and the ionophore A23187 in another study (Emanuel 2015).

Since exosomes originate from endosomes, they contain proteins involved in membrane transport and fusion (GTPases, annexins, flotillin), tetraspanins (CD9, CD63, CD81, and CD82), heat shock proteins (heat shock cognate [Hsc70], heat shock protein [Hsp 90]), proteins involved in the biogenesis of MVBs (Alix and TSG101), and lipid-related proteins and phospholipases (Simpson et al. 2008). Despite the fact that there are significant differences in proteins among exosomes from various sources, these proteins are utilized as positive "labels." The most commonly employed markers of exosome presence include TSG101, Alix, flotillin, and Rab5b; these markers are identified using antibody-based methods including western blot and ELISA (Abak et al. 2018). Over 4400 different proteins have been identified that are typically identified by mass spectrometry; these proteins are associated with exosomes and serve as a means of cell-to-cell communication in addition to the membrane proteins. Exosomes are composed of a lipid-rich matrix, and the different types of exosomes are composed of different amounts of lipid. Phosphatidylcholine, phosphatidylethanolamine, phosphatidylserine (PS), lysobisphosphatidic acid, ceramide, cholesterol, and sphingomyelin are just a few of the lipids that have been documented to produce exosomes (Lucotti et al. 2022). The stiffness and effectiveness of the delivery of exosomes are both affected by the lipid composition of these structures, including sphingomyelin and N-acetylneuraminyl-galactosylglucosylceramide (GM3). By binding the outer proteins, phosphatidylserine is released on the exosome membrane by the enzymes floppase, flippase, and scramblase; this protein has a role in signaling and attaching to the plasma membrane (van der Koog et al. 2022). As a result, the capacity of exosomes to communicate may be impaired by variations in the PS concentration on the exosomal membrane. Also, reports have been made of saccharide-containing groups on the exterior membranes of exosomes. Mannose, polylactosamine, α2,6-sialic acid, and complex N-linked glycans have all been identified in exosomes, as per Batista et al. Additionally, nucleic acids including miRNA, mRNA, and other noncoding RNAs are present in exosomes, along with proteins and lipids. Several studies have documented that exosomal RNA is different from that of the parent cell

(Zhang et al. 2015). Other researchers have demonstrated that the miRNA content of cancer cell-derived exosomes is identical to that of their parent cells, which makes them a useful tool as biomarkers. Because of the expense and time required for EM validation of exosomes, miRNAs may be a superior method for confirming the existence of exosomes and disease markers. While the miRNAs and noncoding RNAs that are carried by exosomes can regulate gene expression, the mRNAs they carry can be translated into proteins in the recipient cells (Lai et al. 2022). Also present in exosomes is DNA, but the purpose of this genetic material is still unknown. The software ExoCarta was created to document the protein, RNA, and lipid content of exosomes via web-based databases, as the amount of exosomal content has increased significantly over the past few years (http://www.exocarta.org/). Overall, a wide variety of cells secrete exosomes, which are made up of components similar to proteins, lipids, and nucleic acids from their original cells. Exosomes are considered to function as conduits because they allow cells to communicate with each other by transferring materials from their primary cells to other cells. Experimental research can corroborate the neuroprotective effects of exosomes in neurodegenerative diseases by looking at the contents and examining potential biomarker candidates. This will gain credence with more explanation of the supposed role exosomes play in cell-to-cell communication (Westergard and Stefeson. 2018).

11.3 Neuronal Communication Via Exosome

Exosomes are utilized as a conduit for communication between cells; they also interact with nearby cells to expedite the release of active substances. The research of Skog et al. demonstrated the transfer of micro-RNAs and proteins from glial cells to axons. Neurons in the hippocampus and cultured cortical cells release exosomes into the extracellular environment in response to glutamatergic synaptic activity (Kalani et al. 2014). These exosomes include receptor molecules called GluR2/3 subunits. Galactocerebrosides, sulfatides, and cholesterol, the defining myelin lipids, are released into the exosomes by oligodendrocytes, which are essential for myelin sheath and neuronal conduction (Fauré et al. 2006). Microglial cells, which are macrophages in the central nervous system, become activated in disease states and convert into antigen-presenting cells by releasing exosomes. The cytokine IL- is pro-inflammatory and contains the exosomes released from the plasma membranes of microglia and astrocytes as a result of ATP stimulation and sphingomyelinase activation (Turola et al. 2012). In reaction to heat and oxidative stress, cultured astrocytes produce exosomes that contain the proteins synapsin-1 and heat shock protein 70. Exosomes carrying immunosuppressive and oncogenic factors are released by glioblastoma and astrocyte-derived brain tumor cells. Therefore, exosomes act as carriers for molecules of many origins, enabling cell-to-cell communication. The process of biogenesis, types of exosomes, role of exosomes, and signaling molecules are represented in Fig. 11.1.

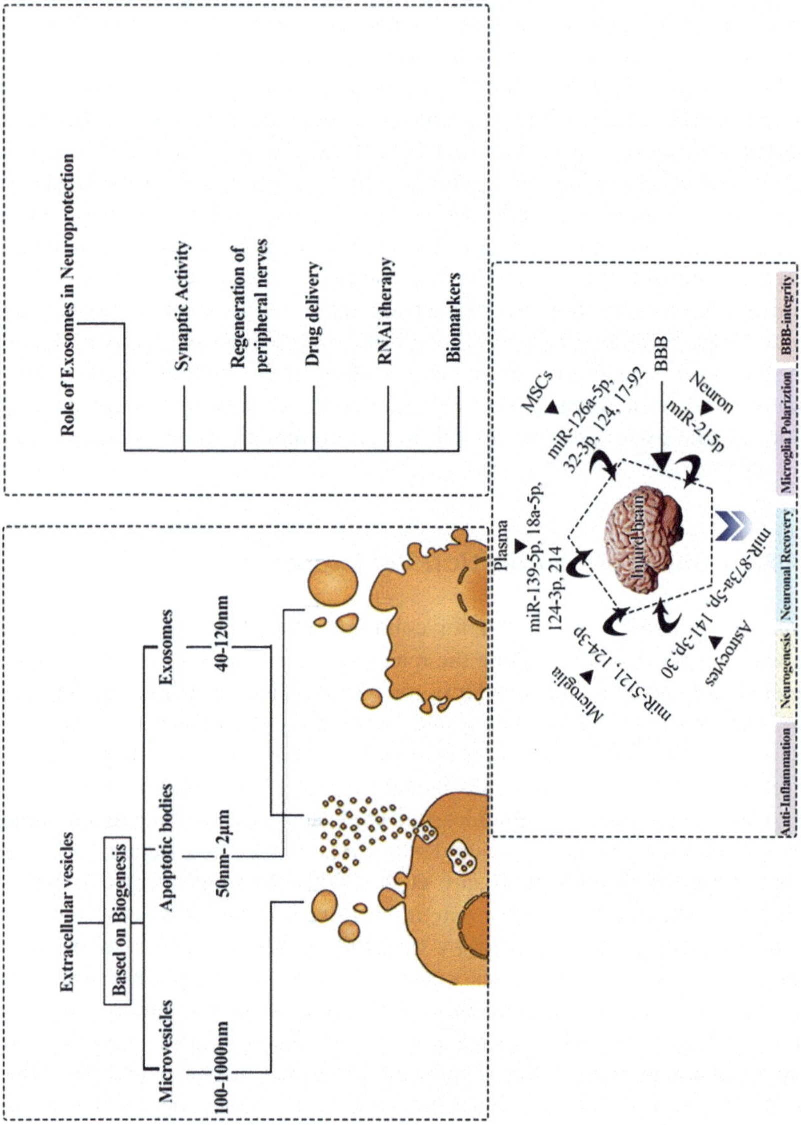

Fig. 11.1 Figure representing the biogenesis of extracellular vesicles and their roles and applications in neuroprotection

11.4 Exosomes and Neuropathology

Exosomes are implicated in the development of many disorders of the nervous system that are inflammatory or degenerative (Zappulli et al. 2016). Neuroinflammation is an innate immune response that is triggered by microglia and astroglia, both of which are resident macrophages of the central nervous system, when they are stimulated by different types of insults or damage. Neuroinflammation causes the production of cytokines, chemokines, reactive oxygen species, and additional messengers. Astrocyte-derived exosomes can also transport proteins that are misfolded or have been overexpressed into neurons, which then have the ability to initiate or propagate inflammation in the brain, causing death of neurons and degeneration. Exosomes that contain inflammatory molecules like IL-1 and other cytokines that are associated with the promotion of neuroinflammation can also be released by glial cells (Cadoni et al. 2020). Their scavenging activities are crucial to the removal of toxic substances. Glial exosomes can also transmit endocrine messages from hematopoietic cells to the brain; this phenomenon is increased in an inflammatory environment (Traina 2021). It is intriguing to observe that extracellular vesicles can easily pass over the BBB, giving a new route for systemic inflammation to influence CNS physiological functions. MiRNAs, which have the ability to dysregulate the gene expression of nearby cells, are also present in exosomes. For example, exosomes that contain pro-apoptotic proteins (e.g., prostate apoptosis response 4 and ceramide) and tau proteins are released by the astrocytes; these proteins are transferred to other cells in order to induce cell death and neurodegeneration in a human cell line that has similar levels of tau protein as those observed in the postmortem brains of patients with Alzheimer's disease (Hagey et al. 2023). Neuronal exosomes contain precursors of amyloidogenic proteins as well as enzymes that facilitate the conversion of other precursors (Leong et al. 2020). These findings support the hypothesis that exosomes from neurons promote the formation of amyloid plaques. Neuronal exosomes are implicated in the release of toxic oligomers of alpha-synuclein to the extracellular space in PD by broadcasting this protein to surrounding neurons and glia (Xia et al. 2019). This, in turn, causes an inflammatory response and cell death. Neurons and astrocytes are capable of releasing exosomes that contain the mutant form of Cu/Zn superoxide dismutase 1 (SOD1) into the extracellular space; this misfolding can be propagated in both human and murine cell models, which results in motor neuron damage; this is what happens in amyotrophic lateral sclerosis (Grad and Cashman 2014). Additionally, both oligodendrocytes and microglia simultaneously take in antigens that are associated with the pathogenesis of autoimmune disorders in the CNS. In an experimental mouse model of autoimmune encephalomyelitis, for instance, pro-inflammatory cytokines encourage immune cells to produce exosomes, which then trigger additional pro-inflammatory chemicals by spreading inflammation. Because of the glutamate's toxicity, the amount of glutamate in this condition increases, which causes oligodendrocytes to release exosomes, which may lead to demyelinated areas (Jiang et al. 2014). Several tissues from autoimmune diseases and multiple sclerosis have also been observed to have altered levels of almost

100 miRNAs. Exosomes have an effect on the onset and progression of negative effects on neuronal survival; this makes them relevant to viral infections as well. Exosomes that contain certain miRNAs (mir-29b, mir-128a, and mir-146a) of mouse cells infected with neurons that participate in the dysfunction of neurons by altering key genes have an effect on the quantity of misfolded prion protein that accumulates in neurons and causes neuronal loss in prion disorders (Bellingham et al. 2012). In primary rat astrocyte cultures, the use of drugs during infection increases the amount of exosomes that are enriched with mir-29b; this may lead to neurotoxicity in neurons. Infectious diseases of the HIV are typically associated with neurological impairments. Numerous studies demonstrate that alcohol misuse causes inflammation and degeneration of the nervous system via the innate immune receptor TLR4. Recent studies on mouse astrocytes have demonstrated that ethanol increases the release of astrocyte-derived EVs and their content to inflammatory proteins (such as NF-κB-p65, NLRP3, caspase-1, and IL-1) and miRNAs (mir-146a, mir-182, and mir-200b) in a TLR4-dependent manner. By promoting neuronal apoptosis, astrocyte-derived EVs would cause naive cortical neurons to have more inflammatory proteins (cyclooxygenase 2) and miRNAs (mir-146a) imported. The purpose of the regulatory functions of the expressed miRNAs in specific genes has been uncovered by the functional analysis of these miRNAs. These functions are associated with various inflammatory pathways (Nassar et al. 2022). These instances suggest that glial EVs may facilitate and augment the neuronal inflammatory response, which could lead to brain damage or dysfunction. It is intriguing to note that mental illnesses have been associated with exosomal participation in neuroinflammation. For example, certain types of psychopathology, particularly depression, are associated with chronic neuroinflammation (Pascual et al. 2020). The release of microglial exosomes is disrupted in a variety of mental illnesses, including depression, anxiety, bipolar disorder, and schizophrenia, as a result of serotonin pathway dysfunction. The same investigation also demonstrated that human microglial cells in culture can be primed to release more inflammatory cytokines, such as IL-1, via EVs derived from the patient's serum that were collected with the autism spectrum disorder. Overall, new research demonstrates that the exosomal origin of neural tissue has a significant role in the propagation of neuroinflammation in neurological and neurodegenerative diseases. However, these microscopic vesicles can also be used to create engineered EVs that have therapeutic applications as their intended purpose (Fayazi et al. 2021). In fact, cutting-edge methods are currently being developed to utilize EVs as intended.

11.4.1 Neurotrophic Factors

The central nervous system contains many different cell types, including oligodendrocytes, microglia, and astrocytes, all of which have been demonstrated to be essential in controlling both healthy and pathological conditions. Glia have a significant role in the degeneration of the nervous system in conditions including Alzheimer's disease, Parkinson's disease, and ischemic stroke, the latter of which is initiated by microglia and culminates in the formation of a glial scar (Peferoen et al. 2014). Past co-culture investigations have explored the significance of these cells to neuronal survival, and the pursuit of neurotrophic factors (NTFs) was initiated by the discovery that the addition of astrocyte-conditioned media would increase the viability of primary neurons in vitro. The discovery of glial cell line-derived neurotrophic factor, a powerful dopamine neurotrophic factor, was thus made (GDNF) (Barbacid 1995).

11.4.2 BDNF

The data from in situ hybridization experiments that utilize RNA indicate that BDNF is first expressed in the embryonic mouse brain at E11.5, and its expression persists through later developmental stages. High levels of BDNF mRNA are present in the cortex, hippocampus, and olfactory regions of the adult mouse brain, while lower amounts are present in the thalamus, hypothalamus, midbrain, and medulla. The analysis of the cell-type-specific transcriptome of the mouse cerebral cortex by RNA sequencing identified about a twofold increase in BDNF mRNA in astrocytes compared to neurons, though the levels of expression are low and are decreasing with age in astrocytes derived from the hippocampus and striatum, but not from the cortex (Peng et al. 2011). Contrasted with mice, the mRNA levels of BDNF in human astrocytes and neurons are minimal. The expression and release of BDNF from glial cultures has been specifically documented in multiple investigations. Rat basal forebrain astrocytes that were cultured on postnatal day 1 expressed mRNA for BDNF. After a spinal cord injury (SCI), immunohistochemistry demonstrated that the levels of BDNF in the rat's astrocytes and microglia were increased at the protein level (Parkhurst et al. 2013). Similar to this, activating rat basal forebrain astrocytes in vitro with KCl, a cholinergic agonist carbachol, and glutamate increased BDNF mRNA levels and protein release. The administration of a medication that is used to treat multiple sclerosis that has a relapsing-remitting pattern increased the expression and secretion of BDNF by around a twofold in cultures of astrocytes in the mouse's cortex, and immunofluorescence labeling of these cultures and the analysis of the media samples revealed a twofold increase in BDNF secretion (Efstathopoulos et al. 2015).

11.4.3 CNTF

Although ciliary neurotrophic factor (CNTF) has long been recognized to affect astrocyte development, there are not many studies that discuss how it affects glial cell expression. Its presence is minimal and confined to particular brain regions in the growing mouse embryo. At E14.5, in situ RNA hybridization in the striatum, basal ganglia, thalamus, and hypothalamus detects moderate levels (Scharfman 2005). CNTF is present in the adult mouse brain at low levels, with the greatest levels of expression found in astrocytes and OLGs that are developing or myelinating.

However, a different investigation that examined the alterations in gene expression in astrocytes from mice of various ages found no evidence of CNTF mRNA expression at any point (Pöyhönen et al. 2019). Similarly, the expression of CNTF in the human brain was not detected by RNA sequencing. However, despite this finding, studies that focused on the detection of CNTF mRNA were able to demonstrate its expression in astrocytes under typical circumstances. In case of stroke and entorhinal cortex injury, CNTF mRNA was increased similarly to BDNF in reactive GFAP-positive astrocytes. This has been hypothesized to occur because, under typical circumstances, interactions with neurons may inhibit the expression of CNTF in astrocytes by means of the integrin signaling pathway (Abrous et al. 2005).

11.4.4 NGF

Astrocytes from rats were observed to have NGF mRNA and protein during development (E20) and postnatal day 3, with the amount of the protein decreasing to around 50% by postnatal day 5. These declines are greatly reduced in mature animal astrocytes (around 10% of the E20 level) (Molteni et al. 2001). Similar to this, astrocytes in the basal forebrain of rats on postnatal day 1 express NGF mRNA, its levels increase in response to glutamate stimulation, peroxynitrite stimulation, or inflammatory polypeptides. Oligodendrocyte precursor cells (OPCs) derived from embryonic mouse brain tissue also possessed NGF mRNA and protein (Mead 2012). In situ hybridization also demonstrated low levels of NGF mRNA in the midbrain and hindbrain of E18.5 mice, but not in the adult brain, which is consistent with these findings. Neurons, OPCs, and newly formed OLGs in the adult mouse cortex have low NGF mRNA levels, according to a study of their RNAs. NGF mRNA was not found in microglia by RNA sequencing, although it could be expressed in rat microglial cells when adenosine A2a-receptors and lipopolysaccharide were present (Pöyhönen et al. 2019).

11.4.5 NT-3

At E18.5, in the telencephalic vesicle, in situ hybridization can begin to reliably detect low levels of NT-3 expression in the developing mouse embryo (Pöyhönen et al. 2019). By postnatal day 4, all of the brain regions are populated with NT-3 mRNA; this mRNA increases dramatically prior to a gradual decline. In the adult mouse cerebral cortex, it is barely detectable by RNA sequencing in any cell type; however, it is present at a low level in adult human neurons. NT-3 mRNA is not stimulated by KCl, glutamate, or the cholinergic agonist carbachol, in contrast to BDNF and NGF (Onorati et al. 2011).

11.4.6 GDNF

Early studies that used in situ hybridization to detect GDNF mRNA expression found that the protein was present in embryonic brain regions as early as E7.5 in mice; its concentration increased at E9.5 and then decreased after E10.5. Only in the ventral midbrain at E13.5 is GDNF mRNA present through subsequent embryonic stages (Cortés et al. 2017). However, it starts to be broadly expressed in the brain at E18.5 and continues through postnatal day 14 and adulthood. Semiquantitative PCR was employed to detect GDNF mRNA in cultures of astrocytes from human fetuses at 12–15 weeks of gestation and from early postnatal mice, where lipopolysaccharide is also capable of increasing its expression. Despite low levels of GDNF mRNA being present in neurons, OPCs, and newly formed OLGs, recent studies that focused on cell-type-specific RNA sequencing in the adult mouse cortex have found no GDNF mRNA in astrocytes, myelinating OLGs, microglia, or endothelial cells. Only human brain OLGs have low levels of GDNF mRNA (Yan et al. 2022).

11.4.7 NRTN

The information on NRTN mRNA's expression during development is not entirely clear. Despite the fact that in situ hybridization suggested that NRTN was present in the developing brain of E11.5 mice, its expression pattern then became restricted to midbrain and hindbrain regions at E14.5, and subsequent experiments failed to detect NRTN mRNA at any subsequent developmental stage, including postnatal day 28. Similar results were obtained from two separate sets of adult mouse brain slices; the first showed widespread expression of NRTN mRNA, while the second failed to detect any. However, RNA sequencing revealed high levels of NRTN mRNA in the brain, specifically in OLGs that myelinate, in a different study, but not in the human brain. These values were lowest in neurons and microglia, the lowest of these three cell types (Pöyhönen et al. 2019).

11.4.8 CDNF and MANF

CDNF and MANF are novel NTFs and might not function like "traditional" NTFs. Under typical circumstances, these proteins localize to the ER and are primarily intracellular. Only after ER calcium depletion neurons secrete them, and proliferating cells do the same in response to ER stress. Furthermore, MANF or CDNF has no known cell surface receptors despite years of extensive investigation (Danilova 2020). According to information now available, MANF regulates the ER stress and UPR pathways and is involved in the migration, extension of neurites, and differentiation of neural progenitor cells in the cerebral cortex (Tseng et al. 2018). The theory that MANF and CDNF function as intercellular signaling molecules that promote neuronal survival in various neurodegenerative scenarios is not entirely proven by our research; however, MANF has already been shown to upregulate markers of phagocytosis and attract phagocytic cells in a rat model of stroke. After inflammation and injury, MANF is also implicated in the polarization of retinal macrophages and microglia. In situ hybridization has been used to find a strong MANF mRNA signal in the adult mouse brain and in the developing mouse brain at all examined embryonic stages (E11.5, E15.5, and postnatal day 7) (Kaiser et al. 2019). As a result, RNA sequencing revealed that astrocytes, OPCs, newly formed OLGs, microglia, and endothelial cells all had high levels of MANF mRNA, whereas neurons and myelinating OLGs had relatively lower levels. Up to 2 years of age, levels in mouse astrocytes are rather steady during the course of the animal's life.

Similar to this, the mRNA for MANF is highly expressed in the human brain, with fetal astrocytes and microglia exhibiting the greatest levels of expression. MANF mRNA is observed in mouse microglia throughout life, except for postnatal day 7 (7 days old). Additionally, LPS, ischemia, and ER stresses all increase the expression of MANF in microglia, astrocytes, OLGs, and, to a lesser degree, neurons (Pöyhönen et al. 2019).

Unlike other NTFs, the expression of CDNF has not been thoroughly investigated; however, only semiquantitative PCR identified CDNF mRNA in the brains of mice from E12 to 21 days after birth (Llana 2013). Very low mRNA levels were observed by RNA sequencing in mouse cortical neurons. In contrast, astrocytes and OLGs in the human brain have CDNF mRNA expressed. It is intriguing to note that the expression of CDNF mRNA in astrocytes from the mouse cortex and hippocampus increases with age, reaching a peak around 9.5 months and remaining elevated through age 2 years. In contrast, striatal astrocytes do not exhibit CDNF mRNA expression. The findings mentioned above demonstrate a typical pattern of NTF expression in glial cells. All NTFs, with the exception of MANF, have relatively modest basal expression levels in glial cells under normal circumstances, but they can be dramatically increased in response to injury, ischemia, and/or stressful conditions (Huttunen and Saarma 2019). This is in agreement with the theory that they have a significant role in promoting neuronal survival during stressful situations. Importantly, the frequency of NTF expression in human brain cells was frequently different than that in mouse cells. This underscored the importance of research utilizing human brain tissues and models, such as the rapidly developing field of human brain organoids (Yadav et al. 2021).

11.5 Signaling Cascade Involved in Neuroprotection

A transcription factor called NF-κB (nuclear factor kappa B) controls many processes, including immune cell activity, cancer cell growth, neuroprotection, and long-term memory (Mincheva-Tasheva and Soler 2013). Two of the most exciting areas of NF-κB research are its role in stem cells and development. The group of Nobel laureate David Baltimore identified NF-κB as a potential transcription factor with inducible binding activity (Kaltschmidt et al. 2022). He then determined that this latent form was caused by a family of arrestins called inhibitor B (IB), which likely interact with NF-κB in the cytoplasm. The neuroprotective properties of PACAP38 in neuron/glia co-cultures were examined by preventing LPS-induced neuronal death. Neuronal death was calculated based on the amount of NSE released into the medium after treatment (Yang et al. 2006). Free radical-induced stress is critically controlled by Nrf2 (nuclear factor erythroid 2-related factor 2). Typically, the protein KEAP-1 (Kelch-like ECH connexin 1) reconstitutes NRF2 with the cullin 3-based E3 ligase in the cytoplasm, leading to protein ubiquitination, and its proteasomal degradation may stimulate the NRF2-ARE signaling system to provide neuroprotection against oxidative damage and cell death (Romani et al. 2018). In addition, the current approach suggests that the NRF2-ARE pathway controls the synthesis of misfolded protein aggregates found in various NDDs (Huntington, PD, and AD) (Grewal et al. 2021). Major brain disorders, including age-related NDD, are affected by phosphoinositide 3-kinase (PI3K). Akt is a regulatory protein in the PI3K/Akt system that controls neuronal survival and plasticity. Phosphatidylinositol-3,4,5-triphosphate (PIP3) activates the downstream molecule AKT (Oudit et al. 2004; Rai et al. 2019). As a result, Akt is phosphorylated, affecting how different target proteins activate or inhibit their function. As such, it controls a wide range of cellular processes, the most important of which are the cell cycle, development, metabolism, proliferation, protein synthesis, and apoptosis. One of the major signals downstream of the mitogen-activated protein kinase (MAPK) signaling cascade is c-Jun N-terminal kinase (JNK). JNK is a member of the threonine protein kinase family consisting of three genes (JNK1, JNK2, and JNK3) that collectively translate ten variants (Oudit et al. 2004). While JNK3 is mainly distributed in the CNS and has shown therapeutic promise in NDD and other CNS diseases. JNK1 and JNK2 are widely distributed in various tissues and play important roles in obesity-induced insulin resistance (Bogoyevitch 2006). JNK is activated by phosphorylation on threonine and tyrosine residues and inactivated by a negative feedback process initiated by MAPK phosphatases. Macrophages and monocytes produce inflammatory cytokines during acute inflammation, one of which, tumor necrosis factor (TNF), is responsible for many cell signaling events leading to apoptosis and necrosis (Turner et al. 2014). TNF molecules can be divided into two types: TNF-α and TNF-β. Microglia are critical in triggering the inflammatory cascade leading to neuronal cell death by releasing neurotoxic and inflammatory mediators such as TNF-α, IL-1, IL-6, and NO. PGC-1, also known as peroxisome proliferator-activated receptor gamma coactivator 1alpha, is critical to the brain. While JNK3 is mainly distributed in the CNS and has shown therapeutic promise in NDD and other CNS diseases. JNK1 and JNK2 are widely distributed in various tissues and play important roles in obesity-induced insulin resistance. JNK is activated by phosphorylation on threonine and tyrosine residues and inactivated by a negative feedback process initiated by MAPK phosphatases. Macrophages and monocytes produce inflammatory cytokines during acute inflammation, one of which, tumor necrosis factor (TNF), is responsible for many cell signaling events leading to apoptosis and necrosis. SIRT-1 is activated in multiple ways, leading to

A-peptide inhibition, Bax-dependent inhibition of apoptosis, and regulation of pro-apoptotic transcription factors. The cyclic AMP-binding protein response element, or CREB, is a transcription factor that regulates the production of genes involved in neuronal survival, function, and neurogenesis (Ortega and Jastrzebska 2021). It is constitutively expressed in the nucleus, in the brain of animal models of AD, and in amyloid-expressing neuronal cell lines. Serine-threonine kinases, also known as mitogen-activated protein kinases (MAPKs), promote a variety of cellular processes such as cell division, proliferation, development, and apoptosis. Serine/threonine interactions are still strongly maintained by MAPKs, which can be divided into traditional MAPKs and atypical kinases. Typical examples are the protein kinases JNK1–3 and ERK1/2 (SAPK1A, 1B, 1C). Atypical are p38 (p38 and ERK5), ERK3/4, ERK7/8, and Nemo-like kinase (NLK). The protein sequence that amplifies the signal consists of the MAPK cascade, which begins with the MAPKK kinase. MAPKs are ultimately activated by MAPK kinases (MAPKKs), which are further phosphorylated and activated by MAPKKKs. Cell growth, apoptosis, and cell survival are all affected by MAPK phosphorylation and activation (Anjum et al. 2022).

11.6 Application of Exosomes in Neuroprotection

Exosomes are small vesicles that have the same membrane structure as cells, have a range of sizes from 30 to 200 nm, and lack an obvious exterior. Even when exosomes are produced from the same cell type, their sizes differ significantly (Pegtel and Gould 2019). Each exosome sizing method, however, adds unique biases and exosome size projections. For instance, methods that measure the hydrodynamic size of proteins and glycans that protrude from the exosome membrane are susceptible to the size of molecules bound to the exosome surface that are weakly attached. Exosomes contain large amounts of specific transport proteins like CD81, CD9, and CD63. Without knowing how many of these proteins are concentrated in exosomes, more than 3000 proteins have been discovered in exosomes secreted by a single cell type. According to this theory, soluble proteins are rarely incorporated into exosomes and, following exosome-cell union, are free to function in target cells. This idea has been skillfully applied for medicinal purposes by using a light-induced exosomal targeting and release technique. Exosomes carry DNA, which can be either single-stranded, double-stranded, mitochondrial, chromosomal, or complementary to reverse transcription. In contrast to other exosomal payloads, it is not known whether specific DNAs are preferentially organized into exosomes (Ke 2020). Further reports suggest that DNA found in exosome-associated particles can reveal the entire genetic code of the cell that produced the exosomes. Also, it is unknown how much exosomal DNA is affixed to the surface of the organelle versus how much is stored inside of it. Exosome-based DNA release may aid in maintaining DNA integrity, possibly in the regulation of inflammation, and may serve as a valuable indicator of malignancy, viral infection, or chemotherapeutic tolerance.

Recent investigations of exosomes have primarily focused on the functions of the molecules that are carried by the exosomes and the signals that they transmit between cells. Exosomes are particle complexes that are loaded with multiple signals and have the capacity to transmit multiple, combinatorial messages by clustering and activating corresponding receptors on the cell's surface. These activities have been documented for proteins associated with the cytokine, Wnt, and Notch pathways as well as the growth factor (Paskeh et al. 2022). Additionally, target cells can receive functional receptors and, in some circumstances, triggered receptors and effectors through exosome union with them. However, it is entirely appropriate to question whether or not research on the exosome-mediated communication between cells and movement is actually significant for metabolic processes in vivo. Despite this question still being unresolved in many systems, there is clear evidence that exosome-mediated PD-L1 signaling contributes to immunosuppression, that exosome-mediated miRNA reprogramming contributes to viral infections and pregnancy, and that exosome-mediated EGFRvIII transfer contributes to cancer (Wolf-Dennen and Kleinerman 2020). As a result, it is sensible to question whether exosomes are necessary for a particular function, but this questioning should be balanced by the likelihood that multicellular life evolved in the presence of continuous, ongoing exosome synthesis and uptake. Exosomes have been linked to numerous noninfectious diseases, including cancers, inflammation, metabolic disorders, autoimmunity, neurodegeneration, chronic pulmonary obstructive disease, addictions, and other normal physiological processes (e.g., development, tissue homeostasis, aging, metabolic regulation, exercise, stress, circadian rhythms, molecular transfer during pregnancy, breastfeeding, eating, and parasitic interactions), as well as many of these processes (e.g., those caused by infection with viruses, protozoans, fungi, worms, arthropods, etc.) (Blum 2019; Naviaux 2019).

The production of exosomes by neurons, microglia, astrocytes, oligodendrocytes, and neural stem cells has been documented in multiple studies to have a role in neuronal protection, regeneration, growth, and learning of new connections. Small (50–80 nm) and large (600 nm) exosomes are released into the ventricular fluid during the early stages of neurogenesis in the developing brain of the rodent (Janas et al. 2016). Exosomes, which can reengineer their recipient cells, may facilitate the transmission of mRNAs that encode for embryonic transcription factors. It is hypothesized that exosomes function as a conduit between spatial and temporal differences that are important for neuronal development. Exosomes are believed to facilitate the genomic alterations (mutations) of embryonic cells by allowing retroposon sequences to be transmitted between cells; this causes gene expression to be altered (Riva et al. 2019).

The participation of exosomes in diverse embryonic processes has been documented in flies. In the drosophila, exosome-like structures are called argosomes, they transport morphogenic Wnt signaling proteins in a manner similar to an endosome, and this movement is directed by geographic and temporal gradients during wing development. Additionally, argosomes may have proteins that signal, such as Hedgehog, Notch, decapentaplegic, and wingless, which are responsible for embryonic gradients in other tissues. As a result, exosomes have a greater role in the systemic and local communication between neurons than direct cell-to-cell contact (Parchure et al. 2018).

11.6.1 Exosomes in Synaptic Activity

In vitro, undifferentiated neurons in the cerebral cortex produce exosomes and become active following a depolarizing stimulus, according to research by Faure et al. The GluR2/3 components of AMPA receptor subunits and the neural cell adhesion protein L1 found in exosomes demonstrate their involvement in synaptic function (Seyedaghamiri et al. 2022). Another research by Musto et al. corroborated this one by demonstrating the same process occurs in completely differentiated neurons of the cortical plate in culture. In this study, the activity of the GABA A receptor increased the exosome, which, in turn, increased spontaneous neuronal activity (Musto 2018). Tetanus toxin was discovered to be present in large quantities in the exosomes of neurons in another study (Kalani et al. 2014). Also, it was demonstrated that exosomes contained GluR2 subunits and that their release was increased with depolarization, which could have a role in regulating synaptic activity (Kalani et al. 2014).

11.6.2 The Role of Exosomes in Peripheral Nerve Regrowth

Exosomes serve a defensive role in repairing wounds and promoting tissue regeneration. The creation of Ndfip1, which is associated with Nedd4 ligases, mediates the degradation of toxic proteins that collect in the brain after damage (Conway et al. 2022). These two proteins, Ndfip1 and Nedd4, which are present in exosomes secreted by neurons, are thought to play significant roles in eliminating toxic proteins following damage. When nerve tissue is damaged, external ATP levels rise, which triggers the release of exosomes from microglia and astrocytes through a sphingomyelinase-dependent process. The inflammatory protein IL-1 found in these exosomes causes an inflammatory reaction. Synapsin 1, a protein unique to neurons and linked to synaptic vesicles, is present in exosomes produced by astrocytes under stress conditions (Holm et al. 2018). Additionally, oligodendrocytes that contain myelin and stress-resistance proteins produce exosomes. The vesicles that contain multiple ribosomes and are located around a dying or damaged peripheral neuron are transported into the axon, and the contents are then released. As a result, exosomes serve as a conduit for the transportation of mRNA and ribosomes to neurons that are injured; this promotes the local synthesis of proteins that are necessary for the repair of the area. Court and colleagues have demonstrated in this scenario that labelled ribosomes in the nerve are derived from the Schwann cells (Liegro et al. 2022).

11.6.3 Exosomes as a Medication Distribution Method

When compared to liposomes, which were previously used as a nano-delivery method, exosomes have a number of benefits. Exosomes exhibit all the qualities that make liposomes ideal for therapeutic cargo transport, including the ability to avoid immune detection and a prolonged half-life in the circulatory system. Exosomes have been successfully used for therapeutic delivery systems because of their minimal immunogenicity, exceptional delivering qualities, and capacity to penetrate the BBB. Exosome lipid bilayers or mRNA, siRNA, proteins, and medicines can all be genetically packaged into the MVBs that create exosomes. As a result, exosomes are now effectively used in medication delivery, immunotherapy, and RNAi treatment (Kalani et al. 2014).

The promise of exosomes as a drug transport mechanism. According to the research, curcumin was successfully loaded into exosomes, where it was more bioavailable and soluble than curcumin alone. Exosomal curcumin greatly reduced the inflammatory activity brought on by LPS. Instead of 30–45% for unfilled exosomes, the exosomes loaded with curcumin could be fractionated at a sucrose density gradient of 45–60% (Harris 2022). Similar to this investigation, Zhuang et al. reported the successful intranasal administration of exosomal curcumin and JS1124 (a statin inhibitor that activates STAT3) to the mouse brain. Curcumin was delivered by exosomes, which greatly reduced the autoimmune encephalomyelitis brought on by myelin oligodendrocyte glycoprotein and LPS-mediated inflammation. Additionally, the intranasal injection of GL26 inhibited the development of the glioma in the brain (Oskouie et al. 2019).

11.6.4 Exosomes in RNAi Therapy

RNAi has been utilized to treat diseases like cancer, genetic abnormalities, and HIV through ribozymes, aptamers, and siRNAs. Short (21–23 nt) single-stranded RNAs called siRNAs can bind to mRNAs with either a perfect or a defective base pairing; this results in gene silencing posttranscriptionally (Burnett and Rossi 2012). The siRNA is bound to the RISC complex in order to target mRNA, following which it is incorporated; it triggers the endonuclease Argonaute 2 to degrade the target mRNA. The RISC complex shields the siRNA from deterioration and allows it to frequently cleave other mRNAs. As a result, siRNA makes a great option for RNAi treatment. Endonucleases found in blood, cells, and interstitial space have the ability to degrade siRNAs and can make them immunogenic. They need effective transport systems, such as exosomes, that can pass through the BBB, preserve mRNAs and miRNAs, and transfer functional RNAs to the target cells. By injecting specific exosomes into the body, Alvarez-Erviti et al. showed how siRNA could be delivered into the rodent brain. This study employed modified dendritic cells derived from yourself that were genetically altered to express Lamp2b, a protein on the exosomal membrane that is combined with rabies glycoprotein. Exosomes that are pure and contain the protein Lamp2b along with the RNA virus RGV were electroporated with the siRNA and then administered intravenously to mice. To precisely silence a gene, the exosomes transported siRNA (for GAPDH) to the brain's neurons, microglia, and oligodendrocytes. As such, exosomes can be effectively employed as a means to transport RNA to specific cells, but more research is necessary to promote their use in clinical experiments (Li et al. 2020).

11.6.5 Exosomes as Indicators for Damage to Neurons

Exosomes are minuscule replicas of the parent cells and have distinctive signatures, or indicators, that indicate the condition of the cell. Exosomes derived from human glioma cells have been documented by Skog to contain proteins and small RNAs specifically present in the cells; these proteins and RNAs are also present in the blood of patients affected by the disease. The research suggests that exosomes may be used in patients with brain cancers as a means of noninvasive diagnosis and treatment. When glioblastoma cells evolve into malignant brain tumor cells, they release a variety of exosomes into the environment. This has been shown in vitro by Balaj et al. They compared the effects of conditioned media from glioma cells to those from normal cells on different cells. Exosomes are employed by cancer cells to manipulate surrounding noncancerous cells and facilitate their development. These changes comprise (1) promotion of angiogenesis, (2) facilitation of tumor development and invasion, and (3) suppression of immune reaction to tumor. These tumor progression-specific micro-RNAs can be found in the blood and used as indicators (Baig et al. 2020).

11.7 Use of Exosomes in Neurological Diseases

Similar to cerebral apoplexy, exosomes may be a useful diagnostic tool and curative approach for the management of severe brain damage and neurodegenerative diseases. Exosomes have been demonstrated in numerous studies to promote neurogenesis and enhance the physiological localization of neurons. They may also have medicinal promise for certain brain diseases. Although the clinical presentations of these neurodegenerative diseases are identical, many of them share the buildup of insoluble proteins (both external and intracellular), giving rise to the name "cerebral proteinopathies" (Lau and Chin. 2022). Neurological diseases like Alzheimer's disease and Huntington's disease share many similar symptoms and mechanisms, aside from the accumulation of incorrectly folded proteins. Several instances involve overloaded protein elimination pathways, impaired protein homeostasis, and a deficiency or degeneration of specific neural groups. The neurological impairment caused by the gradual, permanent loss of neurons and synapses is the primary difference between this disease and another condition (Li et al. 2015). Many cells in the brain, including oligodendrocytes, neurons, and astrocytes, produce exosomes. Exosomes are capable of carrying disease-causing agents and may contribute to the progression of neurodegenerative diseases. Once released into the public, they also function as an effective and useful peripheral method of noninvasive diagnosis that can be used to assess the severity and progression of many neurodegenerative diseases. Exosomes are capable of being utilized in other circumstances as well as the central nervous system, both of which have therapeutic applications (Jan et al. 2017). Additionally, they serve as ideal biological markers. Exosome analysis can be used securely for prenatal assessment of fetal central nervous system damage and diagnostic testing of elderly people because it is noninvasive.

Neuronal cell loss, a defining feature of neurodegenerative diseases like Alzheimer's disease, Huntington's disease, Parkinson's disease, and Niemann-Pick disease, frontotemporal dementia, and amyotrophic lateral sclerosis are all major causes of neurodegeneration. The molecular and cellular mechanisms that lead to protein aggregation and the formation of inclusion bodies in specific parts of the nervous system are the same in all of these diseases (Herms and Dorostkar 2016). Neurons must have the proper protein sorting and degradation to remain functional. In neurodegenerative disorders, exosomes play a role in the dissemination of "toxic" proteins, which are actually mutated or "misfolded" proteins that act as templates for the creation of oligomers. By processing these collected proteins through the endosomal pathway, which either results in their degradation into lysosomes or their integration into MVBs and discharge as exosomes into the extracellular environment, neurons attempt to get rid of them. Studies have detailed how pathogenic prion protein (PrP) and misfolded pathogenic prion protein (PrPsc) are incorporated into exosomes in this setting (Ridolfi and Abdel-Haq 2017). Researchers have demonstrated that PrPsc, which is connected to exosomes, is transmitted to healthy cells that already carry PrP. Numerous neurodegenerative disorders are caused by this process, whereby proteins tend to seed their own aggregation with "infectious" transport via exosomes. Spatiotemporal pathology propagation in neurodegenerative disorders suggests cell-to-cell transmission, and for non-secreted proteins, it is facilitated by exosomes or nanotubes. Exosome participation in various neurodegenerative diseases has been researched; some of the findings are outlined below.

11.7.1 Alzheimer's Illness and Exosomes

The most prevalent neurodegenerative illness is Alzheimer's disease (AD). Typically, it manifests as dementia and is characterized by a significant loss of cognitive function, mental state, and capacity to participate in everyday activities. The two primary causes of AD are the accumulation of Aβ plaque and neurofibrillary tangles, which are caused by the overactivation of tau protein. Both amyloid plaques and NFTs are considered clinical manifestations of Alzheimer's disease. Early in the course of the evolution of the illness, Aβ builds up in oligomers which can help to cause amyloid plaques. Exosomes taken from plasma or CSF samples of AD patients have been shown in experiments to contain disease-related proteins, suggesting that exosomes may be used as AD indicators. Researchers observed that the amount of

LAMP1 had changed and that autolysosome proteins were present in the plasma-derived exosomes of patients with AD (Liu et al. 2019). Researchers have discovered that some neurogenic exosome-related proteins, including heat shock factor 1, lamp 1, and phosphorylated type 1 IRS, were present in the plasma of AD patients, and their concentrations would increase by 10% prior to an official diagnosis of AD. They are all expected to be dependable markers for recognizing and diagnosing Alzheimer's disease as a result. Serum-derived exosomes from people with Alzheimer's disease (DAT), Parkinson's disease dementia (PDD), moderate cognitive impairment (MCI), and vascular dementia have been investigated for the transcript levels of miR-135a, miR-193b, and miR-384 (VaD). Comparing microRNA-384 to the other two short RNAs, it was discovered to be the most efficient at differentiating AD, PDD, and VaD. For an early AD diagnostic, mir-384, mir-193b, and mir-135a appear to work better together (Liu et al. 2019). These findings demonstrate the exosome's potential as a biomarker for Alzheimer's disease early detection as well as its potential to create novel pathways for disease detection and protection. This demonstrates the complexity of exosomes' function in AD, and it is crucial to comprehend how various EV populations interact with various forms of Aβ as well as how they affect the dissemination of Aβ assemblies between cells in order to develop more specific and effective markers for AD and enhance the quality of life of patients (Reza-Zaldivar et al. 2018).

11.7.2 Parkinson's Illness and Exosomes

The second most common neurodegenerative disease and the most common cause of damage to the central nervous system (CNS) is mobility impairment in Parkinson's disease (PD). The frequency increases with aging. The most typical clinical indications of dyskinesia include static tremor, muscular stiffness, slow movement, postural instability, and other symptoms. These symptoms are brought on by an excess of inhibitory (DA) and excitatory (acetylcholine) chemicals in PD patients' substantia nigra. After an unbalance, these motor signs do not immediately manifest. The first movement signs do not manifest until almost 70% of the DAergic neurons in the SNpc and at least 80% of the DA in the striatum have been lost. Parkinson's patients may have a variety of non-motor symptoms, including depression and sadness. The primary component of Lewy bodies and axons in both hereditary and sporadic forms of PD has been identified as misfolded and aggregated alpha-synuclein (α-syn). The release of exosomes is associated with the intracellular transfer of proteins via the endosomal-lysosomal pathway; this suggests that their biological functions may be associated with PD (Leggio et al. 2017). Disruption of protein trafficking to endosomes and lysosomes is now emerging as a potential common mechanism for the pathogenesis of sporadic PD. Since exosomal release is associated with the transport of intracellular proteins via the endosomal-lysosomal pathway, the biology of exosomes may have relevance to Parkinson's disease. Exosomes are believed to play a role in the aberrant development of PD by promoting apoptosis and moving damaging α-syn across cells. Additionally, it has been found that exosomes may help Parkinson's patients by providing neuroprotection. Exosomes have the ability to remove misfolded proteins that impede both intercellular transmission and brain stem cell differentiation. Neurons and glial cells can expel and reduce harmful compounds and proteins (such as α-syn) into cells via exosomal leakage (Olanow and Brundin 2013). In some studies, exosomes derived from PD patients were observed to reduce the neuronal stress response. For example, they investigated the biological effects of adding microvesicles to rat neurons that were starved but otherwise normal; this was followed by a proteomic analysis of the microvesicle preparations from patients with either sporadic or inherited PD. Exosomes derived from brain cells in PD patients have been studied in order to identify the most effective markers.

11.7.3 Huntington's Illness Exosomes

A hereditary, progressive brain condition called Huntington's disease (HD). Mutant Huntington protein is produced as a consequence of the Huntington's gene's (HTT) aberrant CAG repeat increase (mHTT). The process by which mHTT causes HD is unclear, despite the fact that it displays some neurotoxicity is also one of nine neurodegenerative diseases caused by polyQ (polyQ) expansion (Labbadia and Morimoto 2013). HD is often attributed to a loss of cognitive function and impulsive movements. There is no direct evidence that mHTT is associated with exosomes during transneuronal transport, despite suggestions that "tunneled nanotubes" or vesicle mechanisms are responsible for conveying mHTT across nerves. Using a model system that includes a culture of cells, they overexpressed the HTT-exon 1 polyQ-GFP construct in 293 T cells; they observed increased RNA and protein in the EVs. It is possible that these EVs are taken up by mouse striatal neurons, which increases the amount of polyQ-GFP RNA in the cells, but no significant damage to the cells was observed during the trial (Beatriz et al. 2021). A greater understanding of the fate of the transmitted RNA or protein and its possible harm to receptor cells is required in light of the fact that EVs can convey dangerous amplified trinucleotide repetitive RNA. As such, it is possible that exosomes may serve as scavengers, removing polyQ proteins from cells and lowering toxins. These RNAs that are amplified due to disease in the body fluids of HD patients may serve as a promising means of locating biomarkers that can be used to monitor the effectiveness of treatment and detect the development of the disease.

11.7.4 Multiple Sclerosis Exosomes

The most common cause of nontraumatic neurological impairment in young people, particularly women, is multiple sclerosis (MS), an inflammatory process that affects both gray and white matter in the central nervous system. Other clinical manifestations of the illness, besides demyelination and inflammation in the brain and spinal cord, include a breakdown of the blood-brain barrier (BBB), a loss of oligodendrocytes, reactive gliosis, and degeneration of axons and neurons. However, it is generally acknowledged that the primary cause of the inflammatory and degenerative characteristics of MS is the activation of self-reactive Th1 inflammatory cells in the periphery (Dobson and Giovannoni 2019).

The National MS Society (NMSS) divides MS into four main subtypes: clinically isolated syndrome (CIS), relapsing-remitting MS (RRMS), primary progressive MS (PPMS), and secondary progressive MS. Every MS patient has a different pattern of neurological damage (SPMS). Secondary progressive MS (SPMS) is eventually found in more than 80% of MS patients who initially develop RRMS (Eden et al. 2019). The mainstay of MS treatment today are immunomodulatory and immunosuppressive drugs, which increase the risk of malignancy and infection. When alternative disease-modifying therapies (DMTs) first emerged in the 1990s, interferon (IFN) was the first-line treatment for MS.

At least six different parenteral medications for MS, including interferons, immune-suppressants, corticosteroids, glatiramer acetate, sphingosine-1-phosphate receptor modulators, and monoclonal antibodies, have been approved by the FDA. By attacking the immune system at different levels via different mechanisms, these medications have a significant impact on the frequency and severity of relapses in patients with MS and can prevent the disease from progressing (Baecher-Allan et al. 2018). Contrary to the beneficial effects of avoiding relapses in people with MS that has a relapse component, DMT medications have little to no effect on the degree of axonal injury or progressive MS. Additionally, the effectiveness, tolerability, and safety of DMTs vary greatly, and in some cases, they are extremely harmful, including cardiomyopathy, which prevents their continued use. For the treatment of diseases that are caused by inflammation, new immune-modulating methods, such as stem cell transplantation, have emerged. By blocking immune system components in order to allow the system to regenerate on its own, the treatment of MS utilizing stem cells is frequently referred to as immune reconstitution therapy (IRT). The primary reason for the immunomodulatory effects of stem cells, as described by Liu's team, is the expression of the protein HLA-G, which functions as an inhibitor of the NK cell's response to interferon gamma (IFN), the primary inflammatory mediator associated with multiple sclerosis (MS) (Massey et al. 2018).

According to the findings, there has been a substantial remyelination of the spinal cord that is protected from injury because of an increase in anti-inflammatory cytokines, such as IL-10, in contrast to a decrease in pro-inflammatory cytokines, which delays and reverses the progression of MS. Impaired regulatory T cell activity, which is crucial for regulating the Th1/Th2 equilibrium, is one of the pathological characteristics of MS. The capacity of MSC to elicit T regulatory responses and shift from Th1 to Th2 is what allows them to modulate immune responses in autoimmune disorders like MS (Perrin et al. 2005). In a study conducted by Clark and her coworkers, it was found that large doses of EVs generated from placenta-derived MSCs (PMSCs) were able to produce the same clinical effects as treatment alone in the EAE model. Hepatocyte growth factor (HGF) and VEGF were found by proteomic profiling in the EVs generated from PMSC. By activating the regulatory T cell, PMSCs regulated the immune system by secreting high amounts of these factors (Gugliandolo et al. 2020). This finding demonstrated that PMSC-EV can cause myelin regeneration in the EAE mouse model and immunomodulatory reactions similar to PMSC treatments. Similar to this, intravenous injection of human adipose tissue-derived MSC-EVs reduces the production of pro-inflammatory cytokines, inhibits immune cell infiltration into the site of damage, and suppresses immune cell activation to enhance the score of EAE animals.

The induction and maintenance of immunological tolerance are important goals in the treatment of autoimmune diseases, as documented in the aforementioned papers. One recent strategy for peripheral tolerance is to delay the progression of the disease through the advancement of molecules that regulate the disease, including TGF-α, galectin-1, and PD-L1, through biological means, and the host immune system is disrupted. The tolerogenic properties of the MV generated from MSC were investigated in a study by Mokarizadeh and his colleagues that examined the tolerogenic properties of the MV on splenic MNCs derived from EAE-affected mice (Fayazi et al. 2021). The study's results demonstrated that MSCs-MV triggered an apoptotic response in self-reactive lymphocytes; this response was accompanied by a secretion of anti-inflammatory cytokines like IL-10 and TGF-α, an increase in the expression of regulatory molecules like PD-L1 and TGF on the MV's surface, and the differentiation of regulatory T cells that led to a predominance of immune responses in the periphery.

Polarizing microglia to an M2 signature is another newly discovered approach for causing immune tolerance in MS patients. The resident macrophage in the CNS known as microglia is quickly stimulated by its microenvironment to differentiate into either the M1 phenotype, which causes CNS damage by producing pro-inflammatory cytokines, or the M2 phenotype, which encourages tissue regeneration by producing anti-inflammatory cytokines (Roesch et al. 2018). One of the main causes of tissue injury in the brain in the early phases of MS has been attributed to the imbalance of M1/M2 macrophages and their transition to a pro-inflammatory M1 phenotype. As such, it is believed that encouraging microglia to polarize towards the M2 subtype can have a beneficial effect on alleviating MS patients' neurological symptoms. Li investigated the effects of the BMSC paracrine mechanism, specifically the role of exosomes in the microglial polarization and improvement of motor function in the rat model of EAE, in 2019. After evaluating the effects of the normal and treated EAE models, they observed that exosomes derived from BMSCs had the capacity to augment the behavioral scores in the EAE rodent model by decreasing inflammation and demyelination in the CNS. This was accomplished by altering the polarization of microglia towards an M2 phenotype. Additionally, they demonstrated that the treatment of MSCs with exosomes decreased the amounts of TNF-α and IL-12, which are associated with the M1 phenotype, while increasing TGF-α and IL-10, which are associated with the M2 phenotype (ENGIN n.d.).

Another study by Farinazzo connected the decreased demyelination in the spinal cord following treatment with nanovesicles secreted from adipose stem cells with decreased immune cell activity in the CNS, including a reduction in the microglial and T cell extravasation (ASC). Exosomes can be utilized to deliver drugs to an MS patient because they can penetrate the blood-brain barrier. Additionally, in many instances, attaching various functional groups to the surface of the exosomes is facilitated by the development of novel tools, such as antibodies and aptamers, which increase the specificity of the exosomes towards their targets (Teleanu et al. 2022). For example, Shamili's team created surface-modified MSC-derived exosomes that had an antimyelin aptamer (LJM-3064) intended to combat MS in 2019. Previous studies have demonstrated that LJM-3064 has a high propensity to form myelin protein in the brain, as well as the capacity to penetrate the BBB. The research indicated that the co-administration of the LJM-3064 aptamer and the MSC exosome decreased the severity of the rodent's symptoms via the immunomodulatory effects of the MSC exosome and the remyelination effects of the aptamer. In this system, exosomes actually serve as a conduit for the aptamer, which increases the remyelination effects of the LJM-3064 aptamer while also having anti-inflammatory properties. Based on these preliminary findings, the future of MS therapeutics will be based on SC-derived exosomes for a number of compelling reasons, including safety, ability to penetrate the BBB, and their cargo.

11.7.5 In Prion Disorders, Exosomes

Infectious spongiform encephalopathy (TSE), also known as prion disease, is characterized by the gradual loss of neuronal structure and function and the presence of spongiform vacuoles. It is a fatal neurological condition that can spread from animal to human, primarily affecting the central and peripheral nerves. The primary pathogenic attribute of prion disease is the presence of a detectable misfolded protein that is beta structured (PrPSc) (Sikorska and Liberski 2012). A typical prion protein (PrPC), which has a helical shape similar to that of alpha's, causes it to misassemble. High concentrations of PrPC have been documented in exosomes, and young cells are capable of consuming disease-related PrPC variants. This means that exosomes carrying PrPSc can facilitate the spread of the virus between tissues and accelerate the development of the disease. Elevated levels of cholesterol, sphingomyelin, and sphingomyelin GPI-anchored proteins (which contribute to PrPSc production) in exosomes may also play a role in protein sorting in exosomes (Février and Raposo 2004). According to research, if exosomes infected with prion diseases are injected into animals, they can be transmitted to animals. According to some studies, PrPSc can evolve from different cell types, tissues, and stages of infection, leading to different transmission mechanisms or combinations of prions that are not irreversible. In other words, there may be different ways of spreading prion diseases. However, additional research with higher throughput is necessary to identify specific miRNAs that have the potential to serve as prion disease biomarkers for diagnosis and treatment. Other neurological disorders can also have an effect on the up- or downregulation of miRNA in prion-infected exosomes. For instance, the mir-146a-5p may be downregulated in the blood of individuals with primary progression (PPMS). The treatment of diseases caused by prions will benefit from a comprehension of the differences between PrPSc and other neurotoxic proteins in vivo, as well as their association with the release of exosomes, their transport, and linked miRNAs

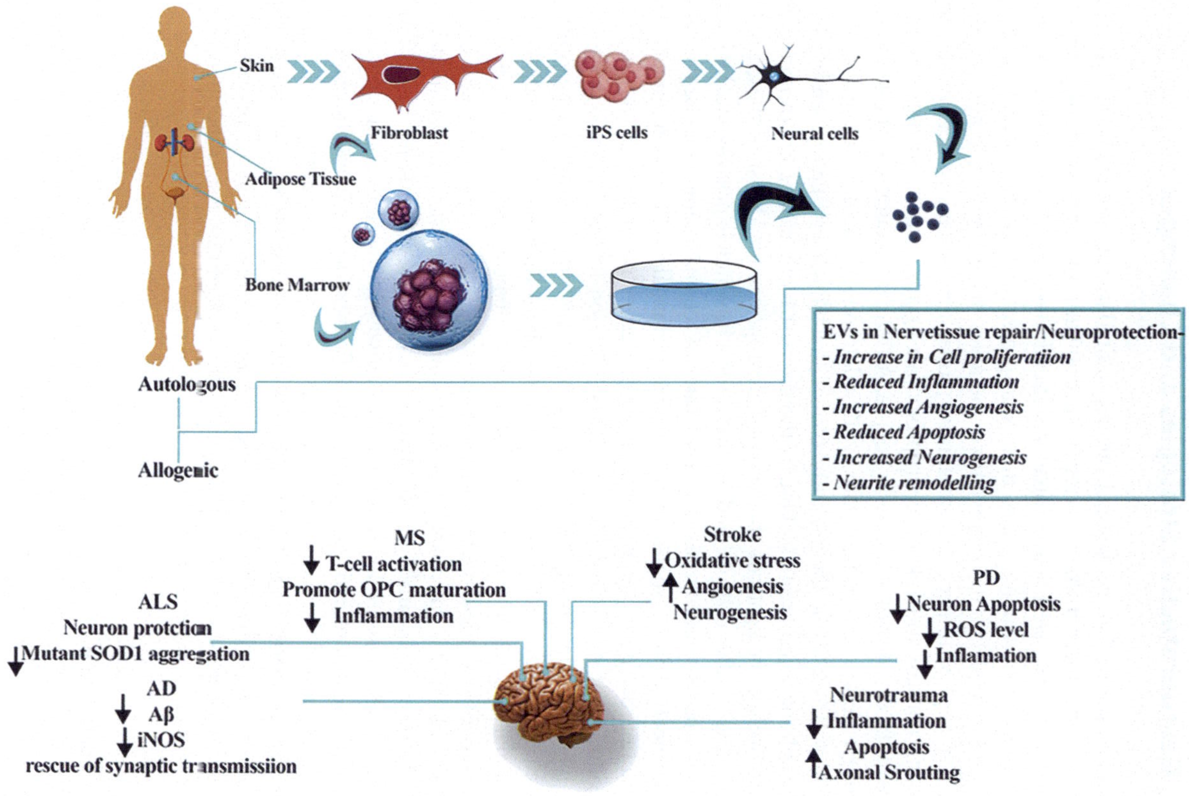

Fig. 11.2 Figure representing stem cell-derived exosomes and their role in neuroprotection

(Liu et al. 2019). The role of neuroprotection by exosomes is represented in Fig. 11.2.

11.8 Neuroprotective Strategies and Challenges

Any agent or agents that can lessen ischemic brain damage by preventing detrimental molecular processes from occurring in the brain as opposed to increasing cerebral blood flow are considered neuroprotectants (Stankowski and Gupta 2011). Preclinical studies have shown for years that neuroprotection may be advantageous in animal stroke models. Such preventive therapies have, however, not yet been successfully applied to people from animal models. Numerous earlier studies only used young animals to evaluate neuroprotective agents, which did not accurately simulate all facets of stroke in elderly people. A frequent occurrence of one or more preexisting comorbidities is associated with a detrimental effect on the prognosis of strokes in patients, even though only 6% of the human population suffers from strokes (without comorbidities). One of the greatest limitations is the timing of the administration of medications that provide neuroprotection. Most of these studies recruited patients after a stroke had occurred for 4 h, which is past the ideal window (4–6 h) for effective neuroprotection. Low rates of recanalization could be the cause of neuroprotective drugs' ineffectiveness in early clinical studies. Co-treatment with drugs that improve nerve function during the process of recanalization may increase the distribution of drugs that improve nerve function to specific brain areas; this would increase the therapeutic effectiveness of stroke during the current thrombectomy period. In the past 5 years, a number of clinical studies to evaluate the beneficial benefits of various neuroprotective medications have been initiated. For the majority of these medications, experimental data were presented on the mechanism of action, functional recovery following treatment in rat models of stroke, effectiveness in stroke models with comorbidities, and extended use following ischemia. The post-stroke inflammatory pathway has been the primary focus of these potential neuroprotective drugs because inflammatory mediators have an effect on the progression of ischemic brain damage, but not all of them have this effect (Paul and Candelario-Jalil 2021). Two of these medications that have anti-inflammatory properties also target the oxidative stress signaling pathway and excitotoxicity. Recent clinical investigations have begun to assess the therapeutic potential of statins in patients at high risk of disease.

11.9 Summary

In terms of cellular and molecular composition, biogenesis, communication mechanisms, regulatory roles, potential as biomarkers, and efficient methods to use exosomes as therapy alone, primed, or packed with therapeutic agents, several studies have significantly contributed to a clear picture of exosome biology. Exosomes could therefore be used as a potential drug transport system, even at modest doses. The success of stem cell banks for potential clinical treatments opens the door to the potential application of comparable, secure, and hopeful technologies for the benefit of people. Exosomes have distinct biochemical functions, neovascularization characteristics, angiogenetic potential, and renewing properties, just like stem cells do. Exosomes can therefore be used to treat specific cases of severe brain diseases. Exosomes have intriguing paracrine properties, and it is possible to create individualized nano-vesicle banks in the future that could support human treatment. Directional research is needed to examine the potential benefits of creating nano-exosomal conservatories, which are comparable to stem cell libraries and may one day be more successful in healing all cerebral illnesses.

Acknowledgement The author would like to thank the Deanship of Scientific Research at Majmaah University for supporting this work.

References

Abak A, Abhari A, Rahimzadeh S (2018) Exosomes in cancer: small vesicular transporters for cancer progression and metastasis, biomarkers in cancer therapeutics. PeerJ 6:e4763

Abrous DN, Koehl M, Moal L, Michel. (2005) Adult neurogenesis: from precursors to network and physiology. Physiol Rev 85(2):523–569

Anjum J, Mitra S, Das R, Alam R, Mojumder A, Emran TB, Aljohani ASM et al (2022) A renewed concept on the MAPK signaling pathway in cancers: polyphenols as a choice of therapeutics. Pharmacol Res 106398:106398

Baecher-Allan C, Kaskow BJ, Weiner HL (2018) Multiple sclerosis: mechanisms and immunotherapy. Neuron 97(4):742–768

Baig MS, Roy A, Rajpoot S, Liu D, Savai R, Banerjee S, Saqib U et al (2020) Tumor-derived exosomes in the regulation of macrophage polarization. Inflamm Res 69:435–451

Barbacid M (1995) Neurotrophic factors and their receptors. Curr Opin Cell Biol 7(2):148–155

Beatriz M, Vilaça R, Lopes C (2021) Exosomes: innocent bystanders or critical culprits in neurodegenerative diseases. Front Cell Dev Biol 9:635104

Bellingham SA, Coleman BM, Hill AF (2012) Small RNA deep sequencing reveals a distinct miRNA signature released in exosomes from prion-infected neuronal cells. Nucleic Acids Res 40(21):10937–10949

Blum K (2019) Neurology & experimental neuroscience

Bogoyevitch MA (2006) The isoform-specific functions of the c-Jun N-terminal kinases (JNKs): differences revealed by gene targeting. BioEssays 28(9):923–934

Bonifacino JS, Traub LM (2003) Signals for sorting of transmembrane proteins to endosomes and lysosomes. Annu Rev Biochem 72(1):395–447

Broughton BRS, Reutens DC, Sobey CG (2009) Apoptotic mechanisms after cerebral ischemia. Stroke 40(5):e331–e339

Burnett JC, Rossi JJ (2012) RNA-based therapeutics: current progress and future prospects. Chem Biol 19(1):60–71

Cadoni MPL, Manetti R, Goswami DA, Galleri DG (2020) VAPB Aggregates in PBMCs and Peripheral Mediators of Inflammation: New Diagnostic Perspectives for ALS. Cells 9(1):164

Conway JA, Kinsman G, Kramer ER (2022) The Role of NEDD4 E3 Ubiquitin–Protein Ligases in Parkinson's Disease. Genes 13(3):513

Cortés D, Carballo-Molina OA, Castellanos-Montiel MJ, Velasco I (2017) The non-survival effects of glial cell line-derived neurotrophic factor on neural cells. Front Mol Neurosci 10:258

Danilova T (2020) MANF as a new regulator of the unfolded protein response and maintenance of pancreatic β-cells in mice. DSHealth Series
Dergunova LV, Filippenkov IB, Limborska SA, Myasoedov NF (2021) Pharmacotranscriptomics of peptide drugs with neuroprotective properties. Med Res Rev 41(2):754–769
Liegro D, Maria C, Schiera G, Schirò G, Liegro D, Italia. (2022) RNA-binding proteins as epigenetic regulators of brain functions and their involvement in neurodegeneration. Int J Mol Sci 23(23):14622
Dobson R, Giovannoni G (2019) Multiple sclerosis–a review. Eur J Neurol 26(1):27–40
Duman RS (2022) Neuronal damage and protection in the pathophysiology and treatment of psychiatric illness: stress and depression. Dialogues Clin Neurosci
Eden D, Gros C, Badji A, Dupont SM, De Leener B, Maranzano J, Ouellette R et al (2019) Spatial distribution of multiple sclerosis lesions in the cervical spinal cord. Brain 142(3):633–646
Efstathopoulos P, Kourgiantaki A, Karali K, Sidiropoulou K, Margioris AN, Gravanis A, Charalampopoulos I (2015) Fingolimod induces neurogenesis in adult mouse hippocampus and improves contextual fear memory. Transl Psychiatry 5(11):e685–e685
Emanuel KM (2015) Alteration of microRNA expression profiles of primary human monocyte-derived macrophage exosomes upon activation. University of Nebraska at Omaha
ENGIN (n.d.) Onur. immunomodulatory effects of stem cells. Prof. Dr. Bedriye Tunçsiper Prof. Dr. Berna Dirim Mete, 389
Fauré J, Lachenal G, Court M, Hirrlinger J, Chatellard-Causse C, Blot B et al (2006) Exosomes are released by cultured cortical neurones. Mol Cell Neurosci 31(4):642–648
Fayazi N, Sheykhhasan M, Soleimani Asl S, Najafi R (2021) Stem cell-derived exosomes: a new strategy of neurodegenerative disease treatment. Mol Neurobiol 58:3494–3514
Février B, Raposo G (2004) Exosomes: endosomal-derived vesicles shipping extracellular messages. Curr Opin Cell Biol 16(4):415–421
Grad LI, Cashman NR (2014) Prion-like activity of cu/Zn superoxide dismutase: implications for amyotrophic lateral sclerosis. Prion 8(1):33–41
Grewal AK, Singh TG, Sharma D, Sharma V, Singh M, Habibur RM, Albadrani GM et al (2021) Mechanistic insights and perspectives involved in neuroprotective action of quercetin. Biomed Pharmacother 140:111729
Gugliandolo A, Bramanti P, Mazzon E (2020) Mesenchymal stem cells in multiple sclerosis: recent evidence from pre-clinical to clinical studies. Int J Mol Sci 21(22):8662
Gurung S, Perocheau D, Touramanidou L, Baruteau J (2021) The exosome journey: from biogenesis to uptake and intracellular signalling. Cell Commun Signal 19(1):1–19
Hagey DW, Andaloussi E, Samir. (2023) The promise and challenges of extracellular vesicles in the diagnosis of neurodegenerative diseases. In: Handbook of clinical neurology, vol 193. Elsevier, pp 227–241
Harris AF (2022) Exploration of Decellularized vegetal scaffolds as new biomaterials for tissue engineering. The University of Arizona
Herms J, Dorostkar MM (2016) Dendritic spine pathology in neurodegenerative diseases. Annu Rev Pathol 11:221–250
Holm MM, Kaiser J, Schwab ME (2018) Extracellular vesicles: multimodal envoys in neural maintenance and repair. Trends Neurosci 41(6):360–372
Hu W, Qiu B, Guan W, Wang Q, Wang M, Li W, Xie G et al (2015) Direct conversion of normal and Alzheimer's disease human fibroblasts into neuronal cells by small molecules. Cell Stem Cell 17(2):204–212
Huttunen HJ, Saarma M (2019) CDNF protein therapy in Parkinson's disease. Cell Transplant 28(4):349–366
Jahan S, Mukherjee S, Ali S, Bhardwaj U, Choudhary RK, Balakrishnan S, Alaidarous M et al (2022) Pioneer role of extracellular vesicles as modulators of cancer initiation in progression, drug therapy, and vaccine prospects. Cell 11(3):490
Jain KK, Jain KK (2019) Neuroprotection in traumatic brain injury. Handb Clin Neurol:281–336

Jan AT, Malik MA, Rahman S, Yeo HR, Lee EJ, Abdullah TS, Choi I (2017) Perspective insights of exosomes in neurodegenerative diseases: a critical appraisal. Front Aging Neurosci 9:317
Janas AM, Sapoń K, Janas T, Stowell MHB, Janas T (2016) Exosomes and other extracellular vesicles in neural cells and neurodegenerative diseases. Biochim Biophys Acta 1858(6): 1139–1151
Jiang Z, Jiang JX, Zhang G-X (2014) Macrophages: a double-edged sword in experimental autoimmune encephalomyelitis. Immunol Lett 160(1):17–22
Juybari KB, Hosseinzadeh A, Ghaznavi H, Kamali M, Sedaghat A, Mehrzadi S, Naseripour M (2019) Melatonin as a modulator of degenerative and regenerative signaling pathways in injured retinal ganglion cells. Curr Pharm Des 25(28):3057–3073
Kaiser J, Maibach M, Salpeter I, Hagenbuch N, de Souza VBC, Robinson MD, Schwab ME (2019) The spinal transcriptome after cortical stroke: in search of molecular factors regulating spontaneous recovery in the spinal cord. J Neurosci 39(24):4714–4726
Kalani A, Tyagi A, Tyagi N (2014) Exosomes: mediators of neurodegeneration, neuroprotection and therapeutics. Mol Neurobiol 49:590–600
Kaltschmidt B, Witte KE, Greiner JFW, Weissinger F, Kaltschmidt C (2022) Targeting NF-κB signaling in cancer stem cells: a narrative review. Biomedicine 10(2):261
Kang X, Zuo Z, Hong W, Tang H, Geng W (2019) Progress of research on exosomes in the protection against ischemic brain injury. Front Neurosci 13:1149
Ke W (2020) Functional nucleic acid nanoparticles and their delivery. The University of North Carolina at Charlotte
Kumar GP, Anilakumar KR, Naveen S (2015) Phytochemicals having neuroprotective properties from dietary sources and medicinal herbs. Pharm J 7(1):1
Kumar GP, Khanum F (2012) Neuroprotective potential of phytochemicals. Pharmacogn Rev 6(12):81
Labbadia J, Morimoto RI (2013) Huntington's disease: underlying molecular mechanisms and emerging concepts. Trends Biochem Sci 38(8):378–385
Lai JJ, Chau ZL, Chen S-Y, Hill JJ, Korpany KV, Liang N-W, Liu Y-C et al (2022) Exosome processing and characterization approaches for research and technology development. Adv Sci 9(15):2103222
Lau HHC (2022) Strains of self-propagating protein aggregates in human neurodegenerative disease. University of Toronto (Canada)
Leggio L, Vivarelli S, L'Episcopo F, Tirolo C, Caniglia S, Testa N, Iraci N et al (2017) microRNAs in Parkinson's disease: from pathogenesis to novel diagnostic and therapeutic approaches. Int J Mol Sci 18(12):2698
Leong YQ, Ng KY, Chye SM, Ling APK, Koh RY (2020) Mechanisms of action of amyloid-beta and its precursor protein in neuronal cell death. Metab Brain Dis 35:11–30
Li C, Donninger H, Eaton J, Yaddanapudi K (2020) Regulatory role of immune cell-derived extracellular vesicles in cancer: the message is in the envelope. Front Immunol 11:1525
Li X, Liu H, Fischhaber PL, Tang T-S (2015) Toward therapeutic targets for SCA3: Insight into the role of Machado–Joseph disease protein ataxin-3 in misfolded proteins clearance. Prog Neurobiol 132:34–58
Liu W, Bai X, Zhang A, Huang J, Xu S, Zhang J (2019) Role of exosomes in central nervous system diseases. Front Mol Neurosci 12:240
Llana ÓC (2013) Tales of Neurotrophism
Lucotti S, Kenific CM, Zhang H, Lyden D (2022) Extracellular vesicles and particles impact the systemic landscape of cancer. EMBO J 41(18):e109288
Massey JC, Sutton IJ, Ma DDF, Moore JJ (2018) Regenerating immunotolerance in multiple sclerosis with autologous hematopoietic stem cell transplant. Front Immunol 9:410
Maxfield FR, Willard JM, Lu S (2016) Lysosomes: biology, diseases, and therapeutics. John Wiley & Sons
Mead EL (2012) Regulation of the microglial NADPH oxidase by neurotransmitters: implications for microglial–neuronal interactions. UCL (University College London)

Mincheva-Tasheva S, Soler RM (2013) NF-κB signaling pathways: role in nervous system physiology and pathology. Neuroscientist 19(2):175–194

Molteni R, Fumagalli F, Magnaghi V, Roceri M, Gennarelli M, Racagni G, Riva MA et al (2001) Modulation of fibroblast growth factor-2 by stress and corticosteroids: from developmental events to adult brain plasticity. Brain Res Rev 37(1–3):249–258

Mummery CL, Zhang J, Ng ES, Elliott DA, Elefanty AG, Kamp TJ (2012) Differentiation of human embryonic stem cells and induced pluripotent stem cells to cardiomyocytes: a methods overview. Circ Res 111(3):344–358

Musto M (2018) Microvesicles: neuroglial unconventional signaling to cortical brain cells

Nassar A, Kodi T, Satarker S, Prasada CG, Upadhya D, Sm F, Nampoothiri M et al (2022) Astrocytic MicroRNAs and transcription factors in Alzheimer's disease and therapeutic interventions. Cell 11(24):4111

Naviaux RK (2019) Metabolic features and regulation of the healing cycle—a new model for chronic disease pathogenesis and treatment. Mitochondrion 46:278–297

Olanow CW, Brundin P (2013) Parkinson's disease and alpha synuclein: is Parkinson's disease a prion-like disorder? Mov Disord 28(1):31–40

Onorati M, Binetti M, Conti L, Camnasio S, Calabrese G, Albieri I, Martynoga B et al (2011) Preservation of positional identity in fetus-derived neural stem (NS) cells from different mouse central nervous system compartments. Cell Mol Life Sci 68:1769–1783

Ortega JT, Jastrzebska B (2021) Neuroinflammation as a therapeutic target in retinitis pigmentosa and quercetin as its potential modulator. Pharmaceutics 13(11):1935

Oskouie MN, Moghaddam A, Nazanin S, Butler AE, Zamani P, Sahebkar A (2019) Therapeutic use of curcumin-encapsulated and curcumin-primed exosomes. J Cell Physiol 234(6):8182–8191

Oudit GY, Sun H, Kerfant B-G, Crackower MA, Penninger JM, Backx PH (2004) The role of phosphoinositide-3 kinase and PTEN in cardiovascular physiology and disease. J Mol Cell Cardiol 37(2):449–471

Parchure A, Vyas N, Mayor S (2018) Wnt and hedgehog: secretion of lipid-modified morphogens. Trends Cell Biol 28(2):157–170

Parkhurst CN, Yang G, Ninan I, Savas JN, Yates III, John R, Lafaille JJ, Gan W-B et al (2013) Microglia promote learning-dependent synapse formation through brain-derived neurotrophic factor. Cell 155(7):1596–1609

Pascual M, Ibáñez F, Guerri C (2020) Exosomes as mediators of neuron-glia communication in neuroinflammation. Neural Regen Res 15(5):796

Paskeh MD, Abad E, Maliheh M, Sepideh Z, Amirhossein S, Hossein N, Javad M, Hushmandi K et al (2022) Emerging role of exosomes in cancer progression and tumor microenvironment remodeling. J Hematol Oncol 15(1):1–39

Paul S, Candelario-Jalil E (2021) Emerging neuroprotective strategies for the treatment of ischemic stroke: an overview of clinical and preclinical studies. Exp Neurol 335:113518

Peferoen L, Kipp M, van der Valk P, van Noort JM, Amor S (2014) Oligodendrocyte-microglia cross-talk in the central nervous system. Immunology 141(3):302–313

Pegtel DM, Gould SJ (2019) Exosomes. Annu Rev Biochem 88:487–514

Peng C, Aron L, Klein R, Li M, Wurst W, Prakash N, Le W (2011) Pitx3 is a critical mediator of GDNF-induced BDNF expression in nigrostriatal dopaminergic neurons. J Neurosci 31(36): 12802–12815

Perrin FE, Lacroix S, Avilés-Trigueros M, David S (2005) Involvement of monocyte chemoattractant protein-1, macrophage inflammatory protein-1α and interleukin-1β in Wallerian degeneration. Brain 128(4):854–866

Pöyhönen S, Er S, Domanskyi A, Airavaara M (2019) Effects of neurotrophic factors in glial cells in the central nervous system: expression and properties in neurodegeneration and injury. Front Physiol 10:486

Rai SN, Dilnashin H, Birla H, Singh SS, Zahra W, Rathore AS, Singh SP et al (2019) The role of PI3K/Akt and ERK in neurodegenerative disorders. Neurotox Res 35:775–795

Reza-Zaldivar EE, Hernández-Sapiéns MA, Minjarez B, Gutiérrez-Mercado YK, Márquez-Aguirre AL, Canales-Aguirre AA (2018) Potential effects of MSC-derived exosomes in neuroplasticity in Alzheimer's disease. Front Cell Neurosci 12:317
Ridolfi B, Abdel-Haq H (2017) Neurodegenerative disorders treatment: the MicroRNA role. Curr Gene Ther 17(5):327–363
Rikhtegar R, Yousefi M, Dolati S, Kasmaei HD, Charsouei S, Nouri M, Shakouri SK (2019) Stem cell-based cell therapy for neuroprotection in stroke: a review. J Cell Biochem 120(6): 8849–8862
Riva P, Battaglia C, Venturin M (2019) Emerging role of genetic alterations affecting exosome biology in neurodegenerative diseases. Int J Mol Sci 20(17):4113
Roesch S, Rapp C, Dettling S, Herold-Mende C (2018) When immune cells turn bad—tumor-associated microglia/macrophages in glioma. Int J Mol Sci 19(2):436
Romani A, Cervellati C, Muresan XM, Belmonte G, Pecorelli A, Cervellati F, Valacchi G et al (2018) Keratinocytes oxidative damage mechanisms related to airbone particle matter exposure. Mech Ageing Dev 172:86–95
Ronaghi M, Erceg S, Moreno-Manzano V, Stojkovic M (2010) Challenges of stem cell therapy for spinal cord injury: human embryonic stem cells, endogenous neural stem cells, or induced pluripotent stem cells? Stem Cells 28(1):93–99
Scharfman HE (2005) Growth factors and epilepsy. Nova Publishers
Seyedaghamiri F, Salimi L, Ghaznavi D, Sokullu E, Rahbarghazi R (2022) Exosomes-based therapy of stroke, an emerging approach toward recovery. Cell Commun Signal 20(1):110
Sikorska B, Liberski PP (2012) Human prion diseases: from kuru to variant Creutzfeldt-Jakob disease. Protein Aggregation and Fibrillogenesis in Cerebral and Systemic Amyloid Disease. Subcell Biochem 65:457–496
Simpson RJ, Jensen SS, Lim JWE (2008) Proteomic profiling of exosomes: current perspectives. Proteomics 8(19):4083–4099
Sinha K, Sun C, Kamari R, Bettermann K (2020) Current status and future prospects of pathophysiology-based neuroprotective drugs for the treatment of vascular dementia. Drug Discov Today 25(4):793–799
Stankowski JN, Gupta R (2011) Therapeutic targets for neuroprotection in acute ischemic stroke: lost in translation? Antioxid Redox Signal 14(10):1841–1851
Teleanu RI, Preda MD, Niculescu A-G, Vladâcenco O, Radu CI, Grumezescu AM, Teleanu DM (2022) Current strategies to enhance delivery of drugs across the blood–brain barrier. Pharmaceutics 14(5):987
Traina G (2021) The role of mast cells in the gut and brain. J Integr Neurosci 20(1):185–196
Tseng K-Y, Anttila JE, Khodosevich K, Tuominen RK, Lindahl M, Domanskyi A, Airavaara M (2018) MANF promotes differentiation and migration of neural progenitor cells with potential neural regenerative effects in stroke. Mol Ther 26(1):238–255
Turner MD, Nedjai B, Hurst T, Pennington DJ (2014) Cytokines and chemokines: at the crossroads of cell signalling and inflammatory disease. Biochimica et Biophysica Acta (BBA)-molecular Cell Res 1843(11):2563–2582
Turola E, Furlan R, Bianco F, Matteoli M, Verderio C (2012) Microglial microvesicle secretion and intercellular signaling. Front Physiol 3:149
van der Koog L, Gandek TB, Nagelkerke A (2022) Liposomes and extracellular vesicles as drug delivery systems: a comparison of composition, pharmacokinetics, and functionalization. Adv Healthc Mater 11(5):2100639
Westergard TR, Stefeson. (2018) Dipeptide repeat proteins associated with C9orf72 ALS/FTD: understanding how they are made and how they potentiate disease. Thomas Jefferson University
Wolf-Dennen K, Kleinerman ES (2020) Exosomes: dynamic mediators of extracellular communication in the tumor microenvironment. Adv Exp Med Biol 1258:189–197
Xia Y, Zhang G, Han C, Ma K, Guo X, Wan F, Huang J et al (2019) Microglia as modulators of exosomal alpha-synuclein transmission. Cell Death Dis 10(3):174

Yadav A, Seth B, Chaturvedi RK (2021) Brain organoids: tiny mirrors of human neurodevelopment and neurological disorders. Neuroscientist 27(4):388–426

Yan, Yan, Herlihy, Bridget, Boyles, Sean, Tao, Junyi, Hulme, William, & Roy, Sabita. (2022). Transcriptomic profiling of mouse amygdala during opioid dependence and withdrawal by single-cell RNA sequencing

Yang S, Yang J, Yang Z, Chen P, Fraser A, Zhang W, Hong J-S et al (2006) Pituitary adenylate cyclase-activating polypeptide (PACAP) 38 and PACAP4–6 are neuroprotective through inhibition of NADPH oxidase: potent regulators of microglia-mediated oxidative stress. J Pharmacol Exp Ther 319(2):595–603

Zappulli V, Friis KP, Fitzpatrick Z, Maguire CA, Breakefield XO (2016) Extracellular vesicles and intercellular communication within the nervous system. J Clin Invest 126(4):1198–1207

Zhang J, Li S, Li L, Li M, Guo C, Yao J, Mi S (2015) Exosome and exosomal microRNA: trafficking, sorting, and function. Genomics Proteomics Bioinformatics 13(1):17–24

Engineered Exosomes as Nano-Vectors against Neurodegenerative Disorders

12

Ghazala Muteeb, Qamar Zia, and Adil Alshoaibi

Abstract

Neurodegenerative disorders are a diverse cohort of ailments marked by gradual neuronal damage. Accumulating evidence suggests the function of exosomes in the pathogenesis of brain illnesses through intercellular communication and the exchange of bioactive molecules. Exosomes, nano-sized membranous vesicles ubiquitously released by brain cells in the cerebrospinal fluid (CSF), are extremely efficient in traversing the blood–brain barrier (BBB). However, innate guiding properties of unmodified vesicles have proven to be insufficient for clinical applications. Researchers have now tailored the exosomes to fine-tune their on-target binding capabilities. Exosomes can be fabricated either by linking target proteins to the exterior or encapsulating bioactives inside their cavities. Bioengineering modifications bestow them with active targeting ability and deliver the payload to specific cell types/tissues. Here, we summarize the current

Ghazala Muteeb and Qamar Zia contributed equally with all other contributors.

G. Muteeb (✉)
Department of Nursing, College of Applied Medical Science, King Faisal University, Al-Ahsa, Saudi Arabia
e-mail: graza@kfu.edu.sa

Q. Zia (✉)
Department of Medical Laboratory Sciences, College of Applied Medical Sciences, Majmaah University, Al- Majmaah, Saudi Arabia

Health and Basic Sciences Research Center, Majmaah University, Al-Majmaah, Saudi Arabia

A. Alshoaibi
Department of Medical Laboratory Sciences, College of Applied Medical Sciences, Majmaah University, Al- Majmaah, Saudi Arabia

Department of Physics, College of Science, King Faisal University, Al-Ahsa, Saudi Arabia

S. Jahan, A. J. Siddiqui (eds.), *Applications of Stem Cells and Derived Exosomes in Neurodegenerative Disorders*, https://doi.org/10.1007/978-981-99-3848-3_12

state of knowledge of the natural targeting abilities of exosomes with special reference to brain diseases. Later, cutting-edge technologies exploited for bioengineering exosomes are critically evaluated. Moreover, the application of customized exosomes in the management of neurological disorders is highlighted. Conclusively, the bottlenecks in harnessing the true power of exosomes for targeting brain therapy are discussed.

Keywords

Brain disorders · Bioengineering · Drug delivery · Exosome · Nanoparticle

Abbreviations

α-Syn	Alpha-synuclein
5-FC	5-Fluorocytosine
5-FU	5-Fluorouracil
AchR	tylcholine receptor
AD	Alzheimer's disease
ADEs	Astrocyte-derived exosomes
ADMSCs	Adipocyte-derived MSCs
AF4	Asymmetric flow field-flow fractionation
AMO-21	Antisense oligonucleotide against miR-21
APP	Amyloid precursor protein
ASD	Autism spectrum disorder
Aβ	Amyloid-beta
BACE1	β-Bite APP-cleaving enzyme 1
BAP	BA-poly(2-(dimethylamino)ethyl acrylate)
BBB	Blood–brain barrier
BDNF	Brain-derived neurotrophic factor
BIM	Bcl-2-interacting mediator of cell death
BM	Bone marrow
BMECs	Brain microvascular endothelial cells
CD	Cytosine deaminase
Cdk5	Cyclin-dependent kinase 5
CendR	C-End rule
CNS	Central nervous system
CR1	Complement receptor type 1
CSF	Cerebral spinal fluid
cur	Curcumin
DS	Down syndrome
EEs	Early endosomes
EGFR	Epidermal growth factor receptor
EMNVs	Exosome-mimetic nanovesicles
ESCRT	Endosomal-sorting complex needed for transport

EVs	Extracellular vehicles
EXO + EDV	Blood-derived exosomes packaged with Edaravone
ExoCAT	Catalase-loaded exosomes
Exo-cur	Cur-primed engineered
Exo-cur	Exosomes encapsulated with curcumin
EXOtic	Exosomal transfer into cells
EXPLORs	Exosomes for protein loading via optically reversible protein–protein interactions
GAC	Glutaminase C
GAP43	Growth-associated protein 43
GBM	Glioblastoma
GDNF	Glial cell line–derived neurotrophic factor
GNP	Gold nanoparticles
HBMVEC	Human brain microvascular endothelial cell
HP	Hypotonic dialysis
ICAM-1	Endothelial intercellular adhesion molecule 1
IGLV133	Immunoglobulin lambda variable 133
IL-1	Interleukin-1
ILVs	Intraluminal vesicles
imDC	Immature dendritic cell
IMTP	Ischemic myocardial targeting peptide
Lamp2b	Lysosome-associated membrane protein 2
LBs	Lewy bodies
LFA-1	Lymphocyte function–associated antigen
LPS	Lipopolysaccharide
MFH	Magnetic fluid hyperthermia
MOR	Opioid receptor mu
MSCs	Mesenchymal stem cells
MVBs	Multivesicular bodies
Mφ	Naïve macrophage
NALP3	NACHT, LRR, and PYD domain–containing protein 3
Ndfip1	Nedd4 family–interacting protein 1
Nluc	Nano luciferase
NO	Nitric oxide
NRP-1	Neuropilin-1
OA	Okadaic acid
PD	Parkinson disease
PEG	Polyethylene glycol
PFF	Preformed fibril
P-gp	P-Glycoprotein
PMA	Phorbol myristate acetate
PPIs	Protein–protein interactions
Que.	Quercetin
RBC	Red blood cells
r-BMSCs	Rat bone marrow–derived MSCs

REXO	RVG-modified exosome
REXO-C/ANP/S	RVG peptide–modified exosome-curcumin/BAP/siSNCA
ROS	Reactive oxygen species
RVG	Rabies viral glycoprotein
SCNA	Alpha-synuclein
shRNA-MC	Short hairpin RNA-minicircle
SNAP25	Synaptosome-associated protein 25
SphK/S1P	Sphingosine kinase/sphingosine-1-phosphate
SPIONs	Superparamagnetic iron oxide nanoparticles
Stat3	Signal transducer and activator of transcription 3
T7-exo	T7-peptide–Lamp2-decorated exosomes loaded with antisense miR-21
TEM	Tetraspanin-enriched microdomains
TfR	Transferrin receptor
Ub	Ubiquitin
UPRT	Uracil phosphoribosyltransferase

12.1 Introduction

Trams et al. first discovered a cluster of exfoliated vesicular entities, about 40 nm in diameter with 5′-nucleotidase activity, contained within yet another larger vesicle (later named as multivesicular bodies [MVBs]) with size dimensions of 500–1000 nm (Trams et al. 1981). Subsequently, Johnstone et al. isolated these small vesicles from maturing sheep reticulocytes by ultracentrifugation and designated them as "exosomes" (Johnstone et al. 1987; Johnstone 2005). Unfortunately, the existence of exosomes was not realized widely then since these particles were considered to be essentially debris and by-products of aging red blood cells (RBCs).

Extracellular vehicles (EVs), a broad group of membrane-bound vesicles constantly released by every known human cell, take a crucial part in cell-to-cell communication (Wang et al. 2020). Centered on their nano-dimensions and subcellular origin, EVs can be categorized into three major subclasses: exosomes, microvesicles, and apoptotic bodies. Exosomes are petite with a diameter of 30–100 nm. Microvesicles, which are somewhat bigger than exosomes, are formed by spontaneous outward bursting from the cell membrane under normal physiological state or in reaction to particular stimuli, like variations in ATP and cytoplasmic calcium concentration (Fuhrmann et al. 2015a, b; Théry et al. 2009). Apoptotic bodies, the biggest (50–5000 nm) and the most diverse subpopulation of EVs, are produced by cell disintegration during regulated cell demise or apoptosis (Akers et al. 2013). Despite clear variations in size and source of origin, there are a limited number of markers available to differentiate among these subtypes, and cutting-edge purification methods continue to struggle to distinguish these diverse populations of EVs (Kowal et al. 2016; Piffoux et al. 2019).

Exosomes are secreted by every cell in the human body and have been found in practically all bodily fluids (Caby et al. 2005, Pisitkun et al. 2004, Admyre et al. 2007; Adriano et al. 2021, Keller et al. 2011, Ogawa et al. 2011; Zhang et al. 2016). The architecture of donor cells can be reflected in the composition of exosomes (Valadi et al. 2007). During exosome assembly, a variety of active ingredients with varying bioactivities encapsulate into exosomes. Exosomes have been discovered to contain a wide range of bioactives (Vojtech et al. 2014; Xie et al. 2019; Guescini et al. 2010; Sansone et al. 2017; Yang et al. 2017a, b; Tsai et al. 2018; Skotland et al. 2017; Muller 2020; Pegtel and Gould 2019). Typically, exosomes can be identified from their morphology, dimension, content, and unique surface marker (CD9, CD81, CD63, flotillin, and TSG101) (Kim et al. 2022). Recently, Zhang et al. (2018) discovered two subtypes of exosome using the asymmetric flow field-flow fractionation (AF4) technique. They were named large exosomes (exo-L, 90–120 nm) and small exosomes (exo-S, 60–80 nm). They also identified distinct ~35 nm sized non-membrane-bound nanostructures which they termed "exomere" . All three elements displayed unique characteristics as well as distinct functions (Zhang et al. 2018). As such, the definition of an exosome has continually evolved as new kinds of vesicles have been identified (Witwer and Théry 2019; Théry et al. 2018).

Exosomes act as intercellular mediators for cell–cell signaling. Donor cells can transport substances to recipient cells through exosomes (Zhang et al. 2018), directing the biological expression of the receiving cells by altering their gene expression (Cheng et al. 2019). Exosomes can be taken up by cells, move through the bloodstream, and bridge the blood–brain barrier (BBB) (Zhang et al. 2019). When compared to traditional delivery systems, exosomes sequestered from an individual exhibit higher compatibility and fewer hazardous side effects (Quah and O'Neill 2005). Furthermore, an exosome-mediated delivery system avoids the P-glycoprotein (P-gp), an ABC-transporter protein, potentially reducing drug resistance (Kim et al. 2016). These naturally occurring nanocarriers have, therefore, been used for diagnostic, prognostic, and therapeutic purposes (Batrakova and Kim 2015).

Recently, exosome bioengineering has gained tremendous interest to achieve targeted delivery to confer cell specificity. Although exosomes' use as drug-delivery devices has been the subject of several reviews (Barile and Vassalli 2017; Conlan et al. 2017; He et al. 2018; Antimisiaris et al. 2018; Salunkhe et al. 2020), this chapter focuses on the engineering of exosomes for targeting neurological disorders specifically. The capabilities of exosomes as delivery systems, as well as methodologies for modifying the exosomal surface and cargo loading techniques, are discussed. Next, the application of bioengineered exosomes in the management of neurological diseases is highlighted. Finally, the limitations of exploiting engineered exosomes for targeted brain therapy are explored.

12.2 Biogenesis of Exosomes

Exosome originates within the endo-lysosomal pathway when early endosomes (EEs) mature into a late endosomal compartment or multivesicular bodies (MVBs) (Fig. 12.1). The MVBs and its cargo mainly adopt two routes controlled by the protein ubiquitin checkpoint (Wollert and Hurley 2010; Vietri et al. 2020). The MVBs may attach to lysosomes/autophagosomes inside the cell and are destroyed. Alternatively, within the MVBs, membrane invagination of the endosomes may result in the production of intraluminal vesicles (ILVs) (Hessvik and Llorente 2018). MVBs then merge with the inner leaflet of the cell membrane, exocytotically releasing the ILVs as exosomes. This release of exosomes can be initiated by a plethora of signals (Alli 2021) including p53 (Jelonek et al. 2016), lipopolysaccharide (LPS), and phorbol myristate acetate (PMA) (Shimoda and Khokha 2017) as well as nitric oxide (NO) (Nolan et al. 2008), carbon dioxide (Thom et al. 2017), and several other factors.

Exosome sorting occurs both by endosomal-sorting complex needed for transport (ESCRT)-dependent and-independent signals (Barile and Vassalli 2017; Hessvik and Llorente 2018; Kowal et al. 2014). The ESCRT complex comprises of five protein groups, ESCRT-0, ESCRT-I, ESCRT-II, ESCRT-III, VPS4 (AAA ATPase), and other auxiliary proteins including Alix (Murphy et al. 2019; Kalluri and LeBleu 2020). ESCRT-0 localizes ESCRT-I to the MVB membrane (Bowers and Stevens 2005), where ESCRT-I binds to the ubiquitinated cargo via its Vps23 subunit (Katzmann et al. 2001), activating ESCRT-II. ESCRT-II subsequently starts oligomerizing small spiral proteins and folds the MVB membrane, which concentrates the cargo (Babst 2005). This causes the ESCRT-III complex to assemble, which eventually constricts the neck by forming a spiral-shaped structure. The deubiquitinating enzyme (Doa4p) recruited by ESCRT-III eliminates ubiquitin (Ub) from the payload prior to their loading into the ILVs (Amerik et al. 2000). Finally, when cargo selection is over, ATPase VPS4p attaches to ESCRT-III and releases the protein using ATP (Babst et al. 1997, 1998).

It has now been confirmed that deubiquitinylation, although may promote cargo loading into MVBs, is not an essential criterion for protein loading (Huebner et al. 2016). ESCRT-independent pathways do not require ubiquitination for protein sorting to MVBs; rather they are dependent on the presence of tetraspanin, cholesterol, or ceramide-enriched membrane microdomains (Anakor et al. 2021). Tetraspanin-enriched microdomains (TEMs) are associated with the assortment of ILV cargo as well as their release. Lipid raft cholesterol participates in the release of exosomes (Anakor et al. 2021). Ceramide, a cone-shaped lipid abundant in the innermost exosomal membrane (Donoso-Quezada et al. 2021), induces the negative curve in the membranes, promoting the fusion of lipid raft microdomains; thus facilitating ILV generation (Verderio et al. 2018). Moreover, it has now been confirmed that inhibition of sphingomyelinase (nSMNase) II, an enzyme that converts sphingomyelin into ceramide, drastically lowers exosome secretion (Choezom and Gross 2022). Thus, these lipids as well as some enzymes collaborate

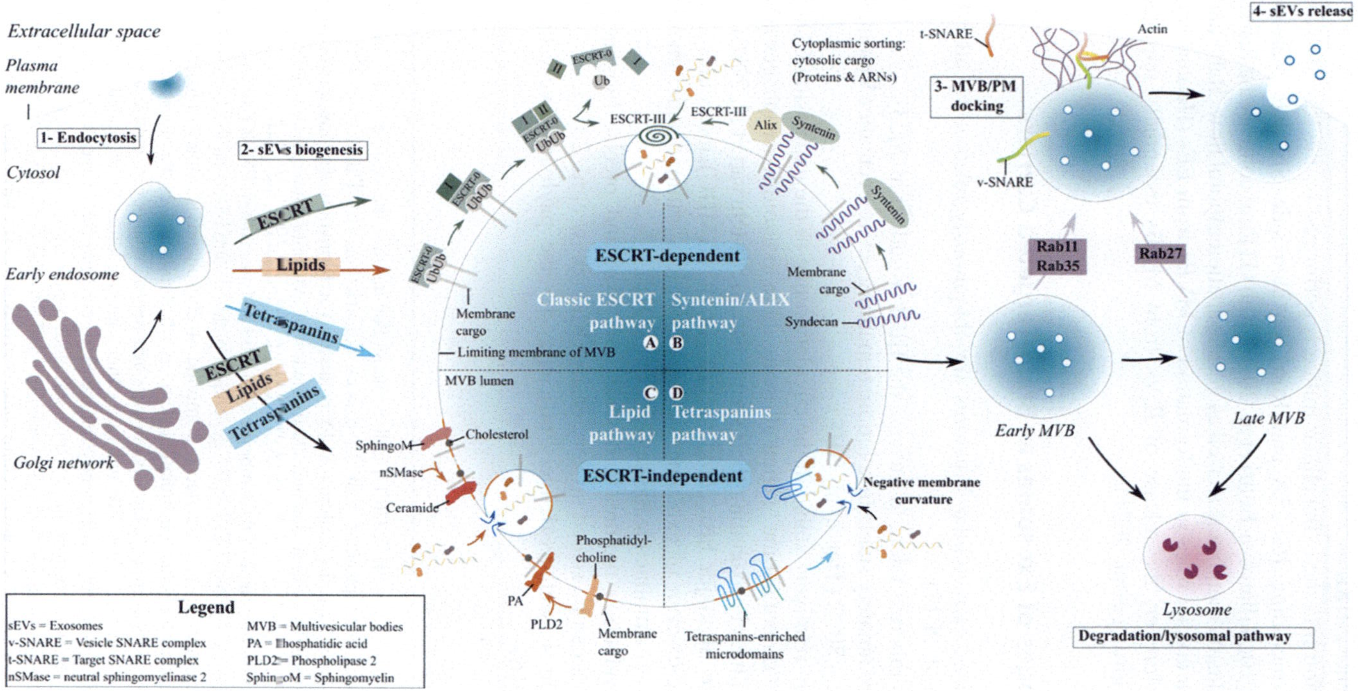

Fig. 12.1 Schematic representation of exosome formation and release in the extracellular space. (1) Exosomes are produced as ILVs by inward budding of the endosomal membrane and accumulate in the lumen of the endosome. (2) Several mechanisms are involved in the biogenesis of exosomes such as ESCRT protein–, lipid raft–, and tetraspanin microdomain–dependent pathways. Whether one or multiple pathways are required simultaneously by one population of MVBs, or if each pathway is specific to one population of MVBs, is still not clear. ESCRT-dependent pathways: (**a**) The most described mechanism involved in the biogenesis of exosomes is the ESCRT-dependent pathway requiring proteins of the ESCRT family. Specific transmembrane ubiquitinated cargo is recruited and clustered at the MVB membrane by the ESCRT-0 complex, subsequently binding to the ESCRT-I structure. The ESCRT-II complex is activated and

closely in MVB membrane invagination, fusion, and release of ILVs as exosomes (Hessvik and Llorente 2018).

Exosome secretion into the extracellular milieu is performed by the Rab class of GTPase proteins (including Rab27A and Rab27B) alongside SNARE proteins (Ostrowski et al. 2010; Villarroya-Beltri et al. 2016). Rab GTPase controls the fusion of MVB with the plasma membrane as well as vesicle spatiotemporal flow. Rab5 and VPS34/p150 play a crucial roles in the formation of MVBs from EEs (Huotari and Helenius 2011). Morever, the release of ILVs from MVBs essentially requires SNARE proteins Gurung et al. 2021).

12.2.1 Advantages of Exosomes as Brain-Targeted Carriers

(i) The small-sized extracellular structures are ubiquitously released by all cell types (Jan et al. 2017).
(ii) EVs possess inherent talent to traverse the BBB (Roudi et al 2023).
(iii) Exosomes are innately stable in solution, remaining active even after 5 months of storage at −80°C (Tian et al. 2018).
(iv) Exosomes are relatively stable in the circulation due to the expression CD47 that relays a "do not eat me" signal allowing them to avoid macrophages and opsonins, ultimately extending their serum half-life (Mehryab et al. 2020).
(v) The presence of major histocompatibility complex (MHC) molecules and CD86 on the exterior of exosomes renders them less immunogenic, greatly improving their immune clearance (Qu et al. 2018).
(vi) Exosome production requires a modest, effective, and precise biogenesis procedure that can be scaled up very easily (Khongkow et al. 2019).
(vii) Encapsulation as well as delivery efficiency for the payload is very high (Ha et al. 2016).

Fig. 12.1 (continued) together with ESCRT-I will create and/or stabilize the vesicle neck. Finally, ESCRT-III and its associated proteins will drive neck constriction. (**b**) The second ESCRT-dependent biogenesis pathway is the syntenin/ALIX pathway. The formation of syndecan-enriched microdomains leads to syndecan cleavage and the formation of syntenin/syndecan complexes that interact with ALIX. The syntenin–syndecan–ALIX complex then favors the recruitment of the ESCRT-III complex to support the MVB membrane curvature and abscission. ESCRT-independent pathways: (**c**) Ceramide- and phosphatidic acid–dependent pathways are based on the formation of lipid rafts where sphingomyelin is converted to ceramide or phosphatidylcholine is converted to phosphatidic acid. The ceramide- and phosphatidic acid–enriched rafts induce the inward curvature of the MVB membrane. (**d**) Similarly, tetraspanin-enriched microdomains can induce a negative curvature in the MVB membrane. (3) MVBs will fuse either with lysosomes for degradation or with the plasma membrane, which will consequently release exosomes into the extracellular space (4). Several proteins have been identified in the transport and fusion of the MVB to the plasma membrane, such as proteins from the Rab GTPase family and SNARE complexes. (Adapted from Anakor et al. 2021 under a Creative Commons Attribution 4.0, https://doi.org/10.3390/cells10112930)

(viii) Bioengineering of exosomes can tailor them to target specific brain cells (Alvarez-Erviti et al. 2011; Liu et al. 2015).
(ix) Any bioactive entity can be loaded or expressed in or on exosomes extending its shelf-life, thereby increasing efficacy by targeted delivery (Yang et al. 2015; Jia et al. 2018).

12.3 Exosomes' Innate Neural Targeting and Homing Abilities

Recently, a number of cutting-edge invasive and noninvasive techniques have been researched with the goal of bypassing the BBB and focusing on the critical disease regions in the brain. The goal of these studies is to develop therapeutic medications that can successfully target the brain. The four main obstacles that hinder standard therapy medications from effectively reaching the brain are the BBB, blood–brain–cancer barrier, blood–CSF barrier, and transporter proteins (Khan et al. 2018). Due to their biocompatibility and nanometer-size range (Arrighetti et al. 2019), exosomes have been the subject of several investigations as potential replacement for conventional agents against neurodegenerative disorders (Lapchak et al. 2018; Qu et al. 2018).

Certain natural exosomal membrane proteins such as Lamp, CD63, CD9, CD81, and CD47 (Raposo and Stoorvogel 2013) can initiate fusion with recipient cells by interacting with the surface receptors, thus allowing the uptake of the exosomal cargo under normal physiological conditions (Mentkowski et al. 2018; Rana et al. 2012). For example, CD47 relays a "do not eat me" signal by interacting with its ligand, signal regulatory protein alpha (SIRPα) (Chao et al. 2012), thus increasing the shelf life of exosomes (Kamerkar et al. 2017).

It has been experimentally proven that exosomes and their payload take part in neuron repair, synapse functionality, cognition, immune responsiveness, and exosome-mediated intercellular interaction, which all promote brain regeneration (Qing et al. 2018; Liu et al. 2019; Branscome et al. 2020). Under CNS disorder, the integrity of the BBB is compromised, which could be used for the passive transfer of bioactives inside the neurons (Shlosberg et al. 2010). For instance, under stroke-like circumstances, luciferase-carrying exosomes may traverse the brain microvascular endothelial cell (BMEC) lining, but not in healthy state. Moreover, the findings reveal that BMECs ingested exosomes via endocytosis. The report also investigated the prospect of exosome targeting in the brain via interactions with BBB cells under inflamed conditions (Chen et al. 2016).

Exosomes have been employed to boost cerebral biodistribution and transport medications to targeted brain cells (Yang et al. 2015; Qu et al. 2018). Nude mice were intravenously inoculated with carbocyanine lipophilic dye, DiD-labeled exosomes, and the results were monitored by acquiring near-infrared fluorescence photographs at various time intervals. Exosomes were discovered to collect preferentially in the brain between one and ten hours of administration, while fluorescence intensity peaked 4 to 8 h after exosomal infusion (Qu et al. 2018). To track exosome movement and its brain-homing capabilities, MSC-derived exosomes were tagged with gold nanoparticles (GNPs) as labelling agent. Brain imaging revealed a comparatively higher localization time of exosomes (up to 96 h) in pathologically significant brain structures of tagged mice than under normal physiological conditions (up to 24 h) (Perets et al. 2019). Microglia cells exclusively ingest exosomes in a mixed brain cell culture via the macro-pinocytosis route (Fitzner et al. 2011). In addition, N2a neuroblastoma cell line–derived exosomes labeled with GFP-CD63 are preferentially taken up by glial cells, while GFP-TTC-labelled exosomes attach primarily to neuron but not glial cells (Chivet et al. 2014). Folate receptor α (FRα) is naturally found in the membrane of choroid plexus epithelial cell–derived exosomes. Researchers exploited FRα-expressing exosomes to deliver bioactive molecules specifically to the brain parenchyma through the choroid plexus (Grapp et al. 2013). Exosomes are therefore appropriate for treating neurological problems since they have a built-in ability to home in on the brain.

12.4 Bioengineering of Exosomes for Brain Targeting

Exosomes can be utilized as a natural vehicle for therapeutic agents due to its innate qualities such as low toxicity, increased circulation time, capacity to load bioactives, robust cargo shelter, and good transport effectiveness to recipient cells, particularly brain cells since they can readily cross the BBB (Rufino-Ramos et al. 2017; Bang and Kim 2019). Additionally, synthetic bioactives may be immunogenic and toxic, whereas exosomes being biological in nature display insignificant untoward reactions (de Abreu et al. 2020). Moreover, exosomes can also be modified by several methods to increase their targeting potential that ultimately enhances their ability to home in specific tissues/organs (Mehryab et al. 2020; Zhang et al. 2020) and ensures targeted delivery of desired payloads (Table 12.1). Thus, numerous reports are available in the literature that have studied the clinical efficacy of cell–derived exosomes in various neurological disorders (Kodali et al. 2020).

Exosomes are presently being engineered to improve biodistribution, renal clearance, and serum half-life. Broadly, two methodologies are preferred in order to bioengineer exosomes (Fig. 12.2). In one scenario, the parent cell is genetically transfected to engineer ligands on the surface or modify drug-loaded exosomes (cell engineering techniques) (Alvarez-Erviti et al. 2011; Liu et al. 2015; Ohno et al. 2013). Practices involving the modification of exosome-producing cells are complex, costly, require a longer time duration, and are essentially non-applicable on pre-isolated exosomes (Kooijmans et al. 2016). Occasionally, the required physicochemical circumstances are unfavorable for live cell lines. Alternatively, the exosome can be loaded with the cargo (bioactive compound) post-isolation (Tanziela et al. 2020).

More recently, exciting nanoparticle modification approaches have been applied to engineer exosomes, wherein direct tailoring of exosomes is achieved by chemical methods (Smyth et al. 2014; Sato et al. 2016). Exosome tailoring is a reliable approach because exosomes are inert entities, making it simple to select the right approach and obtain higher accumulation inside exosomes when compared to cell-

Table 12.1 Engineering strategies for inducing targeted delivery of therapeutic exosomes

Disease	Donor cells	Engineering agent	Target receptor	Cargo	Loading strategy	Target cells/organs	Reference
Parkinson's disease	Exo	–	TfR	Dopamine	Incubation	Hippocampus	Qu et al. (2018)
	HEK293T cells	RVG-Lamp2b	AchR	Aptamer	Incubation	Neurons, microglia, and astrocytes in the midbrain	Ren et al. (2019)
	Murine DC	RVG-Lamp2b	AchR	shRNA-MC	Electroporation	Dopaminergic neurons	Izco et al. (2019)
	Murine DC	RVG-Lamp2b	AchR	α-Syn SiRNA	Transfection	Dopaminergic neurons	Cooper et al. (2014)
	Genetically modified RAW264.7 cells	–	GFRα-1[a]	GDNF	Transfection	Dopaminergic neurons	Zhao et al. (2014)
	RAW264.7 cells	–	–	Catalase	(a) Incubation at RT with or without saponin (b) Freeze–thaw cycles (c) Sonication (d) Extrusion	Microglia, astrocytes, and neuron	Haney et al. (2015)
	HEK293T cells	RVG-Lamp2b	AchR	Catalase mRNA	Transfection	Neurons	Kojima et al. (2018)
	Bone marrow–derived imDC	RVG peptide	AchR	Cur, SCNA	Ultrasonication	Neuron cell	Liu et al. (2020)
Glioma	RAW264.7 cells	RGE peptide	NRP1	Cur	Electroporation	Glioma cells	Jia et al. (2018)
	HEK293T cells	T7-Lamp2b	TfR	AMO-21	Electroporation	Glioblastoma cells	Kim et al. (2020)

(continued)

Table 12.1 (continued)

Disease	Donor cells	Engineering agent	Target receptor	Cargo	Loading strategy	Target cells/organs	Reference
	BM-MSCs	–	–	Anti-miR-9	Transfection	Brain tumor cells	Munoz et al. (2013)
	HEK293T cells	–	–	pCD-UPRT	Transfection	Tumor cells	Erkan et al. (2017)
Ischemic stroke	BM-MSCs	cRGD	Integrin $\alpha_v\beta_3$	Cur	Incubation	Ischemic lesion	Tian et al. (2018)
	BM-MSCs	RVG-Lamp2b	AchR	miR-124	Electroporation	Ischemic brain areas	Yang et al. (2017b)
	Exo	–	TfR	EDV	Sonification	Ischemic brain region	Guo et al. (2021)
Drug addiction (morphine relapse)	HEK293T cells	RVG-Lamp2b	AchR	MOR siRNA	Transfection	Neuron cell	Liu et al. (2015)
Neuroinflammation	RAW264.7 cells	LFA-1	ICAM-1	BDNF	Incubation	Brain-inflamed region	Yuan et al. (2017)
	GL26	–	–	Cur, JSI124	Incubation	Inflammatory microglial cells	Zhuang et al. (2011)
Alzheimer's disease	DC	RVG-Lamp2b	AchR	BACE1 siRNA	Electroporation	Brain neurons, microglia, and oligodendrocytes	Alvarez-Erviti et al. (2011)
	BM-MSC	DOPE-RVG	AchR	–	Incubation	Cortex and hippocampus	Cui et al. (2019)
	r-BMSC and HEK-293T	–	–	miR-29	Transfection	Hippocampus region	Jahangard et al. (2020)
	ADMSCs	–	–	miR-22	Transfection	Microglia	Zhai et al. (2021)
	Exo	Quercetin	HSP70[a]	TLR4[a]	Incubation	Brain neurons	Qi et al. (2020)

	RAW264.7 cells	LFA-1	ICAM-1	Cur	Incubation	Hippocampal microglial cells	Wang et al. (2019)
	HBMVECs	–	–	P-gp	–	Neuronal cells	Pan et al. (2020)

AchR acetylcholine receptor, *ADMSCs* adipose-derived mesenchymal stem cells, *AMO21* antisense miRNA oligonucleotides against miR-21, *BACE1* β-site APP cleaving enzyme 1, *BDNF* brain-derived neurotrophic factor, *BM-MSC* bone marrow mesenchymal stem cells, *Cur* curcumin, *DC* dendritic cells, *DOPE-RVG* dioleoyl phosphatidylethanolamine rabies viral glycoprotein, *EVA* edaravone, *Exo* blood exosomes, *GDNF* glial cell line-derived neurotrophic factor, *GFRα-1* GDNF receptor alpha-1, *GL26* glioblastoma cell line, *HBMVECs* human brain microvascular endothelial cells, *ICAM-1* intercellular adhesion molecule, *1JSI124* Stat3 inhibitor, *LFA-1* lymphocyte function-associated antigen, *MOR* opioid receptor mu, *NRP1* neuropilin-1, *P-gp* P-glycoprotein, *pCD-UPRT* cytosine deaminase-uracil phosphoribosyl transferase, *r-BMSC* rat bone marrow mesenchymal stem cells, *RVG-Lamp2b* rabies viral glycoprotein-lysosome-associated membrane protein 2, *SCNA* alpha-synuclein, *shRNA-MC* short hairpin RNA-minicircle, *TfR* transferrin receptor

[a] Speculated by authors

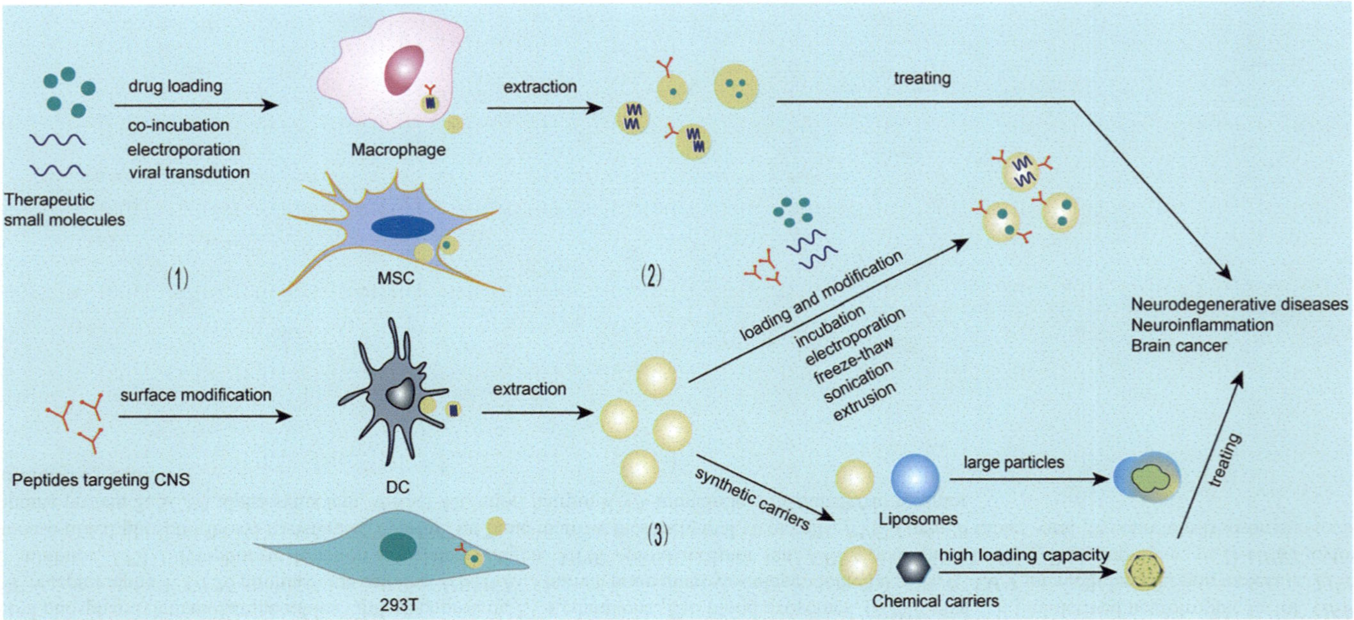

Fig. 12.2 The strategies for preparing targeted exosomes. (1) Preloading the therapeutic agents into the donor cells and modifying exosomes with targeting peptides during their formation. (2) Drug loading and surface modification of the purified exosomes after their extraction. (3) Hybridizing with synthetic vectors to improve drug loading. (Reprinted with permission from Zheng et al. 2019. Bioconjugate Chem. 30, 4, 994–1005, Copyright © 2019 American Chemical Society. https://doi.org/10.1021/acs.bioconjchem.9b00085)

based modification methods. Exosome perfusion, selectivity, and retention time may be improved via surface manipulation techniques. As a result, researchers are often drawn towards post-isolation exosome alteration approaches. The treatment strategies mainly include electroporation (Gilligan and Dwyer 2017), chemical transfection (Chen et al. 2014; Katakowski et al. 2013), co-incubation (Sun et al. 2010), sonication, freeze-thawing cycle, saponin penetration, or extrusion (Haney et al. 2015) (Fig. 12.3). Researchers use these procedures to alter either the surface or the contents of exosomes (Tanziela et al. 2020).

12.4.1 Exosome Surface Modification

Despite recent advances in revealing the intricacies of exosomes, improvements in surface alteration approaches are required for translating it into clinical settings. Surface functionalization can increase exosome specificity. Coating exosomes with a hydrophilic polymer such as polyethylene glycol (PEG) enhances its in vivo retention time and lowers nonspecific contact (Suk et al. 2016). PEGylation was used to improve Neuro2a (N2a) cell–derived exosome selectivity and circulation time by incubating PEG with exosomes via a "post-insertion" mechanism (Kooijmans et al. 2016).

For targeted modification of exosomes, peptides are often attached to exosomal surface proteins due to their low MW, negligible immunogenic nature, and specificity with the target protein (Salunkhe et al. 2020), employing bioengineering methodologies (such as genetic engineering, lentiviral technology, orthogonal chemistry) to construct fusion peptides. Incidentally, a transmembrane protein, lysosome-associated membrane protein 2 (Lamp2b), has been exploited to decorate EVs with targeting ligands as fusion proteins through cell engineering approaches (Alvarez-Erviti et al. 2011; Tian et al. 2014). The N-terminal of Lamp2b protein has been conjugated with neuron-targeting rabies viral glycoprotein (RVG) peptide (Alvarez-Erviti et al. 2011), internalizing cyclic peptide IRGD (CRGDKGPDC) targeting integrin α_v (Tian et al. 2014), and ischemic myocardial targeting peptide IMTP (CSTSMLKAC) (Wang et al. 2019) or modified with glycan residues (Hung and Leonard 2015) to improve circulation time and ensure targeted delivery. Similar research used RVG-modified exosomes to transport opioid receptor mu (MOR) siRNA across the BBB into the neural cells (Liu et al. 2015). The transmembrane domain of epidermal growth factor receptor (EGFR) that binds strongly with the GE11 peptide (YHWYGYTPQNVI) (Li et al. 2005) can act as a substitute to Lamp2b for displaying fusion proteins on the exosomal surface (Ohno et al. 2013).

Click chemistry, first described in 1999, is a dependable, modest, rapid, and greatly effective approach for the bio-conjugation of therapeutic agents to the exterior of exosomes (Hein et al. 2008; Smyth et al. 2014). Exploiting this, researchers integrated cilengitide-like cyclopentapeptide, cyclo(Arg-Gly-Asp-D-Tyr-Lys) or cRGDyK, which displays a strong association for integrin $\alpha_v\beta_3$ (Arosio and Casagrande 2016) into the exosomal surface via the rapid bio-orthogonal chemical method (Tian et al. 2018). Click chemistry was also utilized to conjugate

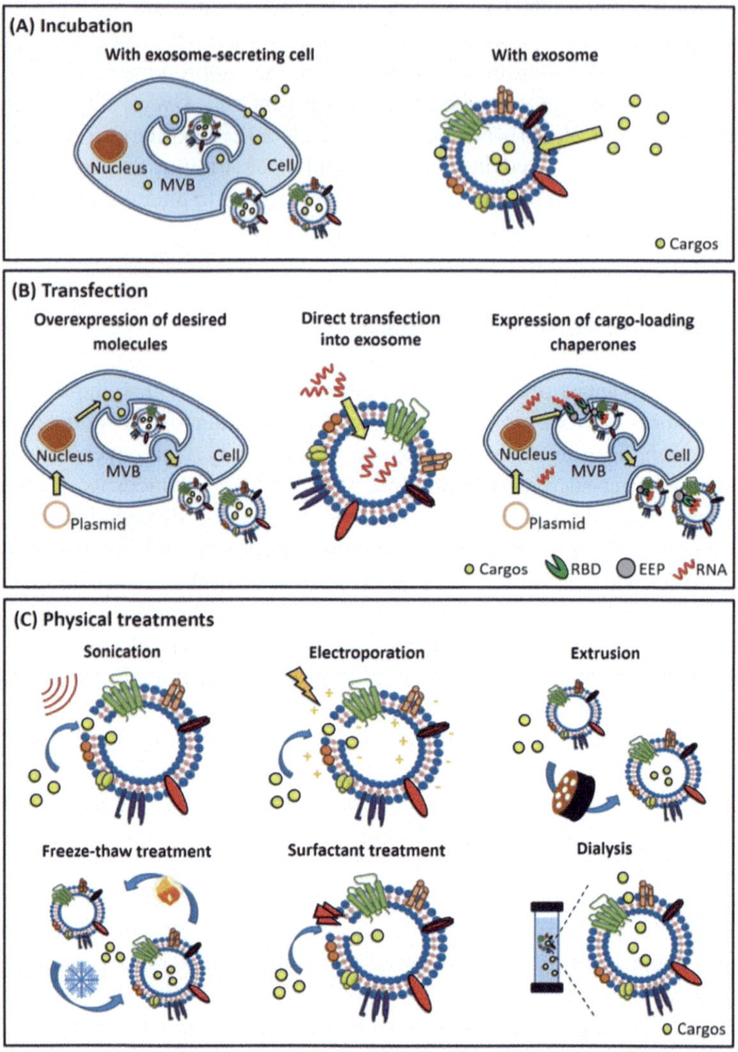

Fig. 12.3 Common methodologies for loading cargos into exosomes. (**a**) Exosome-secreting cells or exosomes are incubated with desired cargo. Cargos diffuse across the cell and exosomal membrane, therefore packaged within exosomes. (**b**) Desired nucleic acids can be loaded into exosomes via a transfection-based strategy. Transfected with vectors, the donor cell generates RNAs/proteins and packages these products into exosomes using endogenous expression and sorting machinery of the donor cell, respectively. Exosomes can be directly transfected with small RNAs for cargo loading purposes. (**c**) Cargos can be loaded into exosomes directly through physical treatments. Electroporation, sonication, and surfactant treatment generate pores on the exosomal membrane that facilitate cargo loading. Freeze-thaw treatment, extrusion, and dialysis enhance cargo loading into exosomes during membrane recombination processes (Adapted from Fu et al. 2020 under the CC BY-NC-ND license, https://doi.org/10.1016/j.impact.2020.100261)

the exosomal surface with neuropilin-1 (NRP-1)-targeted RGE peptide (RGERPPR) to specifically target glioma cells (Jia et al. 2018).

12.4.2 Exosome Cargo Modification

Exosomes can be modified before or after isolation to carry the desired payload. Physical modification methods are now applied to load exosomes with particular cargo for brain targeting (Alvarez-Erviti et al. 2011; Haney et al. 2015). Incubation is the most basic and simplest procedure for packing materials into exosomes, just requiring incubation of exosomes with DNA/RNA/protein. This procedure was utilized to deliver catalase (Haney et al. 2015), curcumin (cur) (Tian et al. 2018), or quercetin (Qi et al. 2020). Another research group modified exosomal payload for cellular distribution by loading it with proteins and ribonucleoproteins (RNA) (Zhang et al. 2020). Similarly, the incubation procedure was exploited to load miR-210 inside exosomes for its targeted delivery to the brain ischemia nodules (Zhang et al. 2019).

Another approach to load bioactives inside exosomes is to incubate exosome-secreting cells with therapeutic agents. To cure AD, a cur-laden exosome was produced by Wang et al. (2019) by incubating macrophages with cur that can bypass the BBB. Yim et al. (2016) utilized an opto-genetically engineered exosome system, EXPLORs (exosomes for protein loading via optically reversible protein–protein interactions), to load proteins inside exosomes through precise, reversible protein–protein interactions (PPIs) that greatly boosted cargo levels in the recipient cells.

Transfection is commonly employed for the stable packing of DNA, RNA, peptides, and polypeptides into exosomes utilizing transfection agents (Fu et al. 2020). Vectors containing an RVG peptide–expressing plasmid as well as mRNA for catalase (Kojima et al. 2018) or MOR siRNA (Liu et al. 2015) were successfully transfected into HEK293 cells. Similarly, exosomes produced from bone marrow (BM)-MSCs genetically manipulated with pre-miR-214 possessing lentivirus upregulated levels of miR-214 in the isolated exosomes (Shi et al. 2020). An alternative technique, sonication, involves applying sound waves to loosen up the exosomal membrane that facilitates cargo loading inside exosomes (Haney et al. 2019; Li et al. 2020). Incidentally, catalase (Haney et al. 2015) and gold nanoparticles have been loaded into exosomes utilizing this approach (Sancho-Albero et al. 2019).

Electroporation is yet another procedure utilized for packaging RNA molecules into exosomes. This technique requires an electrical field to generate small pores in the exosomal membrane that enhances the permeability of the payload. siRNA (Alvarez-Erviti et al. 2011), miRNA (Ohno et al. 2013; Liang et al. 2018), and shRNA (Izco et al. 2019) have been loaded into exosomes employing this technique. Additionally, electroporation was employed to generate superparamagnetic iron oxide nanoparticles (SPIONs) and cur-laden exosomes (RGE-Exo-SPION/cur) for better therapeutic action against glioma (Jia et al. 2018).

Extrusion entails the simultaneous passage of exosomes along with their payload inside an extruder to stimulate the blending of membranes to generate cargo-loaden exosomes after several expulsions (Narayanan 2020). This tactic was utilized by Haney et al. (2015) to produce catalase-loaded exosomes. Further, extrusion can also be applied to engender exosome-mimetic nanovesicles (EMNVs) (Jang et al. 2013). Fast freeze-thaw treatment, a well-known technique used in liposome synthesis (Oku and MacDonald 1983; Pick 1981), was applied to load catalase solution (Haney et al. 2015) inside reconstituted exosomes by adding exosomes to the catalase solution and treating the mixture with repetitive freeze-thaw cycles. Another strategy utilizes surfactants (e.g., saponins, tritons) that permeabilize the membrane by creating pores (Podolak et al. 2010) to facilitate the entry of catalase inside exosomes (Haney et al. 2015). Recently, cargo loading inside exosomes has been achieved by dialysis methods (Fuhrmann et al. 2015a, b; Wei et al. 2019). To achieve this, exosomes and the cargo mixture is placed inside the dialysis membrane/tube, which is then dialyzed by agitating to attain a drug-laden exosome. This hypotonic dialysis (HP) led to a several-fold increment in the cargo encapsulation capacity (Fuhrmann et al. 2015a, b).

A unique vesicle trafficking mechanism comprising late-domain (L-domain) proteins, ubiquitin, and the ESCRT apparatus for the precise transfer of proteins was utilized by Sterzenbach et al. (2017). They targeted a plethora of brain cells including the olfactory bulb, cortex, striatum, hippocampus, and cerebellum. They utilized an adaptor protein Ndfip1 (Nedd4 family–interacting protein 1) that helps in exporting target proteins inside exosomes outside of the cell through interaction with the Nedd4 family of ubiquitin ligases (Putz et al. 2008, 2012). Binding of Ndfip1 with the WW domain of the Nedd4 protein occurs via three L-domain motifs (PPxY) (Harvey et al. 2002). Proteins intended to be transported outside were given the WW tag, which caused them to bind with Ndfip1 that in turn resulted in ubiquitination before being released in exosomes (Sterzenbach et al. 2017).

Hybrid nanocarriers have recently been fashioned that combines the intrinsic attributes of EVs with the drug loading potential of artificial delivery systems. As such, Cheng et al. (2018) created a bioinspired metal–organic nanostructure for targeted therapeutic delivery. The nano-assembles exhibited high drug loading capacity, protected its cargo from extracellular enzymes, evaded the phagocytosis by immune cells, and ensured selective targeting of homotypic tumor sites, autonomously releasing the payload inside the cell. Further, lipid-based nanoparticles (NPs) can be exploited to enhance the drug-loading power of exosomes. Lin et al. (2018) hybridized exosomes with liposomes to generate NPs capable of encapsulating bulky plasmids, for example, CRISPR/Cas9 expression vectors.

12.5 Engineered Exosomes as Therapeutics against Neurological Disorders

Brain parenchyma cells and the blood vessels are connected through an interface called the BBB that shields the brain from potential threats (Daneman and Prat 2015). This layer also acts as a barrier for therapeutic agents restricting their entry into the brain with the help of the non-fenestrated stratum of brain parenchyma cells (Almutairi et al. 2016). A reliable and effective transport mechanism is essential for the management of CNS ailments since difficulties with delivery to the brain constitute a barrier to many effective therapeutic medicines. Exosomes have strong biocompatibility, which allows them to easily traverse the BBB. Further, tailoring of exosomes presents a novel targeted approach for handling brain disorders.

12.5.1 Alzheimer's Disease

Amyloid beta (Aβ) plaques and hyperphosphorylated tau tangles are hallmarks of Alzheimer's disease (AD). Aβ is generated when amyloid precursor protein (APP) is cleaved by either β-site APP-cleaving enzyme 1 (BACE1) or γ-secretase. Persistent loss of cognitive functions disrupts neuronal connections and function. Furthermore, there is no proven cure for this complicated neurological disorder.

Exosomes are being researched as potential carriers for particular medications or therapeutic compounds in the treatment of AD. Cur-primed engineered exosomes (Exo-cur) generated by Wang et al. (2019) retained lymphocyte function–associated antigen 1 (LFA-1) from the parent cell, a protein with strong association with endothelial intercellular adhesion molecule 1 (ICAM-1), that helped them to bypass the BBB. Exo-cur, found to be widely distributed in microglial cells in the hippocampus, activated AKT (a serine/threonine-specific protein kinase) and lowered the phosphorylated tau via the AKT/GSK-3β pathway in okadaic acid (OA)-induced AD mice (Wang et al. 2019). In a similar approach, quercetin (Que) was packed inside plasma exosomes (Exo-Que) to effectively improve cerebral dysfunction in AD mice (Qi et al. 2020). The data of the study established enhanced bioavailability as well as brain targeting of Que. Further, Exo-Que greatly improved the cognitive function by restraining cyclin-dependent kinase 5 (CDK5)-mediated-phosphorylation of tau, preventing the deposition of insoluble twisted fibers (Qi et al. 2020), advocating its potential for treating AD patients (Fig. 12.4).

Exosomes are now recognized to transport miRNA that may well function at various stages in AD. In this regard, Jahangard et al. (2020) transfected mir-29 into rat bone marrow–derived MSCs (r-BMSCs) to obtain miRNA-29-loaded exosomes. A considerable upregulation of miR-29 and underexpression of target proteins, BACE1 and BIM (Bcl-2-interacting mediator of cell death) were established in the transfected cells. They proved that miRNA-29-laden exosomes, when injected in an animal model of AD, protects against Aβ amyloid etiology (Jahangard et al. 2020). Zhai et al. (2021) conducted a similar investigation. They employed adipocyte-derived MSCs (ADMSCs) encapsulated with miRNA-22 (Exo-miRNA-22) to treat

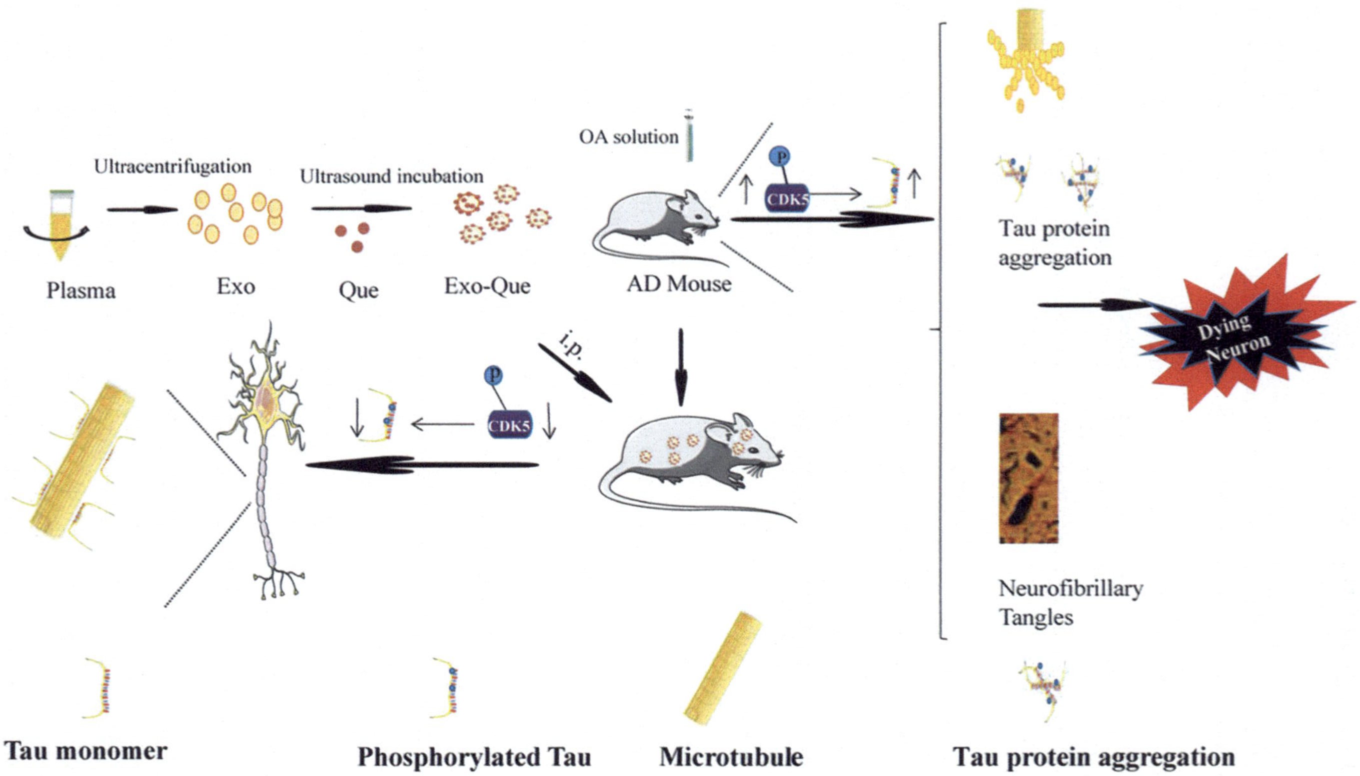

Fig. 12.4 Exo-Que enhanced the bioavailability and achieved brain targeting in vivo. Exo-Que improved cognitive and functional symptoms in OA-induced AD mice by inhibiting CDK5-mediated phosphorylation of tau and reducing NFTs (Adapted from Qi et al. 2020 under the terms of the Creative Commons Attribution License, https://doi.org/10.1080/10717544.2020.1762262)

and restore nerve physiology in an AD mouse model. Nerve cell survival was highly elevated by Exo-miRNA-22, while levels of inflammatory factor lowered drastically, demonstrating that Exo-miRNA-22 may considerably diminish neuro-inflammation (Zhai et al. 2021).

One strategy to enhance exosome therapy is to functionalize their surface. Researchers coupled neuro-specific RVG peptide (YTIWMPENPRPGTPCDIFTNSRGKRASNG) that selectively binds to the acetylcholine receptor (AchR) (Kumar et al. 2007) with the extra-exosomal N-terminus of murine Lamp2b (Alvarez-Erviti et al. 2011). The RVG-exosomes were encapsulated with siRNAs by electroporation and transfected into neuronal N2a cells in order to target BACE1 protease. The siRNA-loaded RVG-targeted exosomes achieved significant delivery efficiency and effective dose-dependent gene knockdown with negligible toxic and immunogenic reactions, localizing mainly to neuronal, microglial, and oligodendrocyte cells (Alvarez-Erviti et al. 2011). Likewise, the RVG peptide conjugated to MSC-Exo through a DOPE-NHS linker (Kajimoto et al. 2013) was used to specifically target intravenously injected exosomes to the cortical and hippocampal region of transgenic APP/PS1 mice. Consequently, a substantial increase in learning and memory capacities, as well as lower plaque formation and Aβ levels, was observed. The targeted exosomes decreased pro-inflammatory components and increased anti-inflammatory elements (Cui et al. 2019).

AD is a complicated illness caused by the interplay of several factors. P-glycoprotein (P-gp), an ABC efflux transport protein, is involved in binding and pumping Aβ into the blood. $A\beta_{42}$ has been linked with the downregulation of P-gp on the BBB leading to the deposition of Aβ in AD (Brenn et al. 2011; Shubbar and Penny 2018). Pan et al. (2020) exploited human brain microvascular endothelial cell (HBMVEC)-derived exosomes (HBMVEC-Ex) containing P-gp to transport Aβ outside neuronal cells, contributing to an efficient improvement in cognitive behavior in Aβ-induced AD animals.

Reports have established the action of MSC-derived exosomes in restoring memory impairments in animal models of AD. Chen et al. (2021) intended to see how MSC-derived exosomes performed in AD models. MSC exosomes lowered Aβ deposition while upregulating neural memory and synaptic plasticity-related genes. Wang and Yang (2021) discovered that BM-MSC-derived exosomes decrease Aβ expression while rescuing brain functions in vivo through the sphingosine kinase/sphingosine-1-phosphate (SphK/S1P) signaling pathway. Another study found that intra-cerebroventricular injection of BM-MSC-derived exosome dwindled levels of NF-κB in astrocytes and reduced cerebral impairment by restoring astrocyte function. They discovered that increased expression of miR-146 and downregulation of NF-κB in the hippocampus prevent cognitive damage (Nakano et al. 2020).

12.5.2 Parkinson's Disease

PD is a neuro-degradative disease illustrated as dopaminergic neuronal loss complemented by the deposition of Lewy bodies (LBs) containing alpha-synuclein (α-syn) fibrils. Decrease in the brain's dopamine levels leads to impaired motor response in PD. Thus, DiD-labeled blood-derived exosomes incubated with dopamine were targeted to the brain through the transferrin–transferrin receptor (TfR) contact (Qu et al. 2018). TfR, overexpressed on the exterior of blood-derived exosomes and brain capillary parenchyma cells (Pan and Johnstone 1983; Qi et al. 2016), is responsible for the transport of exosomes past the BBB through TfR-mediated endocytosis (Qian et al. 2002). DiD-labeled EVs effectively transported dopamine into the brain, striatum, and substantia nigra, while imparting negligible toxic reactions in the neuroblastoma cells, SH-SY5Y (Qu et al. 2018).

Modified exosomes have been utilized as carriers to treat PD. Exosomes bioengineered with RVG peptide were exploited to relay DNA aptamer targeting α-syn fibrils. In brain cells, the RVG-exosome-loaded aptamers drastically lowered the preformed fibril (PFF)-induced aggregation of α-syn and restored dysfunctional motor loss and neuronal death (Ren et al. 2019). In a similar study, RVG-modified exosomes were exploited to specifically deliver short hairpin RNA-minicircle (shRNA-MC) constructs to the brain. This novel approach resulted in the long-term decrease in α-syn expression in an animal model of PD, while improving the clinical symptoms (Izco et al. 2019). A precisely targeted study involved fusing RVG with the N-terminus of Lamp2b protein of exosomes. These modified EVs laden with α-syn siRNA lowered the levels of α-syn protein and its fibrils in intraneuronal space (Cooper et al. 2014). Similarly, ex vivo loading of catalase inside exosomes (ExoCAT) using different methods offered noteworthy neuroprotective properties in PD (Haney et al. 2015).

Liu et al. (2020) engineered a core–shell hybrid system termed RVG peptide–modified exosome curcumin/BA-poly(2-(dimethylamino)ethyl acrylate) siRNA targeting SNCA (REXO-C/ANP/S) as a "nano-scavenger" for clearing α-syn aggregates and reducing their cytotoxicity in PD neurons. A polymer, BA-poly (2-(dimethylamino)ethyl acrylate) (BAP), was mixed with cur and α-syn siRNA (siSNCA) to obtain C/ANP and C/ANP/S, respectively. Ultrasonication was then utilized to encapsulate C/BNP/S and C/BNP/S inside engineered stearoyl-RVG-derived exosomes from the immature dendritic cell (imDC) (REXO) to obtain RVG-modified exosome (REXO-C/ANP/S). The motor performance of animals improved markedly post-therapy. Release of the gene drug siSNCA inside neurons resulted in the inhibition of α-syn fibrillation by diminishing the production of α-syn protein, while cur promptly inhibited α-syn fibrils (Liu et al. 2020). In particular, REXO-C/ANP/S also acted as a "nano-scavenger" by providing immunotherapy because of its imDC innate EXO exterior (Fig. 12.5).

Kojima et al. (2018) developed EXOsomal transfer into cell (EXOtic) devices enabling the efficient customizable production of designer exosomes. Exosomes containing the RNA packaging device, the archaeal ribosomal protein L7Ae which binds to the C/D_{box} RNA conjugated to the C-terminus of CD63 (CD63-L7Ae),

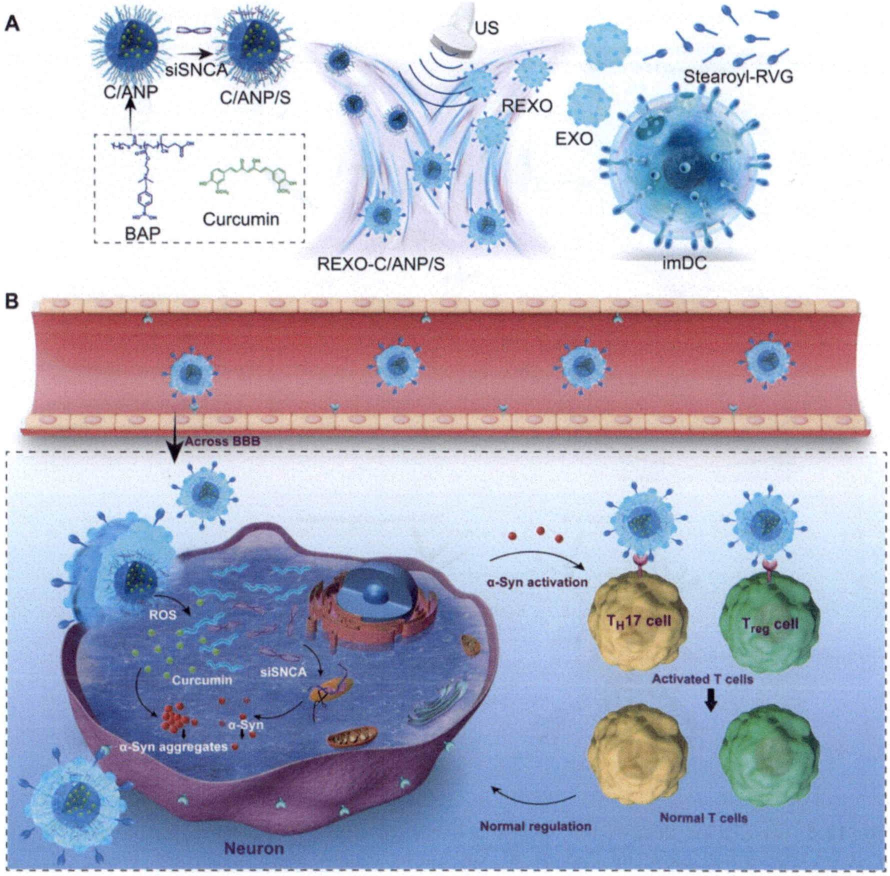

Fig. 12.5 (**a**) The amphiphilic polymer BAP self-assembled and encapsulated the hydrophobic drug curcumin to form curcumin/BAP NP (C/ANP). The final C/ANP/siSNCA (C/ANP/S) nano-complex was formed via electrostatic interaction. Ultrasonication was then utilized to encapsulate C/BNP/S inside engineered stearoyl-RVG-derived exosomes from the imDC to obtain RVG-modified exosome REXO (REXO-C/ANP/S). (**b**) Efficient delivery of siRNA and chemical drugs by the target exosomes reduced the α-syn aggregates in diseased dopaminergic neurons by dual mechanism. Release of gene-drug siSNCA inside neurons resulted in the inhibition of α-syn aggregation by reducing the synthesis of α-syn, while curcumin directly inhibited α-syn aggregates. Further, cellular reactive oxygen species (ROS) content also decreased drastically. This gene-chem nano-complex can further provide a functionalized vector for immunotherapy of neurodegenerative diseases by regulating $T_H{}^{17}$ and T_{reg} cell balance via inhibition of $T_H{}^{17}$ differentiation and promotion of T_{reg} production, thus promoting immune tolerance (Modified from Liu et al. 2020 under Creative Commons Attribution NonCommercial License 4.0, CC BY-NC, https://doi.org/10.1126/sciadv.aba3967)

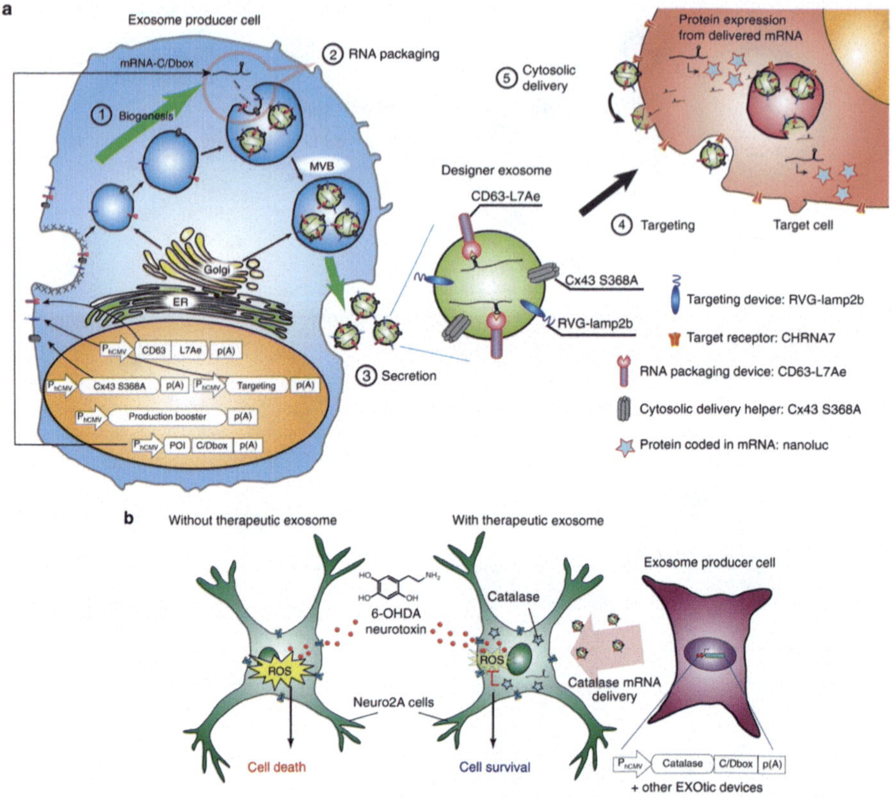

Fig. 12.6 (**a**) EXOtic devices for mRNA delivery. (**b**) Protection against neurotoxicity in an in vitro experimental model of Parkinson's disease by catalase mRNA delivery. Please refer to the text for a description. (Modified with permission from Kojima et al. 2018. under a Creative Commons Attribution 4.0, https://doi.org/10.1038/s41467-018-03733-8)

targeting module (RVG-Lamp2b to target nicotinic acetylcholine receptor, CHRNA7) cytosolic delivery helper (gap junction protein, connexin 43, Cx43 S368A) and mRNA (e.g., nluc-C/D_{box} or catalase) were efficiently produced from exosome producer cells by the exosome production booster. The engineered exosomes were delivered to target cells (HEK-293 T cells expressing CHRNA7) and the catalase mRNA was delivered into the target cell's cytosol with the help of the cytosolic delivery helper, Cx43 S368A. Finally, protein encoded in the mRNA was expressed in the target cells (Kojima et al. 2018). These EXOtics delivered catalase mRNA into the brain and attenuated neurotoxicity and neuroinflammation in the in vitro and in vivo models of PD (Fig. 12.6).

Normal blood exosomes can be utilized for the simplistic management of PD. Upregulation of miR-188-3p (a key player involved in PD pathogenesis) in exosomes exhibited beneficial outcomes on PD by subduing autophagic and inflammatory responses by targeting NACHT, LRR, and PYD domain–containing protein 3 (NALP3) inflammasome and cyclin-dependent kinase 5 (CDK5) in vitro and in vivo (Li et al. 2021). Alternate research exploited RAW 264.7 macrophage genetically modified to express glial cell line–derived neurotrophic factor (GDNF), a survival protein for the dopaminergic neuron cells that deteriorate in Parkinson's (Ramaswamy et al. 2009), to ameliorate neurodegeneration and neuroinflammation in PD mice (Zhao et al. 2014).

12.5.3 Glioma

Gliomas are extremely aggressive and poorly treated malignant brain cancer as conventional therapies failed to deliver drugs to the tumor sites across BBB. Exosomes have emerged as the vehicle of choice to transport genes/drugs for anticancer therapy in the nervous system. Jia et al. (2018) added superparamagnetic iron oxide nanoparticles (SPIONs) and Cur to NRP-1-targeting RGE peptide-modified exosomes (RGE-Exo-SPION/Cur) by electroporation and click chemical methods. The transmembrane glycoprotein, NRP-1, is either underexpressed or not produced at all in healthy brain cells, but is upregulated in glioma cells and the tumor vascular endothelium. The cancer-targeting peptide, RGE, possesses a cryptic C-end rule (CendR) site responsible for NRP-1 attachment (Yang et al. 2013). Cur exhibits a potent antitumor impact against glioma both in vitro and in vivo (Gersey et al. 2017). The engineered RGE-Exo-SPION/Cur exosomes with imaging and therapeutic functions can successfully pass through the BBB making it easier to recognize gliomas accurately. Due to its enhanced targeting capability, RGE-Exo-SPION/Cu in conjunction with magnetic fluid hyperthermia (MFH) and Cur slowed tumor resurgence and prolonged life in model animals (Jia et al. 2018).

The TfR, a potential target for glioblastoma (GBM) (Kang et al. 2015), is upregulated in brain tumors (Recht et al. 1990) and is abundantly expressed in brain capillary endothelial cells as well. TfR aids exosomes in traversing the BBB effectively (Jefferies et al. 1984). T7 (HAIYPRH), a TfR-binding peptide (Lee et al. 2001; Han et al. 2010), has been exploited for GBM-targeted delivery. Kim et al. (2020) designed T7-peptide–Lamp2-decorated exosomes (T7-exo) loaded with antisense oligonucleotide against miR-21 (AMO-21) by electroporation. T7-exo exhibited superior cellular uptake as well as better accumulation in brain tissue, directing higher downregulation of miR-21. T7-exo also demonstrated better BBB traversing ability and tumor targeting via TfR binding, thus effectively reducing tumor size by 50% in GBM models (Fig. 12.7).

Erkan et al. (2017) developed bioengineered EVs encapsulating suicide gene mRNA and protein—cytosine deaminase (CD) fused to uracil phosphoribosyltransferase (UPRT)—as a mechanism to disrupt DNA replication and induce GBM killing by catalyzing the conversion of 5-fluorocytosine (5-FC) to 5-fluorouracil (5-FU). Another study by Munoz et al. (2013) utilized MSC-derived exosomes as a delivery vehicle for anti-miR-9 for successful inhibition of drug efflux transporter P-gp to induce susceptibility to anticancer agent, temozolomide.

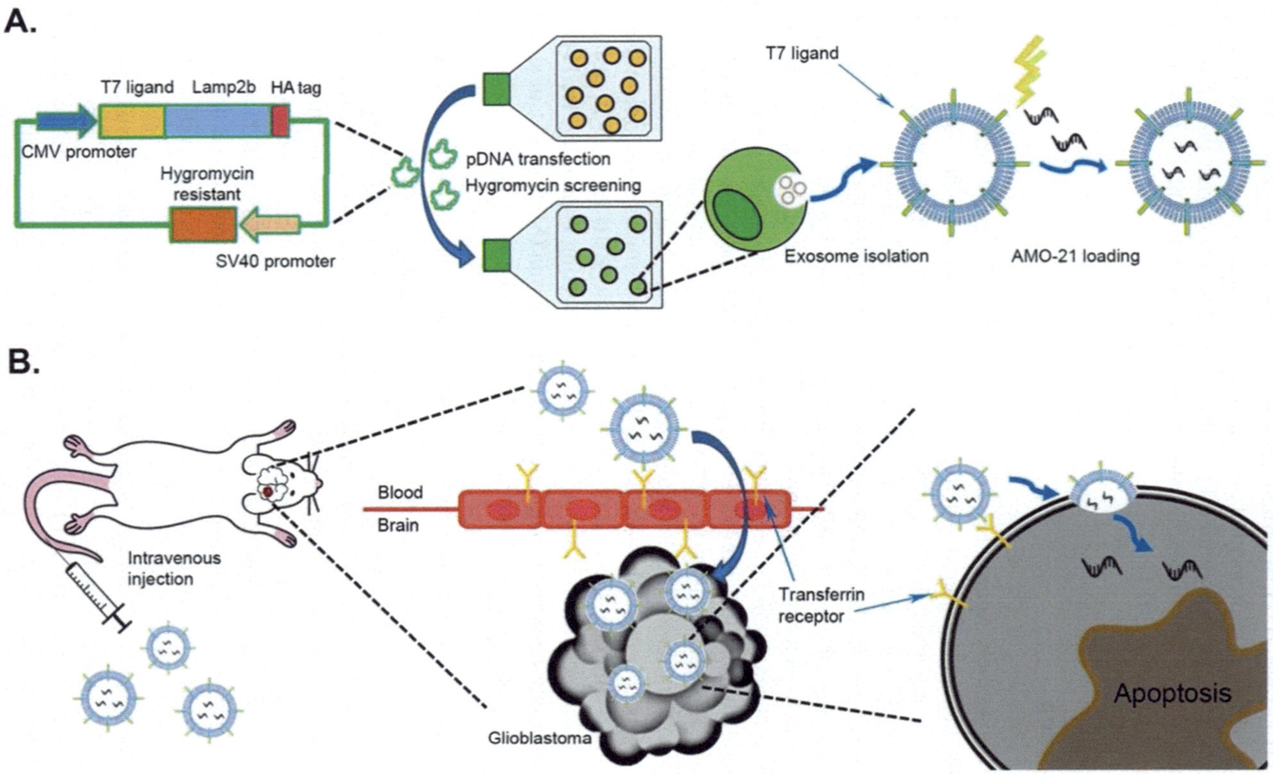

Fig. 12.7 Systemic delivery of microRNA-21 antisense oligonucleotides to the brain using T7-peptide-decorated exosomes (**a**) Isolation of T7-exosomes from a 293 T stable cell line and preparation of AMO-21-loaded T7-exosomes. (**b**) Glioblastoma therapeutic mechanism by systemic injection of T7-exo/AMO-21 (Reprinted with permission from Kim et al. 2020, J Control Release, 317, 273–281. Copyright © Elsevier 2020. https://doi.org/10.1016/j.jconrel.2019.11.009)

12.5.4 Ischemic Brain Stroke

Ischemia is caused by the rapid loss of blood circulation to an area of the brain and is the second leading cause of all global deaths (Mira and Andreas 2018). Exosomes have been exploited as a delivery vehicle for managing stroke. Tian et al. (2018) utilized a bio-orthogonal copper-free azide-alkyne cyclo-addition technique for conjugating functional ligands onto exosomal surfaces. The intravenous administration of c(RGDyK)-coupled MSC-derived exosomes burdened with cur (cRGD-Exo-cur) effectively targeted the ischemic region of the cerebrum, depositing the cargo inside neurons, microglial, and astrocyte cells. cRGD-Exo repressed the inflammatory immune response as well as apoptotic cell death, thus validating its potential as a targeted delivery carrier (Tian et al. 2018). The successful delivery/tropism of cRGD-Exo could be attributed to the overexpression of integrin $\alpha_v\beta_3$ on reactive endothelial cells post-ischemic stroke (Abumiya et al. 1999; Guell and Bix 2014), while its expression is minimal in the nonischemic area (Li et al. 2012). In a separate report, intravenous injection of RVG-exosome-mediated miR-124 successfully delivered the nucleic acid to the stroke site and improved brain damage by stimulating neural progenitor cells to attain normal phenotype at the lesion position (Yang et al. 2017b). Alternatively, blood-derived exosomes (EXO) packaged with edaravone (EXO + EDV), a neuroprotective medication, reversed cerebral ischemia damage in neural tissue (Guo et al. 2021).

12.5.5 Neuroinflammation

Yuan et al. (2017) exploited the association between LFA-1 and ICAM-1 to successfully deliver brain-derived neurotrophic factor (BDNF) inside naïve macrophage (Mφ)-derived exosomes across the BBB into the brain parenchyma. This uptake was heightened in the inflamed regions, when ICAM-1 was overexpressed on capillary parenchyma cells. Similarly, mouse lymphoma cell line–derived exosomes encapsulated with cur (Exo-cur) or a signal transducer and activator of transcription 3 (Stat3) inhibitor, JSI124 (Exo-JSI124), delivered their merchandise to microglial cellss. The intranasal delivery of exosomal cargo protected the brain against LPS-stimulated inflammatory response and induced microglial apoptosis (Zhuang et al. 2011).

12.5.6 Drug Addiction

Drug addiction, a degenerative neuropsychological disorder, is described by harmful effects on the brain due to the tenacious use of drugs. The primary target of the diagnostic and therapeutic use of opioid analgesics/drugs is the opioid receptor mu (MOR). To address morphine obsession, Liu et al. (2015) developed MOR siRNA-loaded, RVG-modified exosomes that effectively transported MOR siRNA to neurons, reducing MOR mRNA and protein levels, and proficiently hampering morphine recurrence.

12.6 Conclusion and Future Perspectives

Targeted delivery to the brain still remains the most exhilarating challenge in the arena of medicinal chemistry. Exosomes, nano-sized vesicles released by a plethora of cells, play a key role in intracellular communications. They have shown great promise as vectors to transport therapeutics into and within the CNS owing to their strong biocompatibility. Nano-size scale, ability to traverse the BBB, biocompatibility, and low endogenous toxicity impart them superiority over traditional carriers. The exosomes are now artificially bioengineered to incorporate unique cargos with precision targeting.

Nevertheless, there are numerous issues that must be addressed. Understanding how exosomes are involved in intercellular communication inside the brain and how cargoes are delivered across the BBB are the foremost questions that warrant prompt resolution. Besides, large-scale production, cargo loading techniques, cell type selection, stability, and bioengineering strategies are the crucial aspects that impact the translational significance of exosomes as targeted delivery agents. Additional systematic in vivo experiments on the efficacy and pharmacokinetics of exosomes are vital to bring this breathtaking advancement near to palliative treatment. A multidisciplinary approach involving collaborative efforts from relevant disciplines is indispensable to discover solutions to the problems listed.

Acknowledgements/Financial Assistance This work was supported by the Deanship of Scientific Research, Vice Presidency for Graduate Studies and Scientific Research, King Faisal University, Saudi Arabia (grant no. 1342), through its KFU Research Summer Initiative.

Author Contribution GM drafted the article and improved the accuracy of the language. QZ carried out the literature search. GM and QZ wrote the first draft of the manuscript and modified the overall structure of the chapter. AA revised the article and incorporated additional material. All authors have read and approved the final manuscript.

References

Abumiya T, Lucero J, Heo JH, Tagaya M, Koziol JA, Copeland BR et al (1999) Activated microvessels express vascular endothelial growth factor and integrin alpha(v)beta3 during focal cerebral ischemia. J Cereb Blood Flow Metab 19:1038–1050

Admyre C, Johansson SM, Qazi KR, Filén J-J, Lahesmaa R, Norman M et al (2007) Exosomes with immune modulatory features are present in human breast milk. J Immunol 179:1969–1978

Adriano B, Cotto NM, Chauhan N, Jaggi M, Chauhan SC, Yallapu MM (2021) Milk exosomes: Nature's abundant nanoplatform for theranostic applications. Bioact Mater 6:2479–2490

Akers JC, Gonda D, Kim R, Carter BS, Chen CC (2013) Biogenesis of extracellular vesicles (EV): exosomes, microvesicles, retrovirus-like vesicles, and apoptotic bodies. J Neuro-Oncol 113(1): 1–11. https://doi.org/10.1007/s11060-013-1084-8

Alli AA (2021) Mechanisms of extracellular vesicle biogenesis, cargo loading, and release. In: Paul MK (ed) Extracellular vesicles–role in diseases, pathogenesis and therapy [internet]. IntechOpen.; (cited Nov 022022), London. https://doi.org/10.5772/intechopen.100458

Almutairi MM, Gong C, Xu YG et al (2016) Factors controlling permeability of the blood-brain barrier. Cell Mol Life Sci 73(1):57–77

Alvarez-Erviti L, Seow Y, Yin H, Betts C, Lakhal S, Wood MJ (2011) Delivery of siRNA to the mouse brain by systemic injection of targeted exosomes. Nat Biotechnol 29:341–345. https://doi.org/10.1038/nbt.1807

Amerik AY, Nowak J, Swaminathan S, Hochstrasser M (2000) The Doa4 deubiquitinating enzyme is functionally linked to the vacuolar protein-sorting and endocytic pathways. Mol Biol Cell 11: 3365–3380

Anakor E, Le Gall L, Dumonceaux J, Duddy WJ, Duguez S (2021) Exosomes in ageing and motor neurone disease: Biogenesis, uptake mechanisms, modifications in disease and uses in the development of biomarkers and therapeutics. Cells 10:2930. https://doi.org/10.3390/cells10112930

Antimisiaris SG, Mourtas S, Marazioti A (2018) Exosomes and exosome-inspired vesicles for targeted drug delivery. Pharmaceu 10:218

Arosio D, Casagrande C (2016) Advancement in integrin facilitated drug delivery. Adv Drug Deliv Rev 97:111–143

Arrighetti N, Corbo C, Evangelopoulos M, Pasto A, Zuco V, Tasciotti E (2019) Exosome-like nanovectors for drug delivery in cancer. Curr Med Chem 26:6132–6148. https://doi.org/10.2174/0929867325666180831150259

Babst M (2005) A protein's final ESCRT. Traffic 6:2–9

Babst M, Sato TK, Banta LM, Emr SD (1997) Endosomal transport function in yeast requires a novel AAA-type ATPase, Vps4p. EMBO J 16:1820–1831

Babst M, Wendland B, Estepa EJ, Emr SD (1998) The Vps4p AAA ATPase regulates membrane association of a Vps protein complex required for normal endosome function. EMBO J 17: 2982–2993

Bang OY, Kim EH (2019) Mesenchymal stem cell-derived extracellular vesicle therapy for stroke: challenges and progress. Front Neurol 10:211. https://doi.org/10.3389/fneur.2019.00211

Barile L, Vassalli G (2017) Exosomes: therapy delivery tools and biomarkers of diseases. Pharmacol Ther 174:63–78

Batrakova EV, Kim MS (2015) Using exosomes, naturally-equipped nanocarriers, for drug delivery. J Control Release 219:396–405

Bowers K, Stevens TH (2005) Protein transport from the late Golgi to the vacuole in the yeast Saccharomyces cerevisiae. Biochim Biophys Acta 1744:438–454

Branscome H, Paul S, Yin DZ, El-Hage N, Agbottah ET, Zadeh MA et al (2020) Use of stem cell extracellular vesicles as a "holistic" approach to CNS repair. Front Cell Dev Biol 8:455. https://doi.org/10.3389/fcell.2020.00455

Brenn A, Grube M, Peters M, Fischer A, Jedlitschky G, Kroemer HK et al (2011) Beta-amyloid downregulates MDR1-P-glycoprotein (Abcb1) expression at the blood-brain barrier in mice. Int J Alzheimers Dis 2011:690121

Caby M-P, Lankar D, Vincendeau-Scherrer C, Raposo G, Bonnerot C (2005) Exosomal-like vesicles are present in human blood plasma. Int Immunol 17(7):879–887. https://doi.org/10.1093/intimm/dxh267

Chao M, Weissman I, Majeti R (2012) The CD47-SIRPα pathway in cancer immune evasion and potential therapeutic implications. Curr Opin Immunol 24:225–232

Chen CC, Liu L, Ma F, Wong CW, Guo XE, Chacko JV et al (2016) Elucidation of exosome migration across the blood-brain barrier model in vitro. Cell Mol Bioeng 9:509–529. https://doi.org/10.1007/s12195-016-0458-3

Chen L, Charrier A, Zhou Y et al (2014) Epigenetic regulation of connective tissue growth factor by MicroRNA-214 delivery in exosomes from mouse or human hepatic stellate cells. Hepatology 59:1118–1129

Chen Y-A, Lu C-H, Ke C-C, Chiu S-J, Jeng F-S, Chang C-W et al (2021) Mesenchymal stem cell-derived exosomes ameliorate Alzheimer's disease pathology and improve cognitive deficits. Biomed 9(6):594

Cheng G, Li W, Ha L, Han X, Hao S, Wan Y et al (2018) Self-assembly of extracellular vesicle-like metal-organic framework nanoparticles for protection and intracellular delivery of biofunctional proteins. J Am Chem Soc 140:7282–7291

Cheng WC, Liao TT, Lin CC, Yuan LTE, Lan HY, Lin HH et al (2019) RAB27B-activated secretion of stem-like tumor exosomes delivers the biomarker microRNA-146a-5p, which promotes tumorigenesis and associates with an immunosuppressive tumor microenvironment in colorectal cancer. Int J Cancer 145:2209–2224. https://doi.org/10.1002/ijc.32338

Chivet M, Javalet C, Laulagnier K, Blot B, Hemming FJ, Sadoul R (2014) Exosomes secreted by cortical neurons upon glutamatergic synapse activation specifically interact with neurons. J Extracell Vesicles 3:24722–24722. https://doi.org/10.3402/jev.v3.24722

Choezom D, Gross JC (2022) Neutral sphingomyelinase 2 controls exosome secretion by counteracting V-ATPase-mediated endosome acidification. J Cell Sci 135(5):jcs259324. https://doi.org/10.1242/jcs.259324

Conlan RS, Pisano S, Oliveira MI, Ferrari M, Mendes Pinto I (2017) Exosomes as reconfigurable therapeutic systems. Trends Mol Med 23:636–650

Cooper JM, Wiklander PB, Nordin JZ et al (2014) Systemic exosomal siRNA delivery reduced alpha-synuclein aggregates in brains of transgenic mice. Mov Disord 29(12):1476–1485

Cui G-h, Guo H-d, Li H, Zhai Y, Gong Z-b, Wu J et al (2019) RVG-modified exosomes derived from mesenchymal stem cells rescue memory deficits by regulating inflammatory responses in a mouse model of Alzheimer's disease. Immun Ageing 16(1):10

Daneman R, Prat A (2015) The blood-brain barrier. Cold Spring Harb Perspect Biol 7(1):a020412

de Abreu RC, Fernandes H, da Costa Martins PA, Sahoo S, Emanueli C, Ferreira L (2020) Native and bioengineered extracellular vesicles for cardiovascular therapeutics. Nat Rev Cardiol 17: 685–697. https://doi.org/10.1038/s41569-020-0389-5

Donoso-Quezada J, Ayala-Mar S, González-Valdez J (2021) The role of lipids in exosome biology and intercellular communication: function, analytics and applications. Traffic 22(7):204–220. https://doi.org/10.1111/tra.12803

Erkan EP, Senfter D, Madlener S, Jungwirth G, Ströbel T, Saydam N, Saydam O (2017) Extracellular vesicle-mediated suicide mRNA/protein delivery inhibits glioblastoma tumor growth in vivo. Cancer Gene Ther 24:38–44

Fitzner D, Schnaars M, van Rossum D, Krishnamoorthy G, Dibaj P, Bakhti M et al (2011) Selective transfer of exosomes from oligodendrocytes to microglia by macropinocytosis. J Cell Sci 124: 447–458. https://doi.org/10.1242/jcs.074088

Fu S, Wang Y, Xia X, Zheng JC (2020) Exosome engineering: current progress in cargo loading and targeted delivery. NanoImpact 20:100261. https://doi.org/10.1016/j.impact.2020.100261

Fuhrmann G, Herrmann IK, Stevens MM (2015a) Cell-derived vesicles for drug therapy and diagnostics: opportunities and challenges. Nano Today 10(3):397–409

Fuhrmann G, Serio A, Mazo M, Nair R, Stevens MM (2015b) Active loading into extracellular vesicles significantly improves the cellular uptake and photodynamic effect of porphyrins. J Control Release 205:35–44

Gersey ZC, Rodriguez GA, Barbarite E, Sanchez A, Walters WM, Ohaeto KC et al (2017) Curcumin decreases malignant characteristics of glioblastoma stem cells via induction of reactive oxygen species. BMC Cancer 17(1):99. https://doi.org/10.1186/s12885-017-3058-2

Gilligan KE, Dwyer RM (2017) Engineering exosomes for cancer therapy. Int J Mol Sci 18(6):1122

Grapp M, Wrede A, Schweizer M, Hüwel S, Galla HJ, Snaidero N et al (2013) Choroid plexus transcytosis and exosome shuttling deliver folate into brain parenchyma. Nat Commun 4:2123. https://doi.org/10.1038/ncomms3123

Guell K, Bix GJ (2014) Brain endothelial cell specific integrins and ischemic stroke. Expert Rev Neurother 14:1287–1292

Guescini M, Guidolin D, Vallorani L, Casadei L, Gioacchini AM, Tibollo P et al (2010) C2C12 myoblasts release micro-vesicles containing mtDNA and proteins involved in signal transduction. Exp Cell Res 316:1977–1984. https://doi.org/10.1016/j.yexcr.2010.04.006

Guo L, Pan J, Li F, Zhao L, Shi Y (2021) A novel brain targeted plasma exosomes enhance the neuroprotective efficacy of edaravone in ischemic stroke. IET Nanobiotechnol 15(1):107–116

Gurung S, Perocheau D, Touramanidou L et al (2021) The exosome journey: from biogenesis to uptake and intracellular signalling. Cell Commun Signal 19:47. https://doi.org/10.1186/s12964-021-00730-1

Ha D, Yang N, Nadithe V (2016) Exosomes as therapeutic drug carriers and delivery vehicles across biological membranes: current perspectives and future challenges. Acta Pharm Sin B 6(4):287–296

Han LA, Huang RQ, Liu SH, Huang SX, Jiang C (2010) Peptide-conjugated PAMAM for targeted doxorubicin delivery to transferrin receptor overexpressed tumors. Mol Pharm 7:2156–2165

Haney MJ, Klyachko NL, Zhaoa YL, Gupta R, Plotnikova EG, He ZJ et al (2015) Exosomes as drug delivery vehicles for Parkinson's disease therapy. J Control Release 207:18–30. https://doi.org/10.1016/j.jconrel.2015.03.033

Haney MJ, Zhao Y, Jin YS, Li SM, Bago JR, Klyachko NL et al (2019) Macrophage-derived extracellular vesicles as drug delivery systems for triple negative breast cancer (TNBC) therapy. J Neuroimmune Pharmacol 15(3):487–500

Harvey KF, Shearwin-Whyatt LM, Fotia A, Parton RG, Kumar S (2002) N4WBP5, a potential target for ubiquitination by the Nedd4 family of proteins, is a novel Golgi-associated protein. J Biol Chem 277:9307–9317

He C, Zheng S, Luo Y, Wang B (2018) Exosome theranostics: biology and translational medicine. Theranostics 8:237–255

Hein CD, Liu XM, Wang D (2008) Click chemistry, a powerful tool for pharmaceutical sciences. Pharm Res 25:2216–2230. https://doi.org/10.1007/s11095-008-9616-1

Hessvik NP, Llorente A (2018) Current knowledge on exosome biogenesis and release. Cell Mol Life Sci 75:193–208

Huebner AR, Cheng L, Somparn P, Knepper MA, Fenton RA, Pisitkun T (2016) Deubiquitylation of protein cargo is not an essential step in exosome formation. Mol Cell Proteomics 15(5):1556–1571. https://doi.org/10.1074/mcp.M115.054965

Hung ME, Leonard JN (2015) Stabilization of exosome-targeting peptides via engineered glycosylation. J Biol Chem 290:8166–8172. https://doi.org/10.1074/jbc.M114.621383

Huotari J, Helenius A (2011) Endosome maturation. EMBO J 30:3481–3500. https://doi.org/10.1038/emboj.2011.286

Izco M, Blesa J, Schleef M, Schmeer M, Porcari R, Al-Shawi R et al (2019) Systemic exosomal delivery of shRNA minicircles prevents parkinsonian pathology. Mol Ther 27:2111–2122

Jahangard Y, Monfared H, Moradi A, Zare M, Mirnajafi-Zadeh J, Mowla SJ (2020) Therapeutic effects of transplanted exosomes containing miR-29b to a rat model of Alzheimer's disease. Front Neurosci 14:564

Jan AT, Malik MA, Rahman S, Yeo HR, Lee EJ, Abdullah TS, Choi I (2017) Perspective insights of exosomes in neurodegenerative diseases: a critical appraisal. Front Aging Neurosci 9:317. https://doi.org/10.3389/fnagi.2017.00317

Jang SC, Kim OY, Yoon CM, Choi D-S, Roh T-Y, Park J et al (2013) Bioinspired exosome-mimetic nanovesicles for targeted delivery of chemotherapeutics to malignant tumors. ACS Nano 7:7698–7710

Jefferies WA, Brandon MR, Hunt SV, Williams AF, Gatter KC, Mason DY (1984) Transferrin receptor on endothelium of brain capillaries. Nature 312:162–163

Jelonek K, Widlak P, Pietrowska M (2016) The influence of ionizing radiation on exosome composition, secretion and intercellular communication. Protein Pept Lett 23(7):656–663

Jia G, Han Y, An Y, Ding Y, He C, Wang X et al (2018) NRP-1 targeted and cargo-loaded exosomes facilitate simultaneous imaging and therapy of glioma in vitro and in vivo. Biomaterials 178:302–316. https://doi.org/10.1016/j.biomaterials.2018.06.029

Johnstone RM, Adam M, Hammond J, Orr L, Turbide C (1987) Vesicle formation during reticulocyte maturation. Association of plasma membrane activities with released vesicles (exosomes). J Biol Chem 262:9412–9420

Johnstone RM (2005) Revisiting the road to the discovery of exosomes. Blood Cells Mol Dis 34: 214–219

Kajimoto T, Okada T, Miya S et al (2013) Ongoing activation of sphingosine 1-phosphate receptors mediates maturation of exosomal multivesicular endosomes. Nat Commun 4:2712. https://doi.org/10.1038/ncomms3712

Kalluri R, LeBleu VS (2020) The biology, function, and biomedical applications of exosomes. Science 367:6478. https://doi.org/10.1126/science.aau6977

Kamerkar S, LeBleu V, Sugimoto H et al (2017) Exosomes facilitate therapeutic targeting of oncogenic KRAS in pancreatic cancer. Nature 546:498–503

Kang T, Jiang MY, Jiang D, Feng XY, Yao JH, Song QX et al (2015) Enhancing glioblastoma-specific penetration by functionalization of nanoparticles with an iron-mimic peptide targeting transferrin/transferrin receptor complex. Mol Pharm 12:2947–2961

Katakowski M, Buller B, Zheng X et al (2013) Exosomes from marrow stromal cells expressing miR-146b inhibit glioma growth. Cancer Lett 335:201–204

Katzmann DJ, Babst M, Emr SD (2001) Ubiquitin-dependent sorting into the multivesicular body pathway requires the function of a conserved endosomal protein sorting complex, ESCRT-I. Cell 106:145–155

Keller S, Ridinger J, Rupp A-K, Janssen JW, Altevogt P (2011) Body fluid derived exosomes as a novel template for clinical diagnostics. J Transl Med 9:86

Khan AR, Yang X, Fu M, Zhai G (2018) Recent progress of drug nanoformulations targeting to brain. J Control Release 291:37–64. https://doi.org/10.1016/j.jconrel.2018.10.004

Khongkow M, Yata T, Boonrungsiman S, Ruktanonchai UR, Graham D, Namdee K (2019) Surface modification of gold nanoparticles with neuron-targeted exosome for enhanced blood–brain barrier penetration. Sci Rep 9(1):1–9

Kim G, Kim M, Lee Y, Byun JW, Hwang DW, Lee M (2020) Systemic delivery of microRNA-21 antisense oligonucleotides to the brain using T7-peptide decorated exosomes. J Control Release 317:273–281. https://doi.org/10.1016/j.jconrel.2019.11.009

Kim MS, Haney MJ, Zhao Y, Mahajan V, Deygen I, Klyachko NL et al (2016) Development of exosome-encapsulated paclitaxel to overcome MDR in cancer cells. Nanomed 12:655–664

Kim J, Li S, Zhang S, Wang J (2022) Plant-derived exosome-like nanoparticles and their therapeutic activities. Asian J Pharmaceu Sci 17(1):53–69. https://doi.org/10.1016/j.ajps.2021.05.006

Kodali M, Castro OW, Kim DK, Thomas A, Shuai B, Attaluri S et al (2020) Intranasally administered human MSC derived extracellular vesicles pervasively incorporate into neurons and microglia in both intact and status epilepticus injured forebrain. Int J Mol Sci 21:181. https://doi.org/10.3390/ijms21010181

Kojima R, Bojar D, Rizzi G et al (2018) Designer exosomes produced by implanted cells intracerebrally deliver therapeutic cargo for Parkinson's disease treatment. Nat Commun 9: 1305. https://doi.org/10.1038/s41467-018-03733-8

Kooijmans SAA, Fliervoet LAL, van der Meel R, Fens MHAM, Heijnen HFG, Henegouwen PMPVE et al (2016) PEGylated and targeted extracellular vesicles display enhanced cell specificity and circulation time. J Control Release 224:77–85. https://doi.org/10.1016/j.jconrel.2016.01.009

Kowal J, Tkach M, Théry C (2014) Biogenesis and secretion of exosomes. Curr Opin Cell Biol 29: 116–125. https://doi.org/10.1016/j.ceb.2014.05.004

Kowal J, Arras G, Colombo M, Jouve M, Morath J, Primdal-Bengtson B et al (2016) Proteomic comparison defines novel markers to characterize heterogeneous populations of extracellular vesicle subtypes. Proc Natl Acad Sci U S A 113(8):E968–E977

Kumar P, Wu H, McBride JL, Jung KE, Kim MH, Davidson BL et al (2007) Transvascular delivery of small interfering RNA to the central nervous system. Nature 448(7149):39–43

Lapchak PA, Boitano PD, de Couto G, Marban E (2018) Intravenous xenogeneic human cardiosphere-derived cell extracellular vesicles (exosomes) improves behavioral function in small-clot embolized rabbits. Exp Neurol 307:109–117. https://doi.org/10.1016/j.expneurol.2018.06.007

Lee JH, Engler JA, Collawn JF, Moore BA (2001) Receptor mediated uptake of peptides that bind the human transferrin receptor. Eur J Biochem 268:2004–2012

Li L, Liu F, Welser-Alves JV, McCullough LD, Milner R (2012) Upregulation of fibronectin and the alphav-beta1 and alphavbeta3 integrins on blood vessels within the cerebral ischemic penumbra. Exp Neurol 233:283–291

Li Z, Zhao R, Wu X, Sun Y, Yao M, Li J et al (2005) Identification and characterization of a novel peptide ligand of epidermal growth factor receptor for targeted delivery of therapeutics. FASEB J 19:1978–1985

Li Q, Wang Z, Xing H, Wang Y, Guo Y (2021) Exosomes derived from miR-188−3p-modified adipose-derived mesenchymal stem cells protect Parkinson's disease. Mol Ther-Nucleic Acids 23:1334–1344

Li YJ, Wu JY, Wang JM, Hu XB, Cai JX, Xiang DX (2020) Gemcitabine loaded autologous exosomes for effective and safe chemotherapy of pancreatic cancer. Acta Biomater 101:519–530

Liang G, Kan S, Zhu Y et al (2018) Engineered exosome-mediated delivery of functionally active miR-26a and its enhanced suppression effect in HepG2 cells. Int J Nanomedicine 13:585–599

Lin Y, Wu J, Gu W, Huang Y, Tong Z, Huang L, Tan J (2018) Exosome-liposome hybrid nanoparticles deliver CRISPR/Cas9 system in MSCs. Adv Sci 5:1700611

Liu L, Li Y, Peng H, Liu R, Ji W, Shi Z, Shen J, Ma G, Zhang X (2020) Targeted exosome coating gene-chem nanocomplex as "nanoscavenger" for clearing α-synuclein and immune activation of Parkinson's disease. Sci Adv 6(50):eaba3967. https://doi.org/10.1126/sciadv.aba3967

Liu W, Bai X, Zhang A, Huang J, Xu S, Zhang J (2019) Role of exosomes in central nervous system diseases. Front Mol Neurosci 12:240. https://doi.org/10.3389/fnmol.2019.00240

Liu Y, Li D, Liu Z et al (2015) Targeted exosome-mediated delivery of opioid receptor mu siRNA for the treatment of morphine relapse. Sci Rep 5:17543. https://doi.org/10.1038/srep17543

Mehryab F, Rabbani S, Shahhosseini S et al (2020) Exosomes as a next-generation drug delivery system: An update on drug loading approaches, characterization, and clinical application challenges. Acta Biomater 113:42–62

Mentkowski K, Snitzer J, Rusnak S et al (2018) Therapeutic potential of engineered extracellular vesicles. AAPS 20:50

Mira K, Andreas L (2018) Global Burden of Stroke. Sem Neuro 38(2):208–211. https://doi.org/10.1055/s-0038-1649503

Muller U (2020) Exosome-mediated protection of auditory hair cells from ototoxic insults. J Clin Invest 130:2206–2208. https://doi.org/10.1172/JCI135710

Munoz JL, Bliss SA, Greco SJ, Ramkissoon SH, Ligon KL, Rameshwar P (2013) Delivery of functional anti-miR-9 by mesenchymal stem cell-derived exosomes to glioblastoma multiforme cells conferred chemosensitivity. Mol Ther–Nucleic Acids 2:e126

Murphy DE, de Jong OG, Brouwer M, Wood MJ, Lavieu G, Schiffelers RM et al (2019) Extracellular vesicle-based therapeutics: natural versus engineered targeting and trafficking. Exp Mol Med 51:1–12. https://doi.org/10.1038/s12276-019-0223-5

Nakano M, Kubota K, Kobayashi E, Chikenji TS, Saito Y, Konari N, Fujimiya M (2020) Bone marrow-derived mesenchymal stem cells improve cognitive impairment in an Alzheimer's disease model by increasing the expression of microRNA-146a in hippocampus. Sci Rep 10(1):10772

Narayanan E (2020) Exosomes as drug delivery vehicles for cancer treatment. Curr Nanosci 16:15–26

Nolan S, Dixon R, Norman K, Hellewell P, Ridger V (2008) Nitric oxide regulates neutrophil migration through microparticle formation. Am J Pathol 172(1):265–273

Ogawa Y, Miura Y, Harazono A, Kanai-Azuma M, Akimoto Y, Kawakami H et al (2011) Proteomic analysis of two types of exosomes in human whole saliva. Biol Pharm Bull 34:13–23

Ohno S, Takanashi M, Sudo K et al (2013) Systemically injected exosomes targeted to EGFR deliver antitumor microRNA to breast cancer cells. Mol Ther 21:185–191

Oku N, MacDonald RC (1983) Differential effects of alkali metal chlorides on formation of giant liposomes by freezing and thawing and dialysis. Biochemist 22:855–863

Ostrowski M, Carmo NB, Krumeich S, Fanget I, Raposo G, Savina A et al (2010) Rab27a and Rab27b control different steps of the exosome secretion pathway. Nat Cell Biol 12:19–30. https://doi.org/10.1038/ncb2000

Pan BT, Johnstone RM (1983) Fate of the transferrin receptor during maturation of sheep reticulocytes in vitro: selective externalization of the receptor. Cell 33(3):967–978. https://doi.org/10.1016/0092-8674(83)90040-5

Pan J, He R, Huo Q, Shi Y, Zhao L (2020) Brain microvascular endothelial cell derived exosomes potently ameliorate cognitive dysfunction by enhancing the clearance of Aβ through up-regulation of P-gp in mouse model of AD. Neurochem Res 45(9):2161–2172. https://doi.org/10.1007/s11064-020-03076-1

Pegtel DM, Gould SJ (2019) Exosomes. Annu Rev Biochem 88:487–514. https://doi.org/10.1146/annurev-biochem-013118-111902

Perets N, Betzer O, Shapira R, Brenstein S, Angel A, Sadan T et al (2019) Golden exosomes selectively target brain pathologies in neurodegenerative and neurodevelopmental disorders. Nano Lett 19:3422–3431. https://doi.org/10.1021/acs.nanolett.8b04148

Pick U (1981) Liposomes with a large trapping capacity prepared by freezing and thawing of sonicated phospholipid mixtures. Arch Biochem Biophys 212:186–194

Piffoux M, Nicolás-Boluda A, Mulens-Arias V, Richard S, Rahmi G, Gazeau F et al (2019) Extracellular vesicles for personalized medicine: the input of physically triggered production, loading and theranostic properties. Adv Drug Deliv Rev 138:247–258

Pisitkun T, Shen R-F, Knepper MA (2004) Identification and proteomic profiling of exosomes in human urine. Proc Natl Acad Sci U S A 101:13368–13373

Podolak I, Galanty A, Sobolewska D (2010) Saponins as cytotoxic agents: a review. Phytochem Rev 9:425–474

Putz U, Howitt J, Doan A, Goh CP, Low LH, Silke J, Tan SS (2012) The tumor suppressor PTEN is exported in exosomes and has phosphatase activity in recipient cells. Sci Signal 5:ra70. https://doi.org/10.1126/scisignal.2003084

Putz U, Howitt J, Lackovic J, Foot N, Kumar S, Silke J, Tan SS (2008) Nedd4 family-interacting protein 1 (Ndfip1) is required for the exosomal secretion of Nedd4 family proteins. J Biol Chem 283:32621–32627

Qi H, Long L, Ren Y, Zhang S, Chang X, Qian X et al (2016) Blood exosomes endowed with magnetic and targeting properties for cancer therapy. ACS Nano 10(3):3323–3333. https://doi.org/10.1021/acsnano.5b06939

Qi Y, Guo L, Jiang Y, Shi Y, Sui H, Zhao L (2020) Brain delivery of quercetin-loaded exosomes improved cognitive function in AD mice by inhibiting phosphorylated tau-mediated neurofibrillary tangles. Drug Deliv 27(1):745–755

Qian ZM, Li H, Sun H, Ho K (2002) Targeted drug delivery via the transferrin receptor-mediated endocytosis pathway. Pharmacol Rev 54(4):561–587. https://doi.org/10.1124/pr.54.4.561

Qing LM, Chen HW, Tang JY, Jia XF (2018) Exosomes and their MicroRNA cargo: new players in peripheral nerve regeneration. Neurorehabil Neural Repair 32:765–776. https://doi.org/10.1177/1545968318798955

Qu MK, Lin Q, Huang LY, Fu Y, Wang LY, He SS et al (2018) Dopamine-loaded blood exosomes targeted to brain for better treatment of Parkinson's disease. J Control Release 287:156–166. https://doi.org/10.1016/j.jconrel.2018.08.035

Quah BJ, O'Neill HC (2005) The immunogenicity of dendritic cell-derived exosomes. Blood Cells Mol Dis 35:94–110

Ramaswamy S, Soderstrom KE, Kordower JH (2009) Trophic factors therapy in Parkinson's disease. Prog Brain Res 175:201–216

Rana S, Yue S, Stadel D et al (2012) Toward tailored exosomes: the exosomal tetraspanin web contributes to target cell selection. Int J Biochem Cell B 44:1574–1584

Raposo G, Stoorvogel W (2013) Extracellular vesicles: exosomes, microvesicles, and friends. J Cell Biol 200:373–383

Recht L, Torres CO, Smith TW, Raso V, Griffin TW (1990) Transferrin receptor in normal and neoplastic brain tissue: implications for brain-tumor immunotherapy. J Neurosurg 72(6): 941–945. https://doi.org/10.3171/jns.1990.72.6.0941

Ren X, Zhao Y, Xue F, Zheng Y, Huang H, Wang W et al (2019) Exosomal DNA aptamer targeting α-synuclein aggregates reduced neuropathological deficits in a mouse Parkinson's disease model. Mol Ther-Nucleic Acids 17:726–740. https://doi.org/10.1016/j.omtn.2019.07.008

Roudi S, Rädler JA, El Andaloussi S (2023) Therapeutic potential of extracellular vesicles in neurodegenerative disorders. Handb Clin Neurol 193:243–266. https://doi.org/10.1016/B978-0-323-85555-6.00017-5

Rufino-Ramos D, Albuquerque PR, Carmona V, Perfeito R, Nobre RJ, Pereira de Almeida L (2017) Extracellular vesicles: novel promising delivery systems for therapy of brain diseases. J Control Release 262:247–258. https://doi.org/10.1016/j.jconrel.2017.07.001

Salunkhe S, Dheeraj BM, Chitkara D, Mittal A (2020) Surface functionalization of exosomes for target-specific delivery and in vivo imaging & tracking: strategies and significance. J Control Release 326:599–614. https://doi.org/10.1016/j.jconrel.2020.07.042

Sancho-Albero M, Rubio-Ruiz B, Pérez-López AM, Sebastián V, Martín-Duque P, Arruebo M et al (2019) Cancer-derived exosomes loaded with ultrathin palladium nanosheets for targeted bioorthogonal catalysis. Nat Catal 2(10):864–872. https://doi.org/10.1038/s41929-019-0333-4

Sansone P, Savini C, Kurelac I, Chang Q, Amato LB, Strillacci A et al (2017) Packaging and transfer of mitochondrial DNA via exosomes regulate escape from dormancy in hormonal therapy-resistant breast cancer. Proc Natl Acad Sci U S A 114:E9066–E9075. https://doi.org/10.1073/pnas.1704862114

Sato YT, Umezaki K, Sawada S, Mukai SA, Sasaki Y, Harada N et al (2016) Engineering hybrid exosomes by membrane fusion with liposomes. Sci Rep 6:21933

Shi J, Jiang X, Gao S, Zhu Y, Liu J, Gu T et al (2020) Gene-modified exosomes protect the brain against prolonged deep hypothermic circulatory arrest. Ann Thorac Surg 111:576–585. https://doi.org/10.1016/j.athoracsur.2020.05.075

Shimoda M, Khokha R (2017) Metalloproteinases in extracellular vesicles. Biochim Biophys Acta Mol Cell Res 1864(11 Pt A):1989–2000

Shlosberg D, Benifla M, Kaufer D, Friedman A (2010) Blood-brain barrier breakdown as a therapeutic target in traumatic brain injury. Nat Rev Neurol 6:393–403

Shubbar MH, Penny JI (2018) Effect of amyloid beta on ATP binding cassette transporter expression and activity in porcine brain microvascular endothelial cells. Biochim Biophys Acta Gen Subj 1862:2314–2322

Skotland T, Sandvig K, Llorente A (2017) Lipids in exosomes: current knowledge and the way forward. Prog Lipid Res 66:30–41. https://doi.org/10.1016/j.plipres.2017.03.001

Smyth T, Petrova K, Persaud I, Redzic JS, Gruner MW, Smith-Jones P et al (2014) Surface functionalization of exosomes using click chemistry. Bioconjug Chem 25:1777–1784. https://doi.org/10.1021/bc500291r

Sterzenbach U, Putz U, Low LH, Silke J, Tan SS, Howitt J (2017) Engineered exosomes as vehicles for biologically active proteins. Mol Ther 25(6):1269–1278

Suk JS, Xu QG, Kim N, Hanes J, Ensign LM (2016) PEGylation as a strategy for improving nanoparticle-based drug and gene delivery. Adv Drug Deliv Rev 99:28–51. https://doi.org/10.1016/j.addr.2015.09.012

Sun D, Zhuang X, Xiang X et al (2010) A novel nanoparticle drug delivery system: the anti-inflammatory activity of curcumin is enhanced when encapsulated in exosomes. Mol Ther 18: 1606–1614

Tanziela T, Shaikh S, Jiang H, Lu Z, Wang X (2020) Efficient encapsulation of biocompatible nanoparticles in exosomes for cancer theranostics. Nano Today 35:100964. https://doi.org/10.1016/j.nantod.2020.100964

Théry C, Ostrowski M, Segura E (2009) Membrane vesicles as conveyors of immune responses. Nat Rev Immunol 9(8):581–593

Théry C, Witwer KW, Aikawa E, Alcaraz MJ, Anderson JD, Andriantsitohaina R et al (2018) Minimal information for studies of extracellular vesicles 2018 (MISEV2018): a position statement of the International Society for Extracellular Vesicles and update of the MISEV2014 guidelines. J Extracell Vehicles 7(1):1535750. https://doi.org/10.1080/20013078.2018.1535750

Thom SR, Bhopale VM, Hu J, Yang M (2017) Increased carbon dioxide levels stimulate neutrophils to produce microparticles and activate the nucleotide-binding domain-like receptor 3 inflammasome. Free Radic Biol Med 106:406–416. https://doi.org/10.1016/j.freeradbiomed.2017.03.005

Tian Y, Li S, Song J et al (2014) A doxorubicin delivery platform using engineered natural membrane vesicle exosomes for targeted tumor therapy. Biomaterials 35:2383–2390

Tian T, Zhang HX, He CP, Fan S, Zhu YL, Qi C et al (2018) Surface functionalized exosomes as targeted drug delivery vehicles for cerebral ischemia therapy. Biomaterials 150:137–149. https://doi.org/10.1016/j.biomaterials.2017.10.012

Trams EG, Lauter CJ, Salem JN, Heine U (1981) Exfoliation of membrane ecto-enzymes in the form of micro-vesicles. Biochim Biophys Acta 645:63–70

Tsai MJ, Hsu YL, Kuo PL (2018) Circulating extracellular vesicles in human disease. N Engl J Med 379:2179–2180. https://doi.org/10.1056/NEJMc1813170

Valadi H, Ekström K, Bossios A, Sjöstrand M, Lee JJ, Lötvall JO (2007) Exosome-mediated transfer of mRNAs and microRNAs is a novel mechanism of genetic exchange between cells. Nat Cell Biol 9:654–659. https://doi.org/10.1038/ncb1596

Verderio C, Gabrielli M, Giussani P (2018) Role of sphingolipids in the biogenesis and biological activity of extracellular vehicles. J Lipid Res 59(8):1325–1340. https://doi.org/10.1194/jlr.R083915

Vietri M, Radulovic M, Stenmark H (2020) The many functions of ESCRTs. Nat Rev Mol Cell Biol 21:25–42. https://doi.org/10.1038/s41580-019-0177-4

Villarroya-Beltri C, Baixauli F, Mittelbrunn M, Fernandez-Delgado I, Torralba D, Moreno-Gonzalo O et al (2016) ISGylation controls exosome secretion by promoting lysosomal degradation of MVB proteins. Nat Commun 7:13588. https://doi.org/10.1038/ncomms13588

Vojtech L, Woo S, Hughes S, Levy C, Ballweber L, Sauteraud RP et al (2014) Exosomes in human semen carry a distinctive repertoire of small non-coding RNAs with potential regulatory functions. Nucleic Acids Res 42:7290–7304. https://doi.org/10.1093/nar/gku347

Wang H, Sui H, Zheng Y, Jiang Y, Shi Y, Liang J, Zhao L (2019) Curcumin-primed exosomes potently ameliorate cognitive function in AD mice by inhibiting hyperphosphorylation of the tau protein through the AKT/GSK-3β pathway. Nanoscale 11(15):7481–7496

Wang S, Khan A, Huang R, Ye S, Di K, Xiong T et al (2020) Recent advances in single extracellular vesicle detection methods. Biosens Bioelectron 154:112056. https://doi.org/10.1016/j.bios.2020.112056

Wang X, Yang G (2021) Bone marrow mesenchymal stem cells-derived exosomes reduce Aβ deposition and improve cognitive function recovery in mice with Alzheimer's disease by activating sphingosine kinase/sphingosine-1-phosphate signaling pathway. Cell Biol Int 45(4): 775–784. https://doi.org/10.1002/cbin.11522

Wei H, Chen J, Wang S, Fu F, Zhu X, Wu C et al (2019) A nanodrug consisting of doxorubicin and exosome derived from mesenchymal stem cells for osteosarcoma treatment in vitro. Int J Nanomedicine 14:8603–8610

Witwer KW, Théry C (2019) Extracellular vesicles or exosomes? On primacy, precision, and popularity influencing a choice of nomenclature. J Extracell Vesicles 8:1648167
Wollert T, Hurley JH (2010) Molecular mechanism of multivesicular body biogenesis by ESCRT complexes. Nature 464(7290):864–869. https://doi.org/10.1038/nature08849
Xie Y, Dang W, Zhang S, Yue W, Yang L, Zhai X et al (2019) The role of exosomal non-coding RNAs in cancer. Mol Cancer 18:37. https://doi.org/10.1186/s12943-019-0984-4
Yang Y, Yan Z, Wei D, Zhong J, Liu L, Zhang L et al (2013) Tumor-penetrating peptide functionalization enhances the anti-glioblastoma effect of doxorubicin liposomes. Nanotech 24(40):405101. https://doi.org/10.1088/0957-4484/24/40/405101
Yang S, Che SP, Kurywchak P, Tavormina JL, Gansmo LB, Correa de Sampaio P et al (2017a) Detection of mutant KRAS and TP53 DNA in circulating exosomes from healthy individuals and patients with pancreatic cancer. Cancer Biol Ther 18:158–165. https://doi.org/10.1080/15384047.2017.1281499
Yang J, Zhang X, Chen X, Wang L, Yang G (2017b) Exosome mediated delivery of miR-124 promotes neurogenesis after ischemia. Mol Ther–Nucleic Acids 7:278–287
Yang T, Martin P, Fogarty B, Brown A, Schurman K, Phipps R et al (2015) Exosome delivered anticancer drugs across the blood-brain barrier for brain cancer therapy in Danio rerio. Pharm Res 32:2003–2014. https://doi.org/10.1007/s11095-014-1593-y
Yim N, Ryu SW, Choi K, Lee KR, Lee S, Choi H et al (2016) Exosome engineering for efficient intracellular delivery of soluble proteins using optically reversible protein-protein interaction module. Nat Commun 7:12277. https://doi.org/10.1038/ncomms12277
Yuan D, Zhao Y, Banks WA, Bullock KM, Haney M, Batrakova E, Kabanov AV (2017) Macrophage exosomes as natural nanocarriers for protein delivery to inflamed brain. Biomaterials 142:1–12
Zhai L, Shen H, Sheng Y, Guan Q (2021) ADMSC Exo-MicroRNA-22 improve neurological. Function and neuroinflammation in mice with Alzheimer's disease. J Cell Mol Med 25(15): 7513–7523
Zhang Y, Yu M, Tian W (2016) Physiological and pathological impact of exosomes of adipose tissue. Cell Prolif 49:3–13. https://doi.org/10.1111/cpr.12233
Zhang H, Freitas D, Kim HS, Fabijanic K, Li Z, Chen H et al (2018) Identification of distinct nanoparticles and subsets of extracellular vesicles by asymmetric flow field-flow fractionation. Nat Cell Biol 20:332–343
Zhang H, Wu J, Wu J, Fan Q, Zhou J, Wu J et al (2019) Exosome-mediated targeted delivery of miR-210 for angiogenic therapy after cerebral ischemia in mice. J Nanobiotechnol 17:29. https://doi.org/10.1186/s12951-019-0461-7
Zhang X, Xu Q, Zi Z, Liu Z, Wan C, Crisman L et al (2020) Programmable extracellular vesicles for macromolecule delivery and genome modifications. Dev Cell 55:784–801. https://doi.org/10.1016/j.devcel.2020.11.007
Zhao Y, Haney MJ, Gupta R, Bohnsack JP, He Z et al (2014) GDNF-transfected macrophages produce potent neuroprotective effects in Parkinson's disease mouse model. PLoS One 9(9): e106867. https://doi.org/10.1371/journal.pone.0106867
Zheng M, Huang M, Ma X, Chen H, Gao X (2019) Harnessing exosomes for the development of brain drug delivery systems. Bioconjug Chem 30:994–1005
Zhuang X, Xiang X, Grizzle W, Sun D, Zhang S, Axtell RC et al (2011) Treatment of brain inflammatory diseases by delivering exosome encapsulated anti-inflammatory drugs from the nasal region to the brain. Mol Ther 19:1769–1779

The manufacturer's authorised representative in the EU is Springer Nature Customer Service Centre GmbH, Europaplatz 3, 69115 Heidelberg, Germany. If you have any concerns regarding our products, please contact ProductSafety@springernature.com

Printed and bound by CPI Group (UK) Ltd, Croydon, CR0 4YY
01/10/2026
02225892-0009